AF614942

HUMAN VARIATION
The Biopsychology of Age, Race, and Sex

Contributors

CYRIL D. DARLINGTON

JOHN L. HORN

DWIGHT J. INGLE

ARTHUR R. JENSEN

ROBERT G. LEHRKE

RICHARD LYNN

CLYDE E. NOBLE

R. TRAVIS OSBORNE

DAVID C. RIFE

AUDREY M. SHUEY

NATHANIEL WEYL

HUMAN VARIATION
The Biopsychology of Age, Race, and Sex

Edited by

R. TRAVIS OSBORNE
Department of Psychology
University of Georgia
Athens, Georgia

CLYDE E. NOBLE
Department of Psychology
University of Georgia
Athens, Georgia

NATHANIEL WEYL
Boca Raton, Florida

ACADEMIC PRESS New York San Francisco London 1978

A Subsidiary of Harcourt Brace Jovanovich, Publishers

ACADEMIC PRESS, INC.
111 Fifth Avenue, New York, New York 10003

United Kingdom Edition published by
ACADEMIC PRESS, INC. (LONDON) LTD.
24/28 Oval Road, London NW1 7DX

Library of Congress Cataloging in Publication Data

Main entry under title:

Human variation.

Includes bibliographies and index.
1. Nature and nurture. 2. Ethnopsychology.
3. Age. 4. Sex differences. 4. Race.
I. Osborne, Robert Travis, Date
II. Noble, Clyde E. III. Weyl, Nathaniel,
[DNLM: 1. Individuality. 2. Genetics,
Behavioral--Psychophysiology. 3. Variation
(Genetics) BF697 H918]
BF341.H8 155.7 77-28049
ISBN 0-12-529050-0

PRINTED IN THE UNITED STATES OF AMERICA

In Memoriam

SIR FRANCIS GALTON, FRS

—founder of behavioral genetics and eugenics,
originator of mental tests and twin studies,
pioneer correlational and psychometric analyst—
for his seminal contributions
to the
biopsychology of human variation

Contents

List of Contributors

Numbers in parentheses indicate the pages on which the authors' contributions begin.

CYRIL D. DARLINGTON (379), Professor Emeritus of Biology, Oxford University, Oxford, England

JOHN L. HORN (107), Professor of Psychology, University of Denver, Denver, Colorado

DWIGHT J. INGLE (5), Professor Emeritus of Physiology, University of Chicago, Chicago, Illinois

ARTHUR R. JENSEN (51), Professor of Educational Psychology, University of California, Berkeley, Berkeley, California

ROBERT G. LEHRKE (171), Clinical Psychologist, Brainerd State Hospital, Brainerd, Minnesota

RICHARD LYNN (261), Professor of Psychology, New University of Ulster, Coleraine, County Londonderry, Northern Ireland

CLYDE E. NOBLE (287), Professor of Psychology, University of Georgia, Athens, Georgia

R. TRAVIS OSBORNE (137), Professor of Psychology, University of Georgia, Athens, Georgia

DAVID C. RIFE (29), Professor Emeritus of Zoology, Ohio State University, Columbus, Ohio

AUDREY M. SHUEY (199), Professor Emerita of Psychology, Randolph-Macon Woman's College, Lynchburg, Virginia

NATHANIEL WEYL (1), Social Scientist, 855 Oleander Street, Boca Raton, Florida

Preface

Contemporary research on cognitive and psychomotor behavior influenced by the three biological factors mentioned in the subtitle of this book is producing a notable effect on the psychology of human differences. In the present work we attempt a broad survey and objective appraisal of some of the key concepts and principles that have emerged from recent investigations of the relationship between organismic factors and human performance.

To write on the subject of human variation in the late 1970s presents certain difficulties not encountered by those who write on other psychological topics. Many of the *phenomena* of intergroup variability in the behavior of human beings classified in terms of age, race, and sex are beyond reasonable doubt, yet whether the *provenance* be mainly that of nature or nurture, to what statistical degree it exists, and for which psychological categories, is a set of issues fraught with controversy. Nor is the controversy merely technical; it has deeply infiltrated the public sector where periodically it erupts in the form of staged media events, campus demonstrations, establishmentarian resolutions, special national legislation, institutional coercion in matters of employment and promotion, and the harassment of individual academicians who decline to embrace the null hypothesis with respect to human abilities. Certain teachers and researchers have paid a high price for academic freedom, even in Western nations that guarantee remarkable liberties to their citizens.

Rather than espousing the majoritarian doctrine of biopsychological uniformity, the contributors to this volume may be said, as a group, to entertain the hypothesis that heritable variations in many human reaction tendencies are significantly associated with the taxa of sex, race, and age. We assume continuities rather than typologies, quantitative intergradations rather than qualitative classes. Our biopsychological orientation represents a position whose origins are more evidential than attitudinal because most of us began our careers as empiricists, and we grew to professional maturity in the context of a pervasive environmentalism. But the dramatic advances in behavioral genetics are compelling, and simplistic Watsonian psychology has yielded to the sophistication of neobehaviorism. Consequently, when one finds *belief* conflicting with *knowledge* it is unreasonable to persist—especially when intuition, emotion, and conation are strongly involved at the expense of cognitive considerations. Notoriously in the domain of race, but increasingly in discussions of sex and age, conformity of opinion is expected in the halls of ivy as it is in legislative and judicial chambers.

Usually such social pressures are justified in the name of humanism, or by invoking an ever-lengthening list of a priori "rights." However, a philosophical pragmatism inclines us to the view that humanitarian goals are achieved most readily and completely when one's society liberates and optimizes the pursuit of empirical and theoretical knowledge. Here we allude to the untrammeled acquisition of scientific information concerning: (1) thc manifold dimensions of human behavior, (2) the genetic, anatomical, and physiological correlates of that behavior, (3) the physical and social environments in which people live, and (4) their biopsychological interactions with those environments. For us, the question of determining the relative proportions of phenotypic variance in selected human attributes that may be ascribed to innate, acquired, interactive, and covariate sources under given conditions is entirely a matter of data and theory—not of ideology, politics, authority, or forensics. Of course, there may be some readers who will draw oversimplified conclusions about complex social problems from the pages that follow, but that is a case of inference rather than implication.

Each chapter is new and was commissioned for this volume. Our general intent has been to proceed from the evidence of controlled studies to quantitative hypotheses, then to state qualified generalizations about behavior, finally to test these conditional propositions by well-established statistical techniques—and frequently to remind ourselves of the provisional nature of scientific evidence. Fortunately, theoretical knowledge is corrigible as well as fallible, so errors of fact or logic will eventually be detected and eliminated. The knowledge that science aspires to is adaptive with respect to a variety of problems, theoretical as well as applied, because it is progressively sensitive to new data and to the relevance of operationally defined concepts imbedded in a system of wide generality. In such an enterprise the future is always relevant, so the evidence for scientific knowledge must be continually growing. One should be suspicious of knowledge that is static or unresponsive to novel facts and interpretations. In the field of human variation the quest for knowledge remains vigorous not only due to the intrinsic interest of the subject matter but also due to its dynamic and responsive nature.

Our editorial labors have been facilitated by a number of colleagues, students, and employees. In particular we wish to acknowledge the contributions of Ted Jaeger, Frank Miele, Jennie Parham, Robert B. Payne, Vickie Rabun, Wilma Sanders, Eileen Totter, and Roger Wilkerson.

Note on Taxonomy

Taxonomy is essential for precision in scientific communication, but many technical problems are encountered in labeling the various categories of human beings. Although age and sex are straightforward enough, the ethnic taxa of mankind have no settled classification. This is because the International Code of Zoological Nomenclature does not specify rules for subspecies (geographical races), nor for smaller taxa (e.g., local races, micro-races).

The names of races used in this book are consistent with the terminology of comparative psychology, physical anthropology, and human genetics. They are intended to achieve maximum clarity without sacrificing parsimony. In the table below, some alternative names are provided in addition to the Latin trinomials. All the living peoples of the world belong to the same genus and species, so our principal distinctions will be those of subspecies.

Taxonomy of the Living Geographical Races[1]

Subspecies	Trinomen	Alternative Names
Australoid[2]	*Homo sapiens australicus*	Australasid, Australid
Capoid	*H. sapiens capitalis*	Khoisanid, Khoisan
Caucasoid	*H. sapiens caucasus*	Europid, Europoid
Mongoloid[3]	*H. sapiens asiaticus*	Mongolid, Asiatic
Negroid[4]	*H. sapiens africanus*	Negrid, Congoid

[1] When necessary, hybrid taxa will be denoted as such (e.g., some Afro-Americans, Ainus, Cape Coloreds, Caribbean islanders, Hawaiians, Hottentots, Indo-Dravidians, Indonesians, Melanesians, Mexicans, Polynesians).

[2] Includes several dwarfed local races of Negritos (e.g., parts of India, Southeast Asia, and certain Pacific islands).

[3] Includes numerous local races of Amerinds (North, Central, and South America), Aleuts, and Eskimos (circumpolar regions).

[4] Includes dwarfed local races of Pygmies (Central Africa).

"Each individual's behavior is caused in part by certain tendencies which he has in common with all members of the species, in part by tendencies peculiar to his sex, in part by tendencies peculiar to his ancestry, in part by the stage of development or maturation which he has reached, in part by tendencies peculiar to the 'culture' of his land and time, and in part by the circumstances which characterize his own peculiar life-history."

EDWARD LEE THORNDIKE
Human Nature and the Social Order (1940)

HUMAN VARIATION
The Biopsychology of Age, Race, and Sex

1

Prologue

NATHANIEL WEYL

Boca Raton, Florida

This volume consists of original scientific essays by specialists commissioned for the project. It is designed for intelligent laymen and for senior and graduate students in behavior genetics, biopsychology, developmental psychology, human skills, individual differences, personality, and psychometrics. With few exceptions, the treatment is nonmathematical and within the grasp of readers with only elementary to modest backgrounds in genetics and psychology. The contributors have sought to present the available data objectively and with minimal reference to those political and moral controversies that have aroused so much intense partisanship and public clamor.

The book is a pioneer contribution to the growing study of human variation. For a people of multiple ethnic and national origins, such as the American, the psychological aspects of racial diversity are of compelling importance. As the interconnections among the peoples of the world become more complex, objective information concerning such behavioral differences becomes increasingly needed to facilitate international understanding.

The study of the psychology of aging and of the changes in mental and perceptual functioning that accompany advancing age is a growing field of interest that assumes particular relevance in the light of governmental programs to integrate better the aging and aged into society. Sexual differences in the frequency distribution of mental abilities and their possible genetic basis are an area of inquiry that has until very recently been systematically avoided. To the best of our knowledge, no other book exists with the scope of this one.

In his chapter on fallacies in arguments concerning human differences, Dwight J. Ingle analyzes examples of bad logic advanced by both proponents and opponents of environmental theories of human variation. He concludes that, although many uncertainties remain concerning the relative importance of heredity and environment in human abilities, heredity appears to be the more powerful factor in most cases.

The chapter on genetic markers, pleiotropy, and linkage, by David C. Rife, contains a detailed analysis of dermatoglyphics as a quantitative gene marker. Ethnic differences in fingerprint indices (FPIs) are stressed. Since there is no evidence of their genetic association with survival-influencing genes, FPIs may be uniquely valuable means of assessing ethnic relationships and the probable ancestry of migrant stocks.

Arthur R. Jensen's chapter on assortative mating is probably the most systematic treatment of the subject available. It covers the Mendelian algebra, observed ethnic and national differences, and the relationship to fertility, inbreeding, and hybridization of nonrandom mating. Jensen finds that assortative mating for intelligence may account for four-fifths of the American and British population with IQs above 145.

John L. Horn's chapter is a theoretical and empirical investigation of the quantitative and qualitative changes in human intelligence associated with the maturation and aging processes. As distinct from child-development studies, the main focus is on the primary mental abilities characteristic of infancy, adolescence, and adulthood. Changes in such factors as intellectual speed, memory, visualization, auditory functions, and productive thinking are evaluated in the context of a hierarchical theory of fluid and crystallized intelligence.

R. Travis Osborne's contribution makes public for the first time a summary of the results of his in-depth heritability studies of large samples of Caucasoid and Negroid monozygotic and dizygotic twins. Primary mental abilities, as revealed by a battery of 12 separate tests, showed a wide range in heritability, suggesting differences in their underlying genetic and environmental causal components. The differences between the observed variance ratios of Negroid and Caucasoid subjects were not statistically significant.

Robert G. Lehrke's contribution on sex linkage explores the biological evidence in favor of greater male variability in intelligence at both the upper and lower ranges of the IQ frequency distribution. The investigation of this hypothesis, to which Lehrke has himself made significant contributions, is a new field in biopsychology with which even graduate students are often unfamiliar. The treatment is both evidential and theoretical.

In the most comprehensive analysis of own-race and self-esteem studies of young Negroid and Caucasoid children ever undertaken, Audrey M. Shuey finds significantly higher scores among Caucasoid subjects in both areas. Neither sex nor the racial composition of the school (when considered independently of its location) significantly affected either. Negroid self-esteem and own-race preference were significantly higher in the North than in the South, and they increased with age.

In the field of racial differences, there is an enormous literature comparing Negroids and Caucasoids. It covers not only IQ in general but the comparative structure of intelligence, sensorimotor differences, and the extent to which variations in any or all of these factors are believed to be genetic or environmental in origin. Quantitative data on the intelligence of peoples other than Caucasoids, Negroids, and citizens of countries of Anglo-Saxon overseas settlement are, however, in most cases fragmentary and, in the case of the Soviet world, nonexistent. In his chapter on ethnic differences in intelligence, Richard Lynn has used culture-fair tests to attempt to outline the geography of human intelligence. Among other interesting findings, he reports that the Mongoloids of Japan are brighter on the average than the Caucasoids of Britain or America.

After tracing the history of research and theory on psychomotor learning and performance, Clyde E. Noble summarizes the available laboratory studies on ethnic, national, age, and sex differences in such perceptual and motor tasks and processes as visual acuity, chemical tasting, optical illusions, pitch and rhythm discrimination, musical talent, color perception, spatial visualization, kinesthetic maturation, athletic skills, eye–hand coordination, maze testing, and multiple-choice learning. He concludes with a treatment of the theory of skill acquisition.

In his Epilogue to this volume, Cyril D. Darlington of Oxford University argues that human evolution is characterized by positive feedback between the growth of intelligence and the environmental changes and transformations it causes. This process has been conducive to the development of individual responsibility and the subordination of instinctual to rational responses to challenge. Natural obstacles to this fruitful transaction, particularly those caused by parasites and endemic diseases, have deprived primitive peoples of the stimulus of the grain revolution.

The chapters of this book deal with the broadest aspects of variation in human behavioral traits. Individual and group diversity provide human society with richness of texture and creative potential. Scientific understanding of the causes and nature of human variety may be a precondition of species survival for *Homo sapiens.*

2

Fallacies in Arguments on Human Differences

DWIGHT J. INGLE
University of Chicago

INTRODUCTION

Debates on the relative importance of heredity and environment in determining the characteristics of individuals and of groups commonly focus on those factors that determine success and status. The question arouses pride and other self-assertive tendencies. Such debates are likely to include misinformation and faulty reasoning, and they often become emotionally charged. Self is identified more with heredity than with environment. It is a threat to self-esteem to acknowledge genetic limitations; it is less so if environment and society can be blamed. I find it paradoxical that those who profess environment to be all important in determining personality and individuality seem to cherish inherited nature more.

Historically, wealth, power, and special privileges were determined mainly by family and social class; this was rationalized by the assumption that traits needed for leadership are inherited and that the lower classes are inferior in those qualities. There was supporting dogma about "royal blood" or "good blood" as opposed to "bad blood" or "ordinary blood." Such dogma was used to justify slave–master and lord–serf relationships.

The hypothesis that heredity is important in determining differences between individuals and between groups is out of fashion today. The view that people are "equal born" may have developed at different times and places and sometimes in response to oppression. It appears in religious, political, social, and scientific writings and teachings. It has been used to support democracy as well as socialism and communism. It is a fallacy to assume that the uses to which ideas are put determine their truth or validity. In what follows, I shall use the word "fallacy" in a broad sense to include weak arguments as well as absolute errors in deduction and induction.

The *fallacists' fallacy* is to assume that fallacious reasoning always leads to error. Some faulty reasoning has led to truth. Logical reasoning sometimes results in factual errors. Arguments vary along a continuum in quality, consistency, and strength of inference. Truth and error are intrinsically probabilistic in the natural world, but the frequency of error can be reduced when we recognize weak arguments and make them more secure. Before reviewing common fallacies, I shall list the requirements for testing claims to empirical knowledge.

REQUIREMENTS FOR PROOF

1. *Methods of identifying and measuring traits must show technical evidence of validity and stability.* In the physical sciences, exact methods of identification and measurement are accepted. Psychometric methods of measuring intelligence and other complex human attributes are less direct, and the meaning of each technique is debated. Logical and statistical criteria are guides to the measurement of human traits.
2. *Any explanation should account for all of the facts and show that strong competing hypotheses are improbable.* If it fails to do this, however, its rejection is not always required; the life of a hypothesis can sometimes be extended by revision.
3. *The evidence for a claim to knowledge must be repeatable by others on demand.*
4. *A claim for discovery should be supported by favorable instance statistics.* This means that taking away or adding the putative causal factor(s) under study should bring about a statistically significant change in results.
5. *The explanation should permit prediction and control of the events or processes under study.* To achieve prediction and control of events or processes adds powerful support to claims that we understand them and have identified the causal factors. However, some incorrect theories and models have correctly predicted some experimental results. Not all scientific knowledge leads to the prediction and control of events under study.
6. *Agreement of different lines of evidence is a sign of probable truth.* This is a sign of truth, not proof of it. Claims are sometimes made that a number of bits of evidence, each weak when taken by itself, represents proof when taken together. If each line of evidence is faulty, then the

combination does not establish proof. The assumption that it does is known as the *fagot fallacy*.

7. *The more implausible the claim, the greater the need for strong evidence*. If a claim does not fit our experience with nature, we are likely to doubt the explanation even when the supporting evidence seems strong.

8. *General agreement by experts*. There are no rules in science that force an individual to accept or reject an idea. The individual is free in this regard, and belief is a personal matter. However, an idea is not commonly regarded as established until there is general agreement among experts that it is true. This does not mean that majority opinion is a trustworthy guide to truth.

It may be helpful to consider the following list of fallacies in relationship to the above requirements for proof. These guides to the acceptance of evidence, taken separately or together, do not establish certainty. It is almost impossible to comply with all of them. Proof is relative, and belief should remain open for reexamination and possible revision.

The taxonomy of common fallacies is not internally consistent. Categories range from absolute errors to insecure deductions and inferences. Ordinary reasoning does not follow the rules of logic and can go awry by novel paths. I do not suggest that it is impossible to develop a logically consistent taxonomy of common fallacies, but no one has done so, and I cannot.

We shall consider fallacies that are common in debates on the relative importance of biological factors and of social environment in causing changes associated with age, sex, and racial differences, especially those debates on the causes of individual and group differences in mental abilities and achievements. Only a few references are given to faulty arguments. My object is to list fallacies to be kept in mind by anyone examining these debates. I have tried to identify fallacies used by both hereditarians and environmentalists, but I am not an impartial author. I believe that heredity probably plays a more important role in causing individual and group differences in mental abilities and achievements.

Some fallacies have been named. Those that have not are identified by description. I have listed five broad families of fallacies: those based on association, those of generalization, verbal fallacies, fallacies involving oversimplification, and those that beg the question. Many specific fallacies will have close relatives in different families.

FALLACIES BASED ON ASSOCIATION

1. *The fallacy of* post hoc, ergo propter hoc *(after this, therefore because of this)*. It has been argued that enslavement of the African ancestors of contemporary American Negroids was a cause of the Negroid–

Caucasoid gap in IQ and in school achievement. The argument assumes that the gap arose after Negroids were brought to America. There is no evidence that this was true. Such studies as have been done show a gap in the average IQs of Caucasoids and African Negroids (Chap. 9). Negroids of today have not lived as slaves, although many have experienced socioeconomic discrimination and other environmental handicaps. Other racial and ethnic groups in America have experienced socioeconomic discrimination. Mongoloids and Jews in this country have an average IQ that is above the national average. Amerinds are lower than American Negroids on the socioeconomic scale but have a higher average IQ (Coleman 1966; Jensen, 1973).

It has also been argued that neglect of Negroid education in the past has caused the present Negroid–Caucasoid gap in test performance and school achievement. Many groups of Negroids have experienced only good schools, but the gap has not been narrowed during several decades of Negroid advancement in educational opportunities (Jensen, 1972). Several extensive studies support the Coleman Report (1966) that differences in educational achievement correlate only slightly with those variables that schools traditionally control.

If a hypothesis cannot be tested, and if the postulated causal factor cannot be shown to have a constant relationship to the trait under study, the hypothesis cannot be accepted as proved. Some have claimed that since sex-role training differs for girls and boys, this is the sole cause of sex differences in interests, drives, and achievements. The hypothesis has not been proved because it has not yet been shown that the differences can be abolished by changes in sex-role training. The alternative hypothesis that there are biological bases of sex roles is supported by data from comparative studies on hormones and behavior and by the facts about sex differences in the neurophysiology of the brain.

2. *The fallacy of* cum hoc, ergo propter hoc *(with this, therefore because of this).* Hereditarians have sometimes argued that the common association between family and achievement confirms the role of heredity in determining success. It does not because closely related individuals tend to have similar environments. The *environmentalists' fallacy* lies in looking for any environmental difference between two populations having a mean difference in IQ or in achievement and assuming that the associated environmental difference is the cause. Each hypothesis requires testing by other forms of inference.

3. *The fallacy of assuming a cause-and-effect relationship between sets of correlated values.* This is a form of the *cum hoc fallacy.* Correlations are merely descriptive. There may or may not be a causal connection. It is difficult to find a correlation between any two variables that is precisely zero. However, correlations can provide useful clues to causal relationships. The inference of cause and effect can be strengthened by

statistical methods (e.g., multiple correlations and path analysis) and by the experimental control and testing of the putative causal factors.

There is an apparent contradiction in two arguments by environmentalists. First, it is claimed that the high correlation between the IQs of identical twins reared apart is due to the similarity of environments for each pair. Second, the low correlation of IQs of unrelated children reared together is explained as due to dissimilar environments. Each claim is possibly correct.

4. *The fallacy of the consequent.* The assumption is that an effect always results from the same cause. Some hereditarians and some environmentalists have separately argued that individuality has either a single cause or a fixed pattern of causes. Causal patterns frequently illustrate equifinality, that is, a general result can be reached by different pathways. Mental retardation can be produced by several diseases and by injuries, some of them genetic and some environmental.

The association between physiological aging and a decline in creativity is sometimes assumed to be a simple cause–effect relationship. Causality is plausible but has not been proved. A decline in creativity and other mental abilities can result from any one of several diseases of the brain, from brain injury, from changes in motivation, or from distractions based on changes in the social environment.

5. *The fallacy of assuming that the cause(s) can always be found at the time the effect is observed.* Hates and fears and other attitudes can last a lifetime beyond the original cause. The bases of some feelings of insecurity and fears among the aged may have been established in childhood. Some biological pathogenic factors have a latent time of years between the times of infection and manifestation of disease. Failure to demonstrate a postulated causal factor does not exclude it as a possible cause.

6. *The fallacy of assuming that the cause of a problem is a deficiency of the remedy.* Lack of water is not the cause of fire. Improvement of the behavior of a child with Down's syndrome by intensive training does not prove that lack of training caused this form of mental retardation.

7. *The fallacy of accident.* Treating a nonessential associated factor as essential for a result illustrates this fallacy. A classic example is the assumption that competence is associated only with "Caucasoid" skins. Another example is the assumption that high intelligence is to be found only among "plain" women.

FALLACIES OF GENERALIZATION

Inductive reasoning allows generalization from a few cases to a group if a representative sample has been drawn from the group. It is logical to generalize only about the group from which the sample was drawn, or to groups that are identical in all relevant factors. This has to do with the strength of the analogy: for example, some nutrition studies using animals may be validly generalized to humans, depending on the similarities of the two species with respect to metabolic, digestive, and natural diet factors.

1. *The fallacy of* secundum quid *or* para pro toto *(overgeneralization)*. Some environmentalists have failed to recognize individuality by assuming that all people are equally endowed with the genetic factors affecting social mobility and achievements. This is the *fallacy of the universal man*. Some hereditarians, especially of an earlier day, have argued that all members of a racial group are alike. They have also ignored individuality and the extensive overlap of abilities between groups.

There are average differences in sex roles, interests, and achievements between the sexes. This has led to the belief that women are inherently inferior to men in certain roles. On the other hand, that some women excel in roles that are customarily assumed by men has led to the assertion that the sexes are equal in the biological bases of all forms of mental achievement.

There is also generalization from age norms, so that people are treated according to age in respect to schooling, voting rights, certain restraints, retirement, and so on rather than according to individuality.

2. *The* per contra *fallacy*. This means drawing a conclusion about an individual on the basis of the general characteristics of the group to which the person belongs—the converse of the *secundum quid* fallacy. It fails to recognize individuality and overlap of group characteristics. Individual Negroids sometimes complain that they are judged solely on the basis of racial identity, not individual traits and achievements. So do certain women with respect to their sex classification.

3. *The fallacy of the faulty sample*. There has been a number of widely published reports of Negroid schoolchildren having an average IQ higher than the national average for Caucasoids. In each case, the children were highly selected either on a socioeconomic basis or by admission standards. One example was the sixth-grade class of the Windsor Hills Elementary School of Los Angeles (Jensen, 1972). The majority of these students were from an affluent neighborhood.

The George Report (George, 1962) reviewed evidence that the average brain weight of Negroids is significantly less than the average brain weight of Caucasoids. The brains of the Negroids were of cadaver origin, either from the slums of Baltimore or from an African village. These brains were compared with the brains of Caucasoids that had been collected in Germany several decades earlier. Some of the more important variables such as age, nutritional status, and cause of death were uncontrolled.

It has been reported that when a human reaches the age of 30 or thereabouts, the brain begins to lose functional cells. Such estimates are based on superficial studies of cadaver brains that are not representative of living healthy brains of that age. The average life expectancy of elderly people who enter homes for the aged is usually less than the average for a control group not placed in homes. Although the groups may be judged to be similar in respect to the incidence of disease, there is typically only a superficial appraisal of health. It should be shown that the two groups are similar in respect to variables known to be relevant.

The theory of psychosexual neutrality at birth is based partly on evidence that when sex roles are assigned to sexually anomalous individuals, many of them accept the gender role assigned to them even when the chromosomal constitution is of the opposite sex. Such sexually anomalous cases represent less than 1% of the population. Gender assignment is made according to the extent that the infant is already masculinized or feminized, and this corresponds to changes already established in the brain by the presence or absence of androgens. Masculinization and feminization are determined by the presence or absence of androgenic hormones rather than by chromosomal constitution (Hutt, 1972). Failure of a hermaphrodite to repudiate an assigned gender role does not mean that normal psychosexuality has been established or that the individual is fully satisfied with it.

4. *The fallacy that within-group measurements can be extrapolated to different groups.* A few writers have used estimates of the heritability of IQ among Caucasoids to draw conclusions about the heritability of IQ among Negroids. It is plausible that if the heritability of individual differences in IQ is high among Caucasoids, it is also high among Negroids, but the extrapolation is insecure.

We have noted studies indicating that the average brain weight of Negroids may be significantly less than the average brain weight of Caucasoids (George, 1962). Critics have replied that the low correlation between brain weight and IQ among Caucasoids implies that racial differences in average brain weight are without significance. This is not a self-validating conclusion. A possible causal relationship between racial differences in brain weight and IQ would need testing by other lines of evidence and inference. Radiological methods could possibly be used to measure brain size in samples of different human populations. Research on the fine structure, the neurophysiology, and the chemistry of the brain might lead to important discoveries about the biological bases of abilities (see Baker, 1974, pp. 429–437).

5. *The fallacy of hasty extrapolation.* Extrapolation can be accurate only when the characteristics of the process under study and its causes

are known. Many processes are so poorly understood that extrapolation of present trends involves risk of error.

It is predicted that medical science will soon extend life expectancy by many decades. At least two types of errors are represented in these predictions. First, gains in life expectancy have resulted almost entirely from public health measures of preventive medicine and from the successful treatment of some of the great killing diseases. Medical scientists do not yet understand how to control the aging process. The upper limit of life span has not increased. Second, some predictions based on the assumption that all of the killing diseases will eventually be controlled involve the *fallacy of spurious replication.* The assumption is made that if cancer can be cured, all cancer patients will then be disease-free until they die of old age. The majority are likely to die of another disease. The same would be true if all cases of heart disease could be cured, and it would be true for every killing disease. Not only do patients sometimes have more than one serious disease at a time, but also the cure of one is commonly followed by other disorders. The gain in life span predicted from the cure of each disease and the assumption of a disease-free existence until old age would be replicated spuriously.

6. *The fallacy of equating individual values with averages.* The average Negroid–Caucasoid difference in IQ is about 15 points (Jensen, 1972). Critics of the hypothesis that there is a genetic basis for this average difference point out that individuals moving from a poor to a good environment sometimes show IQ changes of more than 15 points. (In a minority of cases, there is a decrease in IQ or no change rather than a gain.) A few individual cases of identical twins raised apart have larger differences in IQ. (The higher IQ is sometimes found in the twin raised in the poor environment.) Average changes in IQ are much smaller than the extreme changes in individual values. Averages must be compared with averages, not values for individuals.

It is also a fallacy to assume that similarity in group averages proves that there is no difference in the distribution of individual values. In comparisons of boys and girls and of men and women, it is found that the average IQ is approximately the same, but there is greater variability among individual males; more are dull and more are gifted than is the case for females. There is no complete agreement as to the reasons for sex differences in intellectual variability (see Chapter 7).

7. *The fallacy of ethnomorphism.* The assumption is made that the characteristics of another group are similar to or identical with one's own. It is illustrated by the efforts of some social reformers to solve the problems of minority groups by exhorting them to abandon their culture and adopt that of a majority group, commonly that to which the social reformer belongs. The obverse of this fallacy may be seen within the ranks

of ardent women's liberation advocates who believe that women are free only when forced to assume a male role.

8. *The fallacy of ethnocentrism.* The importance of one's group in relation to other groups is exaggerated. Nationalities, ethnic groups, and the sexes each tend to overemphasize their roles in history.

9. *The fallacy of false analogy.* Although strong inferences can be formed on the basis of similarity between categories of things and processes, reasoning based on superficial or incomplete resemblances is likely to be faulty. The evidence for a heritability of individual differences in IQ of approximately +.80 has been questioned because this is much higher than the heritability of egg production or yield of corn. There are at least two important differences between human intelligence and the other two categories. First, in animal husbandry and in agronomy, there have been many generations of selective breeding, so that breeds of chickens and breeds of corn are much more homogeneous in respect to genetic endowment. The range of genetic differences is relatively small; hence, a much higher proportion of individual differences is due to environmental factors. Second, corn production and egg laying are vegetative functions that are being compared to the marvelously wide range of human intelligence. To assume that the ranges of individual differences in corn yield and egg production are analogous to the range of individual differences in IQ scores is to commit the *fallacy of the variable base.*

Many physiological processes show periodicity. This is true of all hormones that affect sexual functions. It has been argued that because there is periodicity in the secretion of hormones in men, this is analogous to the menstrual cycle of women. The analogy is weak or false. There is similarity in some respects, but menstruation involves the growth and sloughing of uterine tissue, with an accompanying release of toxins and loss of blood. The pattern of changes in the secretion of hormones in women differs basically from the cyclic secretion of hormones in men.

10. *The fallacy of assuming randomness of self-assembly and social mobility.* Social anthropologist Otto Klineberg (1935, 1944) argued that regional differences in the United States caused the notorious Negroid–Caucasoid gap in Army Alpha scores during World War I. Pitting "best" Negroids against "worst" Caucasoids, he found four Northern states in which the average (median) test scores for Negroids were higher than the average scores for Caucasoids in four Southern states. Klineberg assumed that the Negroids who had moved North were representative of the whole population of American Negroids. Yet Caucasoids still scored significantly higher than Negroids in every state. (Incidentally, he failed to note that a higher percentage of Caucasoids than Negroids was capable of being tested.) It is interesting that the "Klineberg effect" did not

appear for the nonverbal Army Beta test; only for the verbal Army Alpha test, which correlates +.70 with years of schooling. Many Negroids could not qualify for military training; large numbers, found to be illiterate, were given the nonlanguage Army Beta test.

The principal weakness of the Klineberg argument, which Jensen (1973) calls the *Klineberg fallacy,* is to assume that migration is independent of the genetic bases of abilities and that each migrating subgroup is representative of its parent racial population. The Negroid–Caucasoid gap in armed forces test scores has not been narrowed by more than 50 years of Negroid advancement, and there is evidence (Jensen, 1973; Loehlin, Lindzey, & Spuhler, 1975; Weyl, 1969) that the gap has increased. Nevertheless, the fallacious Klineberg thesis is still being cited as proof that the average Negroid–Caucasoid IQ difference is due mainly to regional social inequalities (see also Chapter 9).

Certain sociologists claim that if Caucasoids and Negroids were matched well for cultural and socioeconomic status, then their average IQ differences would disappear. It is probable that inherited abilities play a causal role in cultural or socioeconomic success. If this is true, then a sample selected in respect to one variable would be selected in respect to the other; thus, matching doubtless stacks the cards against a genetic view. The assumption that social mobility, self-assembly, and socioeconomic achievements are randomly related to the genetic bases of abilities has, probably unfairly, been termed the *sociologists' fallacy.* Some have even suggested that, owing to racial discrimination, Negroids must be more intelligent than Caucasoids to have achieved the same level of success in society. In the majority of studies in which Negroids and Caucasoids were actually matched for socioeconomic status, however, the IQ gap was not abolished (Jensen, 1973, pp. 235–242).

VERBAL FALLACIES

1. *The fallacy of composition.* This is using a word of more than one meaning to characterize both the whole and the part. A social organization may be made up of strong individuals without being strong as a social organization. A group may be made up of intelligent individuals and still make unintelligent decisions as a group.

2. *The fallacy of division.* This is the converse of the above fallacy. It is using a word of different significance when applied to the individual instead of the group. To describe a neighborhood as wealthy does not mean that every individual in the neighborhood is wealthy.

3. *The synonymic fallacy.* Words of significantly different meanings, such as "intelligence" and "fitness," are used interchangeably. The

words "equal" and "identical" are not the same. People can be equal in rights and freedoms without being identical in biological or psychological terms.

The average life expectancy of human females is greater than for males. The greater average viability of the human female is evident from the time of conception; more males than females die *in utero*. On this basis, the human female is sometimes characterized as the "stronger" or the "superior" sex, although neither word is a synonym for "viable."

4. *The fallacy of* plurium interrogationum *(the fallacy of the false question or of many questions).* The question contains a false assumption, a false implication, or information not in evidence. "Why does training of the hard-core unemployed make them successful workers unless they are of good intelligence?" The question falsely implies that all trainees become successful workers.

5. *The fallacy of adjective conjugation.* Some examples are: "I have ethnic pride." "You are prejudiced." "She is a bigot." "I have pride in the accomplishments of my sex." "You are close minded." "He is a chauvinist pig."

6. *The fallacy of assuming that each name represents an entity having independent existence.* Each race of mankind is sometimes treated as an independent entity, although each is of mixed origins. The converse of this fallacy is to assume that races do not exist; hence, the word "race" has no meaning. These and other fallacies appear in different forms in my list.

7. *The fallacy of using nondescriptive names.* Some self-interest groups describe themselves as "100% American," "democratic," and as having "equality" and "freedom" as goals. Names do not prove the nature of things.

8. *The fallacy of allegation.* Statements of faiths and beliefs are sometimes accompanied by "And that's the truth," "That's a fact," "I tell it like it is," or "That's right." Other examples are "Everybody knows that ———," "Surely you agree that ———," or "It is obvious that ———."

9. *The fallacy of ostensive definition.* This is a form of incomplete definition that points to those parts of a whole that are most in evidence. The concept of environment does not commonly embrace those unobserved and unmeasured nongenetic biological effects that are prenatal and postnatal. During the development of monozygotic (MZ) twins, there is unequal division of the fertilized ovum, and there is some degree of asymmetry in the development of twins. There is a higher incidence of fetal loss, birth defects, and differences in birth weight in MZ than in dizygotic (DZ) twins. Second, the exquisitely complex activities of the brain, conscious and unconscious, are a part of environment but are not commonly thought of as such. Within the complexity called "environ-

ment," there may be many unknown and unmeasured causes of human differences. The ambiguity of words is one of the sources of the discrepancy between truth and ordinary communication.

FALLACIES INVOLVING OVERSIMPLIFICATION

Such fallacies are among the non sequiturs because they base an argument or conclusion on insufficient reasons. This is characteristic of many fallacies.

1. *Fallacies of simplism.* Social anthropologist Ashley Montagu (1969) asserted that human groups do not differ significantly in respect to the genetic bases of intelligence. He stated:

> It appears highly probable that over the long course of man's prehistory the selective pressures on behavioral adaptation have in all human groups been much the same. . . . All the evidence we have indicates unequivocally that the behavioral potentialities of different peoples are everywhere much of a muchness, and that the differences in cultural achievement are not due to genetic factors but to differences in the history of cultural experience which has fallen the lot of each people [p. 88].

This is a strong statement that includes the word "unequivocally." He and other cultural anthropologists have argued that those groups of mankind that were able to survive the stresses of precivilized times, while many races and genera disappeared, would have to represent "much of a muchness" in "intelligence." Herein is the *synonymic fallacy* of confusing intelligence and biological fitness. The lower primates, as well as many forms of life having no capacity for abstract reasoning, also survived.

Because only vestiges of the past remain, because civilizations rise and fall for unclear reasons, and because many causal factors of the past cannot be identified and replicated in the present, any judgments on the causes of past events are largely subjective and intuitive. Relating to these simplistic judgments about the past is the hypothesis that the Negroid race originated before the Caucasoid race; the opposing hypothesis is that the former is younger than the latter (see Baker, 1974).

2. *The fallacy of the faulty criterion.* Many arguments about human psychological differences and their causes focus on the worth of the criteria used to measure complex traits that are important in human affairs. The most heated debates concern the measurement of intelligence. An intelligence test is said to be psychometrically "valid" when there is strong evidence, based on extra-task correlations, that it measures those cognitive abilities that it was designed to measure. Psychological tests are based on several decades of effort to establish their reliability and to "validate" them against outside, practical criteria. High test reliability

means that a test is consistent: that is, there is a high correlation (about +.90) between two forms of the same test, between test and retest, between scores of odd and even items, or between scores on split halves of the test. Intelligence tests developed by psychologists are useful in predicting scholastic achievement and job success. They differentiate between groups selected on the basis of independent judgments of intelligence by educational specialists. They likewise differentiate between different grades of mental deficiency as classified independently by experts in mental retardation. Moreover, IQ scores are responsive to the sort of brain damage that limits abstract reasoning. It happens that IQs are also well correlated with judgments by laymen who rank occupations according to the amount of intelligence required by each (Duncan, Featherman, & Duncan, 1968).

Still, there are no arguments that *compel* the conclusion that IQ is a "true" index of intelligence apart from the specified operations (e.g., ratio of mental age to chronological age). The items that make up the tests are chosen by special statistical methods, and most tests requiring cognitive abilities are significantly intercorrelated, even when they scarcely resemble each other (e.g., spatial versus verbal versus numerical items). Montagu (1969) has made the dogmatic statement that "IQ tests do not provide any measure whatever of intelligence (p. 89)." In order to prove that IQ is not an index of "true" intelligence, it would be necessary to measure the hypothetical trait directly, then show that empirical IQ scores have a zero relationship to "true" intelligence.

There is a more moderate view that intelligence tests are useful in assessing the abilities of middle-class Caucasoid children but are useless for culturally disadvantaged children. Although the most widely used intelligence tests were originally standardized on middle-class Caucasoid children, the tests were extended to and cross-validated on children and adults representing almost the complete range of cultural backgrounds.

Some "culture-fair" tests have been put together ad hoc that contain informational items only and have not been validated as tests of intelligence. When group differences in test performance have been reduced or abolished by the use of faulty criteria, the results are accepted by some critics of standardized intelligence tests. Although no test of intelligence is completely "culture-free," those "culture-fair" tests that have been developed by accepted methods of test construction and standardization show significant average differences between Caucasoids and Negroids. Furthermore, the statistical regression lines are remarkably alike (Stanley, 1971); that means that the tests have about the same validity for predicting the educational performances of the two races. For more evidence, see Jensen (1973).

It is commonly claimed by environmentalists that if "true" intelligence could be measured directly, then the average differences in IQ between ethnic, racial, and social groups would disappear. Random errors of measurement do not affect the average (mean) scores of groups. If the validity of IQ measurement were increased, it is possible that the mean group differences in IQ would increase rather than decrease. There is no logical argument or concatenation of evidence that compels a conclusion in this regard.

3. *The fallacy of grouping dissimilars.* Grouping Negroids, certain Pacific and Mediterranean peoples, Bushmen and Hottentots, and Indo-Dravidians all together as "Negroids" or "Coloreds" ignores their distinctive racial origins.

4. *The fallacy of treating categories that merge as independent.* A common related error is regarding people called "Caucasoid" and people called "Negroid" as completely independent races. Each is of mixed origins. There are no "pure" races. Many mulattoes in the United States and elsewhere have some Caucasoid ancestors but are called Negroids even when they have more Caucasoid than Negroid ancestors. A reverse error is applied to predominantly Negroid mulattoes who are called "whites" by the "blacks" majority in Haiti. It is fallacious to regard sex as a dichotomous concept: Physiologists and psychologists have identified several gradations of sexuality based on anatomical, hormonal, and behavioral evidence.

5. *The fallacy of treating categories that merge as identical.* This is the converse of the above fallacy. It is sometimes argued that because racial groups are of mixed origins, they must be regarded as identical. Although few biological characteristics are associated with but one race, there are important average differences in the gene pools of racial groups (Baker, 1974).

6. *The fallacy of assuming that the interaction of processes must be unidirectional.* Does intelligence determine the kind of environment that a person seeks out and helps to shape, or does the environment mold the level of intelligence? The fallacious assumption that when two processes interact, one must be solely the cause and the other solely the effect has been used to support each view. Each affects the other and to different extents, depending on the person and the environment. Some forms of mental retardation can be caused by injury of environmental origin, such as a blow to the head or a severe protein deficiency during infancy. Regardless of the cause, the mental retardate cannot build a good environment without aid. Inability to perform well on standardized tests can be caused by severe social deprivation. There are other clear examples of both genetic and environmental factors affecting IQ. It seems probable that there is a complex interplay of many factors, some environmental and some genetic, and that the relative effects of each can vary among individuals and groups.

7. *The fallacy of absolute priority.* It is assumed that there must always be the same first cause for every effect. The principle of equifinality—different pathways to the same general result—applies to psychological and sociological events and processes as well as to physical and biological events and processes.

8. *The fallacy of negative proof.* It is concluded that a hypothesis must be true because there is no proof that it is not true. Such arguments in more subtle terms are found among both hereditarians and environmentalists.

9. *The fallacy of the excluded middle.* This fallacy does not allow a position of middle ground, of neutrality, or of indecision. I have had Negroid students say to me, "You are either for us or against us. You either believe that blacks are equal to whites, or you believe that blacks are inferior." The fallacy also excludes the possibility that a statement may be true in one set of circumstances but not in others. It takes several forms and is sometimes called the *fallacy of insufficient alternatives.*

Arguments over the cause of sex differences in vocational interests and success sometimes focus on biological bases or on social pressures, each to the exclusion of the other. The hypothesis that culture can reinforce biologically based tendencies and abilities and that social pressures can effect biological development—especially muscle strength and neuromuscular skills—is excluded.

10. *The fallacy of exclusive particularity.* This is believing that truth in one situation is truth in all other situations. The idea has come into fashion that because grade schools are useful in educating average children, they should be equally useful in educating mental retardates.

As I mentioned earlier, the high heritability of a trait within one population is sometimes taken as evidence that the heritability of that trait will be high in a different population. It is plausible to expect them to be similar, but the argument is not secure.

11. *The fallacy of confusing a proposition with its converse.* It is reasoned that some people of African origin are Negroid, so the word "Negroid" identifies people of African origin. Other racial groups have dark skins but are genetically distinct (e.g., Indo-Dravidians).

12. *The fallacy of argument* ad novitam. It is common to assume that the "last word" is either correct or is more likely to be correct than is a hypothesis that was developed earlier. Thus, M. J. Herskovits (1961) scoffed at the century-old hypothesis that Negroid–Caucasoid differences in average IQ have a genetic basis on the grounds that the hypothesis is outdated and that anthropologists have become bored by the debate.

13. *The fallacy of argument* ad antiquitam. This, the converse of the above fallacy, is not now in fashion, although it is sometimes said, "Our ancestors knew that Negroes are less intelligent than whites."

14. *The fallacy of argument* ad crumenam. Money and material gains

are sometimes used as criteria of success in life, the value of an education, and other social programs. There are other criteria of equal validity, such as the ability to solve problems in the first case or personal enlightenment in the second. It is fallacious to believe that a single criterion exists.

15. *The fallacy of neglecting negative instances.* The proponent of a view on any subject may fail to examine all relevant evidence and to neglect data that fail to support his position. This fallacy sometimes takes the form of withholding from publication those data that fail to support a cherished hypothesis. Some scientific journals have an editorial policy of rejecting manuscripts that report negative results. A review of published evidence is not necessarily a representative sample of all research results.

FALLACIES THAT BEG THE QUESTION *(PETITIO PRINCIPII)*

1. *The fallacy of* circulus in probando *(circular reasoning).* One often finds an element of circularity in psychometric test development and validation because individual test items must tend to agree with the whole test, and also new tests are commonly validated against established tests. The choice of a test is sometimes determined by how well it supports an intuitive judgment as to what the results should be. Protagonists of the view that racism is the cause of racial differences in IQ take data on racial differences in IQ as proof of racism. This is also a form of post hoc reasoning.

It is often argued that education deals solely with environmental variables. Hence, when individuals and groups differ in school achievement, it is concluded that the cause and cure must each be environmental. If the premise is wrong, and if genetic factors affect school achievement, it follows that achievements can be optimized only by attending to individual differences, some having a genetic basis.

2. *The* ad hominem *fallacy.* Until a few years ago, anyone who worked for civil rights and/or equal opportunities was likely to be attacked by some form of character assassination. The same tactics are now used in opposing anyone who claims that heredity is important in human affairs. Some of these attacks are highly organized and are linked to social and political ideologies. Psychologists Arthur R. Jensen (1972) and Richard J. Herrnstein (1973) have given accounts of the pressures to keep them from teaching, speaking, and writing on this topic. Some of the more militant feminists make *ad hominem* attacks on those who believe that there may be biological bases of differences in sex roles, and they ridicule other women who prefer traditional sex roles in society.

3. *The* tu quoque *fallacy.* This is to respond to a criticism by saying, in effect, "You're another." The argument takes the form: *(a)* "Your data are wrong;" *(b)* "So are yours;" and so on.

4. *The* ad populum *fallacy.* A belief may be accepted because it is popular rather than because the supporting evidence has been examined and found to be strong.

5. *The* ad verecundiam *fallacy.* A proposition is accepted because it is supported by authority rather than by strong evidence. In the United States, some organizations, even scientific societies, have issued authoritarian statements, and/or the results of polls and votes by members, to the effect that the question being debated is "settled" and that certain conclusions are not to be questioned. The claim has been made that only members of certain disciplines—such as social anthropology—are qualified to study and make judgments on the origin of racial differences and that relevant data should be kept out of the hands of scientists in other fields.

6. *The fallacy of confusing the origin of an idea with its validity.* This is sometimes called the *genetic fallacy,* the word "genetic" referring to "origin" rather than to "biological heredity." It has been argued that environmentalists' views are invalid because they were developed in a communist society.

7. *The poisoned-well fallacy.* This is to completely discount the data and arguments of anyone who has made one or more mistakes, or to allege "guilt by association" with persons or fragments of ideas. One form of the fallacy is to cry "Hitler" whenever it is suggested that there may be a genetic basis of some social problem. Often the women's rights movement is derogated by an appeal to the apparent lack of femininity of some of its more militant leaders.

8. *The furtive fallacy.* Actions and ideas are regarded as representing a conspiracy or evil plotting. Intelligence tests are sometimes alleged to be devices used by the Establishment to exclude Negroids from the mainstream of America, and eugenic proposals are believed to be aimed at achieving Negroid genocide.

9. *The* argumentum ad ignorantium *fallacy.* During an argument, one disputant may say to the other, "The burden of proof is on you." There are no rules in science or logic that place the burden of proof on one side rather than the other. It is true, however, that the more implausible the claim, the greater the need for strong evidence that a debate should be opened.

10. *The fallacy of retreating behind untestable hypotheses.* When one hypothesis has been shown to be untenable, its proponent may move from one ad hoc hypothesis to another and may show a preference for untestable hypotheses. We should have no strictures on speculation. There are many fields of inquiry having no strongly supported expla-

nations for natural phenomena. It is not unreasonable to allege that subtleties and complexities are important. The fallacy is to accept an untestable hypothesis as true, or as a favored substitute for a plausible hypothesis that is strongly accepted.

11. *The fallacy of sophistic refutations.* To pooh-pooh an idea as a "myth," to exaggerate what someone has asserted, to attack a "strawman," and to allege what has not been denied are all common forms of specious argumentation.

12. *The fallacy of "explaining" by appealing to the unknown.* This is sometimes called the *passe-partout fallacy.* For example, the Negroid–Caucasoid gap in average IQs is explained away as being due to factors of which we know nothing. It implies that all current explanations are false.

13. *The fallacy of changing arguments in response to pressure.* Some scientists are careful to avoid expressing any view on racial or sex differences that might bring social disapproval, handicap their professional advancement, or threaten their safety.

14. *The fallacy of two-valued reasoning.* Different rights and privileges for different ethnic groups have been common in the past. In recent times, a form of reverse racism has become evident. A classic example of two-valued reasoning was that of maintaining different moral standards for men and women.

15. *The prodigious fallacy.* Exciting happenings and claims are judged to be most important. Many crucial points at issue in debates on racial comparisons and sex differences are commonly omitted in news stories.

16. *The fallacy of argument* ad consequentiam. An argument is accepted or rejected on the basis of expected or predicted consequences of its acceptance. Some people urge that possible genetic bases of racial differences not be researched because of the risk of creating mischief.

17. *The fallacy of blaming the messenger for the message.* Anyone who discovers unwelcome evidence and reports it risks disapproval and attacks on one's aims and reputation.

18. *The fallacy of obscurantism.* It is argued that because no race is homogeneous and because the word "race" cannot be clearly defined, the word should be dropped from our vocabulary. Therefore, problems relating to the biology of race do not exist and should not be debated or investigated. Those who oppose the study of biological differences among races do not hesitate to study environmental causes of racial problems, or to recommend social actions based on racial identity rather than individuality. It is argued that "race" is a social concept, but even if this conclusion is accepted, there is no logical reason why groups defined in social terms cannot be studied for biological correlates of social differences. Since there is some degree of error in assigning racial membership to individuals, it seems possible that the role of heredity in causing racial differences may have been underestimated. This would be true unless

the errors of identification are selective, such as identifying dull Caucasoids as "Negroids" and bright Negroids as "Caucasoids." A similar fallacy would apply to labeling awkward boys as "feminine" and coordinated girls as "masculine."

Some writers assert that since "intelligence" cannot be defined to the satisfaction of everyone, it should not be studied. It is also difficult to adequately define "cancer," "gravity," "electricity," and numerous other concepts in scientific research.

The following question can be put to those who try to obscure research on the Negroid–Caucasoid IQ gap and at the same time claim that it has been proved that all races are equally endowed with the biological bases of intelligence: "If 'true' intelligence cannot be measured, then explain how it has been proved that races are equally endowed with the biological bases of intelligence?" Evidently, the proposition has been *assumed*, not demonstrated empirically, by the obscurantists.

DISCUSSION

Most of the data and ideas that have been used in the debate over the relative importance of heredity and environment in causing individual and group differences are available to anyone willing to search the literature. Little if any of the evidence is private knowledge. Some of the relevant evidence is indirect, and many of the inferences are insecure. Why, then, is so much emotion generated in these debates? Why are they not carried out in better humor? I mentioned in the introduction that self-esteem appears to relate more to heredity than to environment. The concept that people are equal born has great appeal. There must be additional factors affecting attitudes toward the issue. It may have occurred to others, as it has to me, that many individuals regard this as a moral issue and are able to ameliorate guilt feelings over racial, sex, and class discrimination by identifying with the side of the angels, regarding their opponents as devils and seeking to exorcise them from society. The cant seems ritualistic, commonly parroted by individuals having no contact with the original literature. At other times, I am only bewildered by the debate, which seems unreal and foreign to our cherished concepts of the methods and freedoms of science.

Peter Urbach's insightful paper, "Progress and Degeneration in the IQ Debate" (1974), concludes that the quality of evidence and argument of those who believe that there may be a genetic basis of intelligence has improved. I agree. The design of experiments, statistical methods, and quality of reasoning have all improved. Although the role of heredity in the average Negroid–Caucasoid gap and in school achievement has not been fully established, there is a remarkable concatenation of supporting evidence. The strongest support for biological bases of intelligence comes from studies of MZ and DZ twins reared together and apart. Is the evidence spurious to some extent because of a genotype–environment correlation? Do the environments of MZ twins tend to be more similar than for DZ twins? Were the environments of separated Caucasoid twins as far apart as those of the average Negroid and the average Caucasoid? Do Negroids experience IQ-limiting factors not found within the range of environments to which Caucasoids are exposed? Unidentified variables are involved in many natural processes. It is reasonable and necessary to examine this possibility. The error is to conclude that because the above questions have not been answered, there cannot be a genetic basis for Negroid–Caucasoid differences in IQ and school achievement.

Urbach (1974) concludes that the position of the environmentalists has degenerated. I do not agree. The best of research and logic supporting this position is superior to the best offered by environmentalists a few years ago. I do not believe that this aspect of the debate should be judged by association with ad hominem appeals, fallacious reasoning, obscurantism, or attempts to block research. However, if environment is all important in causing the Negroid–Caucasoid gap in IQ and in school achievement, then it should be possible to close the gap by manipulation of environmental variables. Efforts to do so have almost always failed (Jensen, 1972). The few claims of success have not yet become repeatable by others on demand, a requirement for proof. Each of these studies has been criticized as poorly controlled.

Many individuals and a number of groups of scientists and other members of the academic community have demanded that research on possible genetic bases of social problems be stopped. I am concerned by such efforts to constrain research, but there are bright spots in recent publications. Baker (1974), Jensen (1973), and Loehlin, Lindzey, and Spuhler (1975) have reviewed the evidence for and against the hypothesis that the genetic potential for intelligence is distributed equally among all racial groups. Jensen has also reviewed the methods that can be used to study this problem.

What can be said in reply to Lewontin's (1970a, b, p. 25) question, "But suppose the difference between the black and white IQ distributions were completely genetic: What programs for social action flow from that fact?" I suggest the following:

1. Negroids as well as Caucasoids would be encouraged to practice selective population control (Ingle, 1973) so that those who are more intelligent, self-sufficient, and best qualified for parenthood would carry the greater part of the reproductive load. Some authorities believe that Negroids in the United States may be drifting toward genetic enslavement. If this be true, it could become the greatest of all barriers to Ne-

groid advancement. Is it an act of friendship to Negroids to block research on this problem?

2. Education would be reoriented toward the abilities and interests of individuals rather than carrying all children through the same methods of education, frequently by unearned promotions. The Negroid–Caucasoid gap in school achievement has not been appreciably narrowed by efforts to treat all children alike without regard for abilities and interests.

3. Our society would stop telling Negroids that all of their problems are caused by racism, a dogma that has fostered racial hatred, social malignancy, and has unjustly blamed many fair-minded and competent teachers for the ethnic achievement gap in our schools.

4. There should be a return to the ideal of treating individuals according to abilities, interests, drives, assumption of duties, and behavioral standards unless it can be shown that preferential treatment of Negroids brings greater benefits than harm.

What of the argument that evidence for a genetic basis of some problems of American Negroids would be used to make mischief? Up to 10 years ago, I knew of individuals who would have used such information to support their efforts to reestablish forced segregation of schools and housing and to deprive Negroids of equal rights and opportunities. I suppose that some such still exist, but they have become ineffective. I do not know of any scientists among them. Some racists prefer the false dogma that all Negroids are inferior to all Caucasoids rather than the recognition of overlap and the acceptance of the principle that all persons should be treated according to individuality. I agree with Jensen (1972) that linking the nature–nurture debate with the issue of racial desegregation and discrimination is a non sequitur. It is my opinion that far more good than harm will come from sound knowledge about the reasons that some persons are disadvantaged.

I have given examples of specious reasoning about the causes of differences in interests, roles, and achievements of men and women. There is a great deal of information on physiological and psychological differences associated with sex. It is not difficult to design salient studies on the relative importance of biological and social causes of these differences, but such research is not a high-priority topic. Although I have some ideas about the causes of individual differences relating to sex, they are neither original nor self-certifying.

More serious barriers exist to the study of aging. Developmental biology and developmental psychology have evolved sophisticated forms of research on growth and maturation, but the study of aging has not progressed very far beyond describing the signs and effects of these processes. There are penetrating investigations of the biochemistry and biophysics of aging in microorganisms and some short-lived animals. Research on man and most laboratory animals is complicated by various diseases that cause debilitation and death before the organism dies of old age. Longitudinal studies of aging require keeping the subjects under controlled conditions from birth until death. Neither scientists nor granting agencies are willing to wait for decades to learn the outcomes of such studies. Much of gerontology is concerned with the diseases of old age and with palliation of the effects of senescence. There are wide individual and group differences in rates of development, maturation, and aging. These and all other parameters of individuality are of key importance in human affairs.

SUMMARY

The relative importance of heredity and environment in causing individual differences and average group differences in human abilities remains uncertain. Measurements of psychological traits are complex; some environmental effects on the phenotypes may not have been identified or cannot be measured directly. The genes involved in psychological traits have not been mapped or otherwise identified. Only in the case of MZ twins can the genetic endowments of any two individuals be assumed to be identical. It is not known that the environments of any two individuals are identical; they can be judged to be similar or to show different degrees of dissimilarity.

As an outcome of these uncertainties, the estimates of heritability of different traits and abilities differ over a wide range. Many views have become strongly polarized and are emotionally charged. Most of the conclusions are insecure because of uncertainties inherent in identification and measurement of causal factors. Fallacies in reasoning can be found abundantly in arguments about the causes of human differences. Some of them are absolute errors in logic, many beg the question, and others are weak inferences; even strong inferences having some claim to validity do not exclude alternative hypotheses. This chapter reviews the kinds of fallacies that have appeared in these debates.

It is the opinion of this author that sophisticated research and arguments can be found on all sides of the debates but that the strongest lines of evidence and logic support the conclusion that heredity is of major importance in determining individual and group differences in human abilities.

REFERENCES

Baker, J. R. *Race*. New York: Oxford Univ. Press, 1974.

Coleman, J. S. *et al. Equality of educational opportunity*. Washington: U.S. Department of Health, Education, and Welfare, 1966.

Duncan, O. D., Featherman, D. L., & Duncan, B. *Socioeconomic background and occupational achievement: Extensions of a basic model.* Final Report, Project No. 5-0074 (EO-191). Washington: U.S. Department of Health, Education, and Welfare, Office of Education, Bureau of Research, May 1968.

George, W. C. *The biology of the race problem.* New York: National Putnam Letters Committee, 1962.

Herrnstein, R. J. *I.Q. in the meritocracy.* Boston: Little, Brown, 1973.

Herskovits, M. J. Rear-guard action. *Perspectives in Biology and Medicine,* 1961, 5, 122–128.

Hutt, C. *Males and females.* Harmondsworth, Eng.: Penguin Books, 1972.

Ingle, D. J. *Who should have children?* Indianapolis: Bobbs-Merrill, 1973.

Jensen, A. R. *Genetics and education.* New York: Harper & Row, 1972.

Jensen, A. R. *Educability and group differences.* New York: Harper & Row, 1973.

Klineberg, O. *Negro intelligence and selective migration.* New York: Columbia Univ. Press, 1935.

Klineberg, O. *Characteristics of the American Negro.* New York: Harper, 1944.

Lewontin, R. C. Race and intelligence. *Bulletin of the Atomic Scientists,* 1970, *26,* 2–8. (a)

Lewontin, R. C. Further remarks on race and the genetics of intelligence. *Bulletin of the Atomic Scientists,* 1970, *26,* 23–26. (b)

Loehlin, J. C., Lindzey, G., & Spuhler, J. N. *Race differences in intelligence.* San Francisco: Freeman, 1975.

Montagu, M. The new ideology of race. *Midway,* 1969, *10* (Autumn), 87–100.

Stanley, J. C. Predicting college success of the educationally disadvantaged. *Science,* 1971, *171,* 640–647.

Urbach, P. Progress and degeneration in the IQ debate. *British Journal of the Philosophy of Science,* 1974, *25,* 99–135.

Weyl, N. Letter to the editor. *Perspectives in Biology and Medicine,* 1969, *13,* 122–124.

3

Genes and Melting Pots

DAVID C. RIFE

Ohio State University

INTRODUCTION

Popular beliefs concerning the results of racial mixture may appropriately be termed the myth of the melting pot. According to this concept, extensive intermarriage between members of different racial groups will eventually result in a uniform blend, analogous to a thorough mixing of paints of different colors, the end product depending on the proportions of each component.

The melting-pot concept is true only insofar as it may be applied to environmentally conditioned traits. It does not apply to biological traits. These are determined by genes, discrete units of heredity that maintain their individuality from one generation to another. The relations and effects of mixing genes from different populations may be compared to mixing marbles of different colors. Suppose we have two bowls, one containing 100 red and the other containing 100 blue marbles. The marbles from both bowls are dropped into another bowl. This would result in a mixture consisting of equal numbers of red and blue marbles. If they should be thoroughly mixed, and a blindfolded person then places them in small bags, two in each, he ends up with 100 bags. Approximately half would contain one red and one blue, one-fourth would contain two red, and one-fourth should contain two blue. If each were reemptied into the bowl and the mixing and bagging process repeated, he should again obtain an approximate ratio of one-half containing one red and one blue, one-fourth containing two red, and one-fourth containing two blue. This process could be repeated indefinitely with the same results.

The results of hybridization between human populations are comparable to the mixing of marbles. Let us suppose two islands. Everyone on one is blood type M, and everyone on the other is type N. People of type M are homozygous, having received gene *m* from each parent. People of type N are also homozygous, while those of type MN are heterozygous, having received gene *m* from one parent and *n* from the other. Neither gene is dominant to the other. Genes *m* and *n* are alleles, that is to say, they occur at the same chromosome locus. Thus, every individual has either two *m*, two *n*, or one of each. Every gamete carries either *m* or *n*.

Suppose several people migrate from the first island and intermarry with inhabitants of the second. Their offspring will receive one gene from each pair of alleles from each parent; thus, all will be of type MN. Their offspring, however, will segregate in an approximate ratio of 1 M : 2 MN : 1N. This ratio will remain constant from one generation to another as long as random intermarriage continues among the descendants of the original migrants. One generation of random mating results in genotypic ratios for each set of allelic genes that will remain constant from one generation to another as long as random mating occurs. Such a situation is known as *genetic equilibrium* (Hardy, 1908; Weinberg, 1908).

Racial hybridization is usually a continuous process. Such populations arise from gradual intermixture covering many generations, among which American Negroids and Amerinds are familiar examples. Interbreeding of Amerinds and Caucasoids has been continuous since the discovery of America, and between Negroids and Caucasoids since the introduction of slavery in the sixteenth century.

Populations of hybrid origin are characterized by greater genetic variability than are those of nonhybrid origin. This is especially true of those into which there has been a continuous flow from ancestral strains or in which the process of hybridization is continuous. Intermixture results in a redistribution of the variability. It tends to eliminate sharp contrasts between parental populations and to increase individual genetic variability. It does not result in a biologically uniform product.

Insofar as members of hybrid populations live under somewhat similar conditions, variations due to environment tend to be reduced. From an environmental standpoint, the trend is toward a uniform product, or a true blend. In other words, populations of hybrid origin provide situations in which genetic traits are brought more sharply into focus.

Evidence that individual variations in a particular trait are hereditary indicates that the parent populations will vary from one another with respect to this same trait. Individual differences are usually all or none with respect to simply inherited traits, whereas those between populations are quantitative. For example, everyone belongs to one of the four ABO blood groups. The great majority of populations throughout the world possess all four groups, but in widely differing proportions. Similarity in blood group frequencies is an important criterion of ethnic origins and relationship. This same principle holds true for other genetic traits.

Maximum genetic variation between populations occurs when the latter derive from different racial groups. It is axiomatic that evidence of individual genetic variations constitutes evidence that populations differ with respect to their incidence.

GENETIC MARKERS

Irrefutable evidence for the inheritance of normal traits in human adaptability, behavior, and intelligence is usually very difficult to obtain. There are several reasons why this is so. It is not feasible to make experimental matings in man, unlike the situation with other organisms, in order to evaluate the roles of heredity. Studies must be made of families and populations as they exist by means of gene frequency analysis and other specialized techniques. Although environmental factors obviously do play important roles in the synthesis of such traits, the possible roles of heredity are frequently masked by environmental factors, a problem not so likely to be encountered in other organisms. As a consequence, standard tests sometimes lack high validity, especially when employed for comparisons between populations having widely different backgrounds. In cases of rare and striking abnormalities, it is sometimes possible to identify genetic factors because of their association with chromosomal abnormalities. But here we are concerned with normal variations in the general population and among different populations. Hence, chromosomal aberrations cannot be expected to be of great assistance.

The conventional method of attacking the problem of inheritance of complex normal traits is through twin research. This consists of comparisons of degrees of intrapair similarity between identical and fraternal twins. The terms *monozygotic* (one-egg) and *dizygotic* (two-egg) are employed to categorize the two types of twins, as they accurately describe their modes of origin. Monozygotic twins have identical genotypes (with the exception of mutations that may have occurred in one member subsequent to embryonic separation), whereas dizygotic twins have the same degrees of genetic similarity as do ordinary siblings.

Intrapair differences in monozygotic twins must be due to environment, whereas those in dizygotic twins can be attributed to both heredity and environment. Greater intrapair differences in dizygotic than in monozygotic twins may logically be attributed to heredity. Estimates of the effects of environment can be made by comparisons of intrapair differences among monozygotic twins reared together with those reared apart. Greater differences among pairs reared apart may be attributed to environment.

The twin method of research has been employed in investigations of the heritabilities of numerous traits. The pioneer studies of Newman, Freeman, and Holzinger (1937) on the heritability of intelligence (as measured by Binet IQ) are well known. Comparisons of 50 pairs of monozygotic twins reared together, 50 pairs of dizygotics reared together, and 19 pairs of monozygotics reared apart indicated that both heredity and environment are important in bringing about variations in IQ. Subsequent studies, analyzed by modern biometric techniques, have amplified these findings (Jinks & Fulker, 1970).

Comparisons of IQ differences between parents and natural children, as contrasted with those between parents and foster children, give higher correlations between parents and natural children, thus adding support to the evidence obtained from twin studies that heredity is of considerable importance in bringing about variations in IQ (Munsinger, 1975).

Numerous comparisons between racial groups with respect to IQ and other abilities show significant differences, thus adding another link to the convincing chain of evidence that heredity is of outstanding importance in the determination of mental capacity. Beyond these broad generalizations, we know little concerning genetic components of intelligence, and as yet nothing concerning number and kinds of genes involved. Recent work has established a relationship between dermatoglyphic patterns and various chromosomal and metabolic abnormalities that also have behavioral consequences (Schaumann & Alter, 1976).

A gene responsible for a simply inherited variation, and that segregates independently from other simply inherited variations, is sometimes referred to as a genetic marker. Here the term genetic marker includes not only specific marker genes but also highly heritable quantitative variations. Genetic markers are employed in tests for associations with normal variations of unknown etiology.

Scores of simply inherited qualitative variations are known in man. These include over two dozen series of blood groups as well as numerous hemoglobin, biochemical, and morphological traits. Tests for associations between qualitative genetic markers and traits whose etiologies are not understood can usually be made without great difficulty. But in cases in which quantitative variations are being tested with simply inherited genetic markers, negative results are likely because, if only one of the markers is associated with the quantitative variation, the association may be too small to be of statistical significance. The obvious way out of this difficulty is to include quantitative genetic markers in the test battery.

Anthropometric variations are among those available for consideration as quantitative genetic markers. These include among others skeletal dimensions, stature, cephalic index, somatotype, pigmentation, features, ear shape, body hair, hair form, and dentition. Ideal genetic markers should have high heritabilities and be of minimal importance from the standpoint of selective mating. Dermatoglyphics fulfill the requirements for quantitative genetic markers. They have high heritabilities and are of no importance from the standpoint of selective mating. Postnatal factors do not alter their configurations. Although scars, calluses, and warts show on prints, the types of configurations and numbers of ridges remain constant. This was first demonstrated by Sir Francis Galton (1892). He compared two sets of prints from each of 15 persons, many years intervening. Later prints of one individual were compared after a 70-year interval.

Dermal ridges on palms are homologous with those on fingers, which are so useful as identification criteria. These configurations are fully established during the first half of fetal development and are not altered thereafter except in size during growth (Cummins & Midlo, 1943). Thus, dermatoglyphics offer several unique advantages as quantitative genetic markers.

All fingerprints fall into three categories: arches, loops, and whorls. Ridges in arches bow toward finger tips and usually go clear across the finger. Loops are patterns in which some of the ridges recurve, emerging on the same side on which they enter. There is always a point in loops from which ridges go in three directions. This is termed a triradius. Whorls are patterns in which some of the ridges form circular or spiral configurations. These always have two and occasionally three triradii. There are several subclasses of arches, loops, and whorls (Figure 3.1).

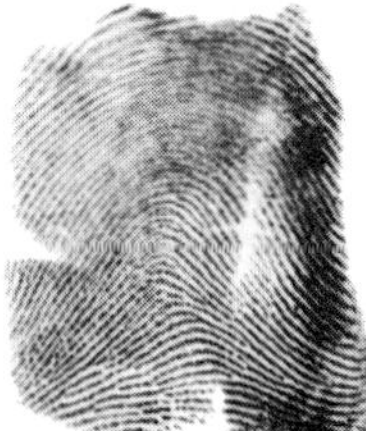

Figure 3.1 From left to right: whorl, loop, and arch. Loop shows line from triradius to core, for ridge counting.

The total number of triradii on the 10 fingers is termed the fingerprint index, or FPI. If a person has arches on all fingers, his FPI is 0; if he has 10 loops, it is 10; and if he has 10 whorls, it is at least 20.

Ridge counts are made by counting the number of ridges transversing a straight line drawn from the triradius to the core or innermost portion of the pattern. Whorl ridge counts customarily include only the one on the side having the largest number. Arches have no triradii; thus, their ridge count is always zero. Ridge counts are a measure of pattern size. Whorls usually have higher counts than loops, whereas arches are patternless configurations. Ridge counts are positively correlated with FPI.

Whorls, loops, and arches do not occur at random. Whorls occur most frequently on ring fingers and thumbs. They also occur more frequently on right than on left fingers, and more frequently in males than in females. Arches occur most frequently on middle and index fingers, and more frequently in females than in males.

Loops and whorls may occur on five different areas of the palm. A triradius is usually present below each of the four fingers. Lines drawn tracing ridges from the triradii to the palmar margin are useful as indicators of transverseness of ridges, which vary from transverse to oblique. Loops or whorls may occur between any two of the triradii in the distal palmar areas. They may also occur below the thumb and on the side of the palm opposite the thumb. Bilateral asymmetry is more pronounced on palms than on fingers. Palmar ridges are usually more transverse on right than on left palms (Figure 3.2).

Although no two hand prints are identical (Newman, 1930), those of monozygotic twins are remarkably similar. In fact, prints of the two right hands or of the two left hands of a pair are usually more alike than those of the two hands of each member (MacArthur, 1938). Hand prints were important criteria in determining that the Dionne quintuplets arose

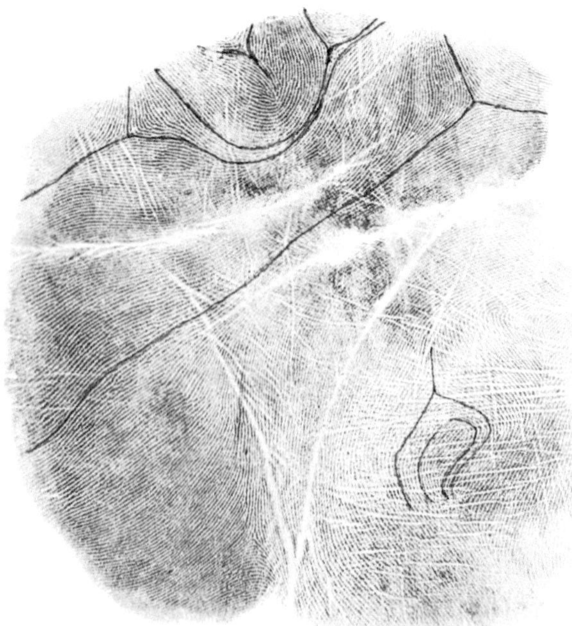

Figure 3.2 Print of a left palm, outlining triradius and loop on thenar/first interdigital area, triradii below each of the four fingers, and direction of palmar ridges. Accessory triradii sometimes occur in interdigital areas in the distal palmar region.

from a single embryo. Total ridge counts for each member of the set ranged from 99 to 102, whereas their three other siblings ranged from 78 to 139. Yet the differences between right and left fingers of each member of the set ranged between 14 and 1, an average of 7 ridges (MacArthur & Ford, 1937).

Newman, Freeman, and Holzinger, (1937) compared the ridge counts among 50 pairs of monozygotic twins with those of 50 pairs of dizygotic twins. Intrapair correlations were approximately .50 among dizygotics and .95 among monozygotics, indicating that 90% of intrapair variation in dizygotic twins is due to heredity. Sarah Holt (1957) obtained a similar estimate of heritability from the prints of 80 pairs of monozygotic and 92 pairs of like-sexed dizygotic twins. This is the highest heritability recorded in man for normal quantitative traits. Like other quantitative variables, the number of genes involved is unknown. Types of patterns and transverseness of palmar ridges are also heritable but to lesser degrees than ridge counts (Van Valen, 1963).

Fingerprint indices vary greatly from one part of the world to another, depending on racial origins (Rife, 1953, 1954c). North American Caucasoids have approximately 27% whorls, 68% loops, and 5% arches on fingers, an FPI of 12.2. This is similar to the indices of European Caucasoids. American and African Negroids have somewhat similar indices, ranging between 11.5 and 12.5. But Amerinds usually have higher indices, those reported for seven tribes ranging from 14 to 15.5. North Carolina Cherokees differ from these in having an index of 12.6. It should be noted that these Cherokees have approximately 40% Caucasoid ancestry, largely from the British Isles (Rife, 1971a, 1972).

The FPIs of American Jews are similar to those of Middle Easterners, averaging around 13.5. Sachs and Bat-Miriam (1957) analyzed the fingerprints of immigrants to Israel from 10 countries, which included North African, Yemenite, Sephardic, and Ashkenazi Jews. FPI averages for the 10 countries ranged between 13.3 and 13.9, an insignificant difference. Asiatic populations have higher indices than Europeans, ranging from approximately 13.5 among Middle Easterners to approximately 15 among Chinese and Japanese.

Australoids in eastern Arnhem Land have the highest indices on record, averaging 17.5 (Cummins & Stetzler, 1951). They have whorls on over 70% and arches on less than 1% of fingers. At the other extreme, Efe Pygmies have an average index of 9.7 and South African Bushmen an average almost as low, 9.9%. These are the only populations on record among which arches occur more frequently than whorls (Cummins, 1955). Although both Efe Pygmies and South African Bushmen are short, their low indices cannot be regarded as reliable criteria of stature. Negritos in New Guinea have an average FPI of 15.7, a relatively high one (Geipel, 1958).

Although Jewish immigrants to Israel have strikingly similar FPIs, significant differences have been found in their blood group distributions. Blood groups are simply inherited and, like other simply inherited variations, genetic drift (random fluctuations in distributions) can occur in small populations over several generations. This is not so likely in quantitative variations, such as fingerprint configurations. A shift in frequencies of particular genes in one direction may be counterbalanced by a shift in the opposite direction by other genes affecting the same trait. In other words, quantitative variations tend to be more stable than simply inherited ones within small populations over long periods of time.

The validity of estimates of genetic relationship between different populations depends to a large extent on the number of genetic markers employed. Hand prints include variations due to several genes, some of which appear to act independently. The information conveyed on a good set of prints is equivalent to that obtained from several blood groups or other simply inherited genetic markers. This is not intended to imply that simply inherited markers are not highly important, but that information from all types of genetic markers is mutually supplementary. For example, FPI and palmar configurations of Amerinds are more similar to those of Mongoloids than to those of either Caucasoids or Negroids. This supports evidence that the ancestors of Amerinds migrated from Asia to America when a land bridge existed over Bering Strait. Yet the ABO blood group distributions of Amerinds are quite different from those of Mongoloids. The ancestors of the Amerinds presumably came to America in small groups and settled in widely separated areas. Genetic drift was likely an important factor in bringing about differences between present-day Amerinds and East Asian Mongoloids.

The story is the same with the criteria used to investigate relationships between Jewish migrants from all parts of the world to Israel. In neither situation did blood group data suggest similar origins, whereas in both situations similar origins were indicated on the basis of dermatoglyphics.

Evidence based solely on fingerprints suggests that Negroids and Caucasoids are more closely related to each other than they are to Amerinds and Mongoloids. But evidence from palm prints suggests Caucasoids and Negroids are more distantly related to each other than either to Mongoloids. Combined evidence from both fingers and palms indicates no close relationship between any of these racial groups.

PLEIOTROPY

The effects of a particular gene on two or more traits is known as pleiotropy. It results in an important type of association in which the traits in question are associated with a genetic marker, both in families and populations.

Biochemical

Phenylketonuria is a disease that is inherited as a simple recessive. Affected individuals excrete large amounts of phenylpyruvic acid in their urine, which in turn becomes bluish-green in color if a ferric chloride is added. Affected individuals have severe mental retardation. The phenylketonuric gene is pleiotropic in that it affects a chemical excretion and results in mental retardation (Lerner, 1968).

A sex-linked recessive gene results in a deficiency of the enzyme glucose-6-phosphate dehydrogenase, commonly referred to as G-6-pd. Affected individuals are sensitive to pollen of the broad bean, which brings on an anemia known as favism. They are also sensitive to the antimalarial drug primaquine and to ingestion of sulfanilamide and napthaline (Price & Clark, 1961).

Gout is an arthritic disease and was formerly considered to result from eating excessive amounts of rich food. It is now known to be caused by a dominant gene, with 80% penetrance in males and only 12% in females. The primary effect of this gene is to produce hyperuricemia, owing to an elevation of uric acid levels in the plasma. This, in turn, often causes renal lesions resulting in high blood pressure (Stern, 1973). There is some evidence to the effect that there is a positive association between hyperuricemia and intelligence, although more data are needed.

Morphological

There are several genetic syndromes that do not appear to be associated with chromosomal aberrations but result from pleiotropic effects of particular genes. Marfan's syndrome results in exceptionally long fingers (spider fingeredness), frequently accompanied by an abnormal position of the eye lens and by heart defects (McKusick, 1960). The Laurence–Moon–Biedl syndrome produces mental deficiency, obesity, polydactyly, and subnormal genitalia (Stern, 1973). Persons affected with the nail–patella syndrome have abnormal fingernails and toenails and defective kneecaps. Still another syndrome, van der Hoeve's, results in fragile bones, otosclerosis (type of deafness), and a thin, bluish sclera (outer wall) of the eye (Bell, 1928). Wardenburg's syndrome is characterized by deafness associated with anomalies in pigmentation of the eye. These syndromes are of particular interest to the medical profession, as they comprise constellations of abnormalities. Although associations are characteristic of each syndrome, they are not always complete. In other words, some pleiotropic effects are variable in expressivity.

Behavioral

There are over 20 million left-handed people in the United States. Yet, in spite of this, most standard equipment is made for right-handers. Chairs for left-handers are seldom found in school and college classrooms. Clothing zippers, strings and pistons on musical instruments, and design of many tools are common examples of conventional conformity to right-handed standards. Although standardization may be more economical, it is also a type of discrimination against the left-handed minority.

If handedness is solely the result of environment and training, it would seem reasonable to train all children to become right-handed. But if handedness is wholly or partially inherited, it is undemocratic to expect left-handers to conform to right-handed standards. Evidence from various sources indicate that handedness is determined by interactions of heredity and environment.

Many left-handed children are taught to write and to perform other unimanual operations with the right hand. But these transfers are much more difficult for some children than for others. Approximately 20% of monozygotic twins are mirror images in handedness; that is to say, one member of the pair is right-handed, and the other is left-handed. These observations provide positive evidence for environmental influence on handedness but do not rule out the possibility that heredity may also be important.

Data taken by Rife (1944, 1959) at Ohio State University over a period of 15 years revealed that approximately 10% of Caucasoid Protestant students were left-handed, whereas the frequency rose to almost 16% among Jewish students. These differences were consistent from year to year and were of high statistical significance. They included 1710 Protestant and 535 Jewish students, all Caucasoids.

There is a positive correlation between handedness of parents and offspring. When both parents are left-handed, about half of their children are left-handed; when only one parent is left-handed, about one child in six is left-handed; and when both parents are right-handed, approximately 1 child in 16 is left-handed. These findings definitely indicate a genetic basis for handedness.

Monozygotic twins possess identical genes, and intrapair differences must be the result of nongenetic factors. It has been found that monozygotic twins showing intrapair differences in handedness or pairs in which both are left-handed have four times as many left-handed relatives among parents and siblings as do those pairs in which both members are right-handed. This indicates that some people are strongly predisposed toward right- or left-handedness by genotype, whereas others are genetically ambidextrous, and their functional handedness depends largely on environmental influences. According to this hypothesis, intrapair differences occur only in those monozygotic twins who are not strongly predisposed toward right- or left-handedness by genotype. In other words, the effectiveness of training and environment on handedness varies from one individual to another, depending on genotype (Rife, 1940, 1950).

Irrefutable evidence of a biological basis for handedness has been obtained from finger and palmar dermatoglyphics. The association between handedness and dermatoglyphics is of a subtle nature and is apparent only when prints of large numbers of right- and left-handers are compared. Patterns occur in different frequencies among right- and left-handers. This holds true for various palmar areas and, to a lesser extent, for fingers (Cummins, 1940).

A pattern area located near the base of the thumb, known as thenar/first interdigital, is especially suitable for comparisons of pattern frequencies. Comparisons of prints of 3916 right-handers with those of 1556 left-handers have revealed higher pattern frequencies in this area among left-handers than among right-handers. The results were consistent throughout, the odds being 199,999 to 1 in favor of the differences being real rather than chance. These studies were conducted independently by various investigators (Bettmen, 1932; Cromwell & Rife, 1942; Lèche, 1933; Newman, 1934; Rife, 1955).

Further analysis of the distribution of these pattern increases on right and left palms was made by Rife (1951, 1955). Approximately 15% of unselected American Caucasoids have patterns on the thenar/first interdigital area of one or both palms. Among those having patterns on only one palm, they occurred on left palms in approximately 8.5% and on right palms in approximately 1% of right-handers. But among left-handers they occurred only on left palms in about 10% and only on rights in 3.5% of cases. The ratio of those having patterns only on lefts to those having them only on rights was 8.5:1 among right-handers, but was only 2.8:1 among left-handers. Differences between pattern frequencies on only right and on only left palms were reduced approximately 67% among left-handers. There was also slightly over 1% higher frequency of people having patterns on both palms among left-handers.

Similar comparisons were made on other palmar areas and fingers. Trends were the same, averaging 30% less difference in bilateral pattern frequencies among left- than among right-handers. Left-handers also manifested less bilateral difference in direction of palmar ridges than did right-handers.

In respect to sex and handedness, left-handed females manifest the least and right-handed males the most bilateral asymmetry, right-handed females and left-handed males being intermediate. Although there are individual exceptions to this rule, it holds true in varying degrees for the populations investigated. The association between handedness and degree of bilateral asymmetry in dermatoglyphics is another example of pleiotropy.

LINKAGE

Linked genes are those located on the same chromosome. Linkage results in two types of association between traits. One of these occurs within families and provides the basis for modern techniques employed in the mapping of human chromosomes. The other type of association occurs only within populations of hybrid origin and has unusual merit as a means for the detection of heritabilities heretofore unknown.

Chromosome Mapping

Within stable populations, linkage results in associations between genetic variations *within families* but in no such associations within the population as a whole. Thus, it can readily be distinguished from pleiotropy, which results in associations in entire populations as well as within families. Linkage results in opposite types of associations from one family to another, canceling associations within the population as a whole. Pleiotropy brings about the same types of association from one family to another and from one population to another.

The classic test for linkage is the conventional backcross. When two sets of alleles are under consideration, offspring should occur in a 1:1:1:1 ratio if they are not linked. If linked, two types of offspring will occur more frequently than the other two. Those occurring in more than 50% of offspring are termed noncrossovers and those occurring in less than 50% of offspring as crossovers. Crossover percentages range from 0 to almost 50, depending on how closely the gene loci are linked. The crossover percentages between any three linked loci enable the investigator to determine their order on the chromosome. Thus, if loci A and B have 15% crossovers, B and C have 25% crossovers, and A and C have 10% crossovers, we know that their order on the chromosome is BAC.

Simple backcross tests for linkage are relatively easy to perform in laboratory plants and animals. Matings may be made according to plan, intervals between generations are frequently periods of weeks or months, and numbers of offspring range from half a dozen or more in mice to hundreds in *Drosophila*. The situation is quite different in man. The interval between generations is usually not less, and frequently much more, than 20 years. Matings are not made for the purpose of studying linkage, so the investigator must test families as they exist. The most efficient test material is found in families in which one parent is heterozygous and the other homozygous for the traits under consideration and in which there are several children. But among individuals who are phenotypically dominant for two pairs of alleles, there are four different genotypes, only one of which is doubly heterozygous. The other three phenotypes include those homozygous for both and those homozygous for one pair and heterozygous for the other. Unless the investigator has prior knowledge of the parents or offspring of the parent with the dominant phenotype, he has no way of knowing whether they are of the desired genotype. For these and other reasons, progress in the mapping of human chromosomes has been slow as compared with that in many experimental organisms.

During the past 40 years, several mathematical techniques have evolved for testing human linkage, some of which enable the investigator to use data from a single generation. Chromosome mapping requires precise knowledge of the inheritance of the traits under consideration prior to testing for linkage. Aside from theoretical interest, linkage maps may prove to be a valuable asset to physicians and marriage counselors. If a particular abnormality or disease susceptibility is known to be linked with a simply inherited normal trait, and if the onset of the deficiency or disease does not occur until the subject is several years of age, prediction may be made on the basis of knowledge of the parents' phenotypes of the likelihood that a child will be affected. Steps can then be taken to prevent or ameliorate the condition.

Detection of Linkage in Racial Melting Pots

This is the type of linkage testing with which we are here particularly concerned. The associations between linked genes within populations of recent hybrid origin, or proverbial "melting pots," are of different types from those occurring within long-established populations in which mating occurs at random. These associations occur throughout the entire hybrid population and are of the same kind. They are found only within those hybrid populations in which certain requirements are met.

Hybridization between populations alters the frequencies of those genes in which the ancestral populations significantly differ. The number of generations required to regain genetic equilibrium in populations of mixed origin depends on the positional relationships of the genes under consideration. These are as follows: *(1)* allelic, *(2)* linked, and *(3)* located on different chromosomes (Robbins, 1918; Li, 1948).

In the nonallelic case, the rate of equilibration depends solely on the percentage of crossovers. If not linked, the deviation from equilibrium is reduced by half each successive generation; if linked, by the percentage of crossovers. The initial degree of deviation from equilibrium depends on the relative sizes of ancestral populations and the frequencies of the genes under consideration. Nonlinked genes attain approximate equilibrium after 8 generations, whereas over 100 generations would be required for genes with less than 1% crossovers. The foregoing calculations are based on the hypothetical assumption that all hybridization occurred during the same generation (Rife, 1954a).

If a statistically significant association is found between a genetic marker and another trait within a population of hybrid origin, whereas no association is found with other genetic markers, linkage is strongly indicated, provided certain conditions are met. These are as follows: Ancestral populations differ significantly in frequencies of genes and traits involved; one of the marker genes or traits showing the association is unimportant in selective mating; the association occurs only within populations of hybrid origin. It is difficult to conceive of any cause other than linkage that satisfies these conditions. This, in turn, indicates a genetic basis for the trait in question.

Populations of hybrid origin, as discussed here, do not conform to the usual concept of hybrids in plants and animals. The hybridization occurs over periods of many generations and even centuries; in other words, there has been a constant gene flow from ancestral populations. The net effect is to retard hypothetical rates of equilibration. In view of the fact that over 100 generations of random mating may be required to attain equilibrium with respect to very closely linked genes, a few centuries is not a long time period. Populations of hybrid origin are racial melting pots.

The objective of research of this type is to test for statistically significant associations between genetic markers and other traits of unknown heritability. If the traits showing the association are of no importance from the standpoint of selective mating or survival, and if no associations are found between these and other genetic markers, linkage is indicated. This, in turn, presents very strong evidence for inheritance. It would be impossible to explain on cultural or environmental grounds.

I conducted a research project of the type under discussion (1954a, b) on American Negroid and northern Sudanese populations. Various estimates agree in finding that American Negroids have 20–30% Caucasoid ancestry (Baker, 1974). Northern Sudanese comprise a hybrid population whose ancestors were southern Sudanese Negroids and Arab slave traders. Thus, both American Negroids and northern Sudanese are derived from mixtures of Negroids and Caucasoids, both of relatively recent origin. There has been a continuous flow of genes from the parent populations since hybridization began.

Two samples of American Negroids and one of northern Sudanese Negroids were tested. The first sample consisted of 35 Negroid students at Ohio State University. Data were collected on ABO, MN, and Rh blood groups; PTC taste reaction; hand prints; handedness; skin color;

anthropometric traits, and photographs. Analysis revealed highly significant associations between shade of skin color and occurrence of an accessory triradius in a distal area of the palm. No other significant associations were found between traits. Accessory triradii in distal regions of the palm occur more frequently among African Negroids than among Caucasoids. They are of no concern in choosing one's mate.

The second sample consisted of 100 northern Sudanese who were college students in Khartoum. Here again, highly significant associations were found between shade of skin color and an accessory triradius in a distal palmar area. No other significant associations were discovered.

The third sample consisted of 167 students at Central State College, Wilberforce, Ohio. This was a Negroid institution. Here, for the third time, significant associations were found between shade of skin color and accessory triradii, whereas no other significant trait associations were observed. Data on associations between shade of skin color and accessory triradii are summarized in Table 3.1.

TABLE 3.1

Summary of Tests for Associations between Presence of Accessory Triradii on Distal Areas of Palms and Shade of Skin Color

	Accessory Triradii			
	Dark shades		Light shades	
	Present	Absent	Present	Absent
Sudanese	11	29	4	56
American Negroids	25	86	8	83
Totals	36 (23.84%)	115	12 (7.94%)	139

Accessory triradii occurred three times as frequently among those with dark as among those with lighter shades of pigmentation. The odds are less than 1 in 1000 that these are chance associations. These differences continued to be highly significant when males and females were compared separately and when darkest and lightest shades (a total of 61 persons) were excluded from the comparisons. Linkage between genes responsible for variations in skin color and accessory triradii in distal palmar areas is strongly indicated.

Subsequent studies of a family of mixed origin gave information that supports the suggestion of linkage between genes affecting skin color and the occurrence of accessory triradii in distal palmar areas. The father is of mixed Negroid and Caucasoid descent, and the mother is a Caucasoid. The father and four children have accessory triradii in distal palmar areas and also have dark skin, hair, and eyes. The mother lacks accessory triradii in distal palmar areas. She has fair skin and auburn hair. Two children resemble the mother in lacking triradii in distal palmar areas and having fair skin and red hair. One child has triradii in distal palmar areas and fair skin. Thus, four children resemble the father both in shade of skin color and triradii in distal palmar areas. Two children resemble the mother in both traits, and only one out of seven children resemble the father in one trait and the mother in another. The mother is homozygous for both traits and the father is doubly heterozygous. Thus, this family represents a typical backcross, yielding six noncrossovers and one crossover. Although it would be hazardous to draw sweeping conclusions on the basis of only one family, the data definitely support evidence for linkage based on that obtained from populations of hybrid origin (Rife, 1971a, b).

I have also tested three African populations for associations between genetic markers and behavioral variations (Rife, 1956). Three population samples were tested as follows: 51 pupils in a Goan school in Kampala, Uganda; 80 prison inmates in Zanzibar; and 135 Capoids in Cape Town, South Africa (Rife, 1956). Although the Goans claimed to be of mixed Indian–Portuguese descent, their appearance and the genetic data obtained from them indicated only a minor portion of Portugese ancestry. The prisoners in Zanzibar were of three types: Indo-Dravidians, Bantu Negroids, and Arabs. The two latter groups were in the majority and were the only ones in Zanzibar of mixed origin. The Capoids were definitely of mixed origin but were derived from various populations, including Caucasoids, Bantu, Malay, Hottentots, and Bushmen. The last two constituted a large proportion of their ancestry, hence predominantly Capoid.

The following data were obtained: ABO blood groups, cell sickling, hand prints, anthropometric variations, performance tests, and photographs. Performance tests included weight discrimination, the test being modeled after one in the Binet series. Three small boxes of identical size, weighing 5, 6, and 8 grams, were employed. The subjects were asked to arrange the boxes in order of weight from light to heavy.

TABLE 3.2

Results of Tests for Associations between Performance of Weight Discrimination Test and Ridge Counts on Fingers among Capoids

Performance on weight test	Number of individuals	Mean number of ridges per person	Differences in mean values
Correct	64	159.09 ± 6.08	27.75 ± 10.956
Incorrect	52	131.34 ± 9.12	

Highly significant associations were found in the Capoid sample between ability to discriminate between small weights and ridge counts on fingers. Ridge counts are highly heritable, and South African Bushmen have among the lowest on record, considerably less than other ancestral populations of the Capoids. Actual data are shown in Table 3.2.

THE MELTING POT AS GENETIC SOURCE MATERIAL

The research on African populations was not conducted under ideal circumstances. The Goans showed slight evidence of hybridization; in Zanzibar only the Arabs were of hybrid origin, and the Capoids were mixtures of more than two ancestral groups. All population samples were smaller than desired. Nevertheless, the association observed between weight discrimination and ridge counts is highly significant among the Capoids and is absent in other populations tested. This type of association is exactly what would be expected between linked genes.

It should be kept in mind that this methodology is to be used essentially for detection of associations characteristic of linkage. When such associations are found, they should be followed by standard methods of analysis. Evidence of heritability obtained through detection of associations characteristic of linkage should be immune to charges that the tests lacked validity.

It may be argued that many genetic variations are quantitative, and therefore the effects of linkage with a particular gene would be so minor as to be unnoticeable. But where more genes are involved, the probabilities of linkage are also increased; in other words, multiple associations also increase the time required to approach equilibrium. It is especially important that quantitative markers be included in testing quantitative variations. Constellations of quantitative variations could have numerous interlinkages that would result in persistent and significant degrees of association for many generations.

The size of sample required for detection of linkage depends on closeness of linkage, difference between ancestral populations in gene frequencies, and proportional contributions of ancestral populations. Since percentages of crossovers are unknown beforehand, and precise data are usually unavailable on other items involved, a minimum size sample of 500 is suggested.

Sources of samples should obviously be from populations of hybrid origin, principally descended from two ancestral populations, between which hybridization began at least a hundred years ago. Such populations are found among Amerinds and American Negroids.

Cherokee Indians in the Cherokee, North Carolina, area constitute a population meeting the foregoing requirements. The tribal council has maintained records of ancestry over a period of seven generations. Pupils in the Cherokee school are of mixed Amerind–Caucasoid descent, in a ratio of approximately 3 : 2. Their Caucasoid ancestors were predominantly of English and Scottish–Irish origin (Rife, 1971a).

Mohawk Indians in the area of Fort Covington, New York, are of mixed Amerind–Caucasoid descent, their Caucasoid ancestors having been predominantly French and Irish. No ancestral records are available for ascertaining actual proportions of Amerind and Caucasoid ancestry (Rife, 1972). Mohawks are famous for their skill in working on high structures and are always in demand as steeplejacks. It is not known whether their lack of fear of high places is of genetic origin or the result of conditioning and training. Assuming availability of valid tests, Mohawk school pupils could provide samples for a project designed to search for evidence of linkage between genetic markers and ability to work efficiently in high places.

Not all Amerind tribes provide usable samples of populations of hybrid origin. Although there is evidence of mixed ancestry among contemporary Seminole schoolchildren, it is seldom observed among their parents. Moreover, degrees of racial intermixture vary considerably among different tribes.

As stated previously, American Negroids have significant proportions of Caucasoid ancestry. This is a broad generalization, actual percentages varying considerably from one portion of the country to another. Negroids in Southern rural areas appear to have comparatively little Caucasoid ancestry, whereas percentages are relatively higher in Midwestern cities. Carib Amerinds in the Caribbean area have high proportions of Negroid ancestry.

In order to be most effective, this type of research should be interdisciplinary, involving geneticists, psychologists, physical anthropologists, physiologists, and biometricians. Both qualitative and quantitative markers should be included, the more the better. The technique is essentially one of testing racial melting pots for associations between genetic markers and traits of unknown etiology. High school and college students are ideal subjects. Family pedigrees, adoptions, and illegitimacies need be of no concern here. Such investigations should be an effective tool for the detection of heritabilities held in dispute or unsuspected.

Importance of and interest in human variability is rapidly increasing, especially in regard to adaptability, behavior, intelligence, and special abilities. Climatic adaptability, adaptability to high and low altitudes, minimal nutritive requirements, differential fertility, disease susceptibilities, alcoholism, pain threshold, components of intelligence, and general adaptability are familiar examples of quantitative traits in which possible roles of heredity are yet largely undetermined.

Racial melting pots offer unique reservoirs of virtually untapped resource material for the detection of heritabilities of complex traits, which would establish beyond reasonable doubt a genetic basis and which could not be explained away on grounds of environmental factors and low validity of tests. Investigations of this type may provide definite answers to some questions that have long been debated.

SUMMARY

Genetic traits occur in widely different frequencies from one population to another. Interbreeding of different populations does not bring about greater uniformity but results in scattering of the variability. Populations of mixed origins tend toward genetic equilibrium, where genetic traits are distributed randomly and remain constant from generation to generation. In the absence of selection, genes located on different chromosomes reach approximate equilibrium in eight generations. Linked genes, those located on the same chromosome, approach equilibrium more slowly and require as much as 100 generations if closely linked.

Statistically significant associations between two nonselective hereditary traits within populations of mixed origins, and lack of such associations with other hereditary traits, indicates linkage. It follows that significant association between a known hereditary trait and a trait of unknown etiology, and the lack of such association between the latter and other known genetic traits within populations of hybrid origin, is indicative of linkage. This presents positive evidence for inheritance. Such associations cannot be attributed to environment or lack of test validity. Numerous simply inherited traits of no importance from the standpoint of selective matings are available for such tests. Hand prints and hair form are useful in testing for associations with quantitative traits of unknown etiology. American Negroids and various tribes of Amerinds are examples of populations of mixed origins.

The technique for linkage tests in populations consists of searching for significant associations between genetic traits and those of unknown etiology. The latter include adaptabilities to climatic variations and high altitudes, alcoholism, susceptibilities to physical and mental diseases, pain thresholds, behavioral traits, and cognitive abilities.

REFERENCES

Baker, J. R. *Race.* New York: Oxford Univ. Press, 1974.

Bell, Julia. Blue sclerotics and fragility of bone. *Treasury of Human Inheritance,* 1928, *2,* 269–324.

Bettman, S. Die Papillarleistenzeichnung der Handfläche in ihrer Beziehung zur Händigkeit. *Zeitschrift für Anatomie und Entwickelungsgeschichte,* 1932, *98,* 149–174.

Cromwell, H., & Rife, D. C. Dermatoglyphics in relation to functional handedness. *Human Biology,* 1942, *14,* 517–526.

Cummins, H. Finger prints correlated with handedness. *American Journal of Physical Anthropology,* 1940, *26,* 151–166.

Cummins, H. Dermatoglyphics of Bushmen (South African). *American Journal of Physical Anthropology,* 1955, *13,* 177–188.

Cummins, H., & Midlo, C. *Finger prints, palms and soles: An introduction to dermatoglyphics.* Philadelphia: Blakiston, 1943.

Cummins, H., & Stetzler, F. M. Dermatoglyphics in Australian aborigines (Arnhem Land). *American Journal of Physical Anthropology,* 1951, *9,* 455–460.

Galton, F. *Finger prints.* London: Macmillan, 1892.

Geipel, G. Die Finger und Handleisten der Ayom-Pygmaen Neuguineas. *Zeitschrift für Morphologie und Anthropologie,* 1958, *51,* 181–187.

Hardy, G. H. Mendelian proportions in mixed populations. *Science,* 1908, *28,* 49–50.

Holt, S. B. Quantitative genetics of dermal ridges on fingers. *Proceedings of the International Congress of Human Genetics, Acta Genetica et Statistica* (Basel), 1957, *6,* 473.

Jinks, J. L., & Fulker, D. W. A comparison of the biometrical genetic, MAVA and classical approaches to the analysis of human behavior. *Psychological Bulletin,* 1970, *73,* 311–349.

Lèche, S. M. Handedness and bimanual dermatoglyphic differences. *American Journal of Anatomy,* 1933, *53,* 1–53.

Lerner, I. M. *Heredity, evolution and society.* San Francisco: Freeman, 1968.

Li, C. C. *An introduction to population genetics.* Peiping: National Peking Univ. Press, 1948.

MacArthur, J. W. Reliability of dermatoglyphics in twin diagnosis. *Human Biology,* 1938, *10,* 12–15.

MacArthur, J. W., & Ford, N. H. C. A biological study of the Dionne quintuplets—an identical set. In *Collected studies of the Dionne quintuplets.* Toronto: Univ. of Toronto Press, 1937.

McKusick, V. A. *Heritable disorders of connective tissue.* (2nd ed.) St. Louis: Mosby, 1960.

Munsinger, H. The adopted child's IQ: A critical review. *Psychological Bulletin,* 1975, *82,* 623–659.

Newman, H. H. The finger prints of twins. *Journal of Genetics,* 1930, *23,* 415–446.

Newman, H. H. Dermatoglyphics and the problem of handedness. *American Journal of Anatomy,* 1934, *55,* 277–322.

Newman, H. H., Freeman, F. N., & Holzinger, K. J. *Twins.* Chicago: Univ. of Chicago Press, 1937.

Price, D. A., & Clarke, C. A. Pharmacogenetics. *British Medical Bulletin,* 1961, *17,* 234–240.

Rife, D. C. Handedness, with special reference to twins. *Genetics,* 1940, *25,* 178–186.

Rife, D. C. A comparison of the frequencies of certain genetic traits among Gentile and Jewish students. *Human Biology,* 1944, *16,* 172–180.

Rife, D. C. An application of gene frequency analysis to the interpretation of data from twins. *Human Biology,* 1950, *22,* 137–145.

Rife, D. C. Heredity and handedness. *Scientific Monthly,* 1951, *73,* 188–191.

Rife, D. C. Finger prints as ethnic criteria. *American Journal of Human Genetics,* 1953, *5,* 389–399.

Rife, D. C. Populations of hybrid origin as source material for the detection of linkage. *American Journal of Human Genetics,* 1954, *6,* 26–33. (a)

Rife, D. C. Distributions of skin pigmentation, dermatoglyphics, tasting ability and blood groups within mixed Negro-white populations. *Acta Geneticae Medicae et Gemellologiae,* 1954, *3,* 259–269. (b)

Rife, D. C. Dermatoglyphics as ethnic criteria. *American Journal of Human Genetics,* 1954, *6,* 319–327. (c)

Rife, D. C. Hand prints and handedness. *American Journal of Human Genetics,* 1955, *7,* 170–179.

Rife, D. C. Associations between weight discrimination and handedness. *Eugenics Quarterly,* 1956, *3,* 213–218.

Rife, D. C. *Heredity and human nature.* New York: Vantage Press, 1959.

Rife, D. C. Palm patterns and pigmentation in a Cherokee Indian population. *Acta Geneticae Medicae et Gemellologiae,* 1971, *20,* 69–76. (a)

Rife, D. C. Simple segregations in two families of mixed origin. *Acta Geneticae Medicae et Gemellologiae,* 1971, *3,* 284–288. (b)

Rife, D. C. Dermatoglyphics of Cherokee and Mohawk Indians. *Human Biology,* 1972, *44,* 81–85.

Robbins, R. R. Some applications of mathematics to breeding problems. *Genetics,* 1918, *3,* 375–389.

Sachs, L., & Bat-Miriam, M. The genetics of Jewish populations. I. Finger print patterns in Jewish populations in Israel. *American Journal of Human Genetics,* 1957, *9,* 117–126.

Schaumann, B., & Alter, M. *Dermatoglyphics in medical disorders.* New York: Springer-Verlag, 1976.

Stern, C. *Principles of human genetics.* (3rd ed.) San Francisco: Freeman, 1973.

Van Valen, L. Selection in natural populations: Human fingerprints. *Nature,* 1963, *200 (4912),* 1237–1238.

Weinberg, W. Über den Nachweis der Vererbung bein Menschen. *Jahreshefte Verein für vaterländische Naturkunde in Württemberg,* 1908, *64,* 368–382.

4

Genetic and Behavioral Effects of Nonrandom Mating

ARTHUR R. JENSEN
University of California, Berkeley

INTRODUCTION

Individual differences among persons in a large number of characteristics, physical and mental, are obvious to everyone. Probably there are many more differences that are less obvious and are revealed only by specialized techniques of observation and measurement. By now, it is a truism that most such differences are a joint result of genetic and environmental factors. But it would be a mistake to view the environment, particularly the cultural environment, as somehow more superficial than the genetic factors or as something that is merely added, like a kind of trimming, after the genes have already laid down the main outlines. True, as far as the individual is concerned, once conception has occurred, his genotype for any characteristic is set once and for all, and the environment will not alter it. Environmental factors may greatly influence its phenotypic expression, however, depending on the particular trait. If it is eye color or blood type, environment will have little or no effect on individual differences; these will reflect genotypic differences only. If it is a personality trait, experiential factors may account for much of the variation among persons. Finding the roles of genetic and environmental factors as they affect the phenotypic expression of individual differences is essentially the problem of heritability analysis. Instead, we shall be concerned here with some of the ways in which cultural–environmental factors can affect the genotypes themselves, by influencing the genetic structure of the population.

One of the principal mechanisms by which culture affects the genetic structure of human populations is the mating system, that is, the cultural and societal determinants of who mates with whom. (Other important factors affecting the genetic structure of populations are migration, contraceptive customs, differential fertility associated with cultural, religious, and socioeconomic differences, and selective mortality resulting from war, famine, and disease.) The strictly genetic effects of different mating systems are great, and their consequences are profound. Yet, in human populations, the determinants of mate selection itself are almost entirely a cultural affair. In the temporal sequence of causality, it may be more accurate to say that culture affects genetics rather than the other way around. In fact, since most external, noncultural selection pressures on populations, such as large-scale famines and disease epidemics, have been virtually eliminated in the industrialized world in the last century, the genetic structure of the population is mainly influenced by systems of mating, and whatever cultural forces, customs, and values influence mate selection in a given population will thereby, indirectly, have definite genetic effects, which may have profound cultural and social consequences, which in turn may affect mating patterns that feed back into the genetical system, and so on (see Eckland, 1972a). Thus, the relationship between genetics and culture is not to be thought of as a unidirectional causal system but as a feedback loop.

The variations in mating patterns and their genetic consequences may all be described in terms of departures from *panmixia* or *random mating*. Completely random mating is a theoretical abstraction in population genetics, but it can be created under laboratory conditions in experimental genetics with plants and infrahuman animals. Random mating, like assortative mating, is character specific; in a given population, mating can be random for some genes and assortative for others. In limited human populations, for example, there is close to random mating for many characters, such as ability to taste phenylthiocarbamide (PTC), blood groups, and serum proteins. Random mating means that among those who reach breeding age (i.e., are eligible to become parents), which male mates with which female is a matter of chance. (Though it is not part of the definition of random mating, in working out simple models based on the assumption of random mating, we usually assume that no individuals are left out in the mating system and that equal numbers of offspring result from each mating.) Few things could be more obvious than the fact that in human populations mating is anything but random for many traits.

The main types of departures from random mating, each with different genetic consequences, are assortative mating (which may be positive or negative), inbreeding, and outbreeding.[1] They are not mutually exclu-

[1] Polygamy is another departure from random mating that, in small populations, has important genetic consequences related to selection and inbreeding.

sive but may coexist in varying combinations and degrees for various traits. In practical importance, however, they differ greatly.

Assortative mating is mate selection based on resemblance in one or more characteristics. Assortative mating can be based on phenotypic similarity, or on genotypic similarity owing to common ancestry. In the latter case, it is called consanguinity, or inbreeding. But phenotypic assortative mating is character specific, while inbreeding is not. (Inbreeding is a special case of assortative mating; the term is used in animal breeding as well as in human population genetics; consanguinity refers specifically to humans who are quite closely related through a common ancestor.) The term assortative mating throughout the present chapter refers strictly to mating on the basis of phenotypic resemblance. The term inbreeding here is reserved for resemblance between mates due to common ancestry.

In theory, assortative mating can be either positive (i.e., greater than chance similarity of mates) or negative (i.e., greater than chance dissimilarity), although, in fact, no one has yet demonstrated statistically significant negative assortative mating for any human trait in any large population.

The psychological, social, and cultural reasons for positive assortative mating, though interesting and important topics in their own right, are not our main concern in this chapter, which deals with the genetic and psychological *effects* of assortative mating rather than with its psychological and cultural origins. Suffice it to note that in modern industrial societies, at least, there is assortative mating as to race, ethnic origin, social class, age, religion, education, intelligence, various personality traits, physical characteristics (height, weight, complexion), values, interests, residential propinquity, and a host of other variables. Tharp (1963) has comprehensively reviewed the literature on the social–psychological determinants of assortative mating. Of the dozens of human traits that have been investigated, those showing the highest degree of assortative mating in European and North American Caucasoid populations are age, amount of formal education, and IQ.

The important thing about assortative mating, from our standpoint, is its effects on heritable traits in the population. Since assortative mating increases the variance of a trait in the population, for example, it has been estimated that the present level of assortative mating for intelligence in England and the United States, assuming that this level of assortative mating has existed for several generations, may account for over half the frequency of persons with IQs above 130 and four out of five of those with IQs over 145, and there are approximately 20 times as many persons above an IQ of 160 as we would find if there were no assortative mating for intelligence (Jensen, 1973a, p. 108). Such effects may greatly affect the character of a population in terms of its intellectual resources. If our society were suddenly to engage in random mating with respect to intelligence, the intellectually most able of the next generation would not be as bright as the same upper $x\%$ of the previous generation.[2] The percentage of the mentally retarded would also be reduced, though perhaps not as much since about one-fifth to one-fourth of mental retardation is attributable to rare genetic abnormalities and nongenetic causes such as brain injury and disease. However, the positive or beneficial effects to society of persons with very high abilities may well be much greater than are the negative effects of an equal number of persons whose abilities are just as far *below* the average. The great cultures of the past are distinguished largely by the number of their creative geniuses, not by their rates of mental deficiency. A population with little intellectual variance but homogeneously centered around an average level of ability, say, with 90% of persons having IQs between 90 and 110 (the range that now contains the middle 50% in the United States Caucasoid population), if left to itself, would probably advance very slowly, if at all, beyond a Stone Age or simple agrarian culture. As a factor in cultural evolution, the amount of variability of talents in a population could be more important than its overall mean.

Positive assortative mating per se does not affect the mean of the population on the trait in question (unless there is directional dominance of the genes involved), but it always increases the variability. (Negative assortative mating decreases the total variance.) Unlike selective mating, assortative mating does not change the frequency of the *genes* for a given trait in the population's gene pool. But it does change the frequency of *genotypes*, that is, combinations of genes. Positive assortative mating increases the frequency of genotypes that make for more extreme phenotypic values in the trait and decreases the frequency of genotypes that make for the more average phenotypes.

As Gordon Allen (1970, p. 186) has pointed out, assortative mating is probably circular and self-reinforcing in its operation; it acts as a positive feedback loop. Since assortative mating for a particular trait increases the population variance, that is, the range of individual differences in the trait, it makes the trait an even more salient basis for assortative mating in the next generation. It has been estimated that in England some 20% of the variance in IQ is attributable to assortative mating (Burt, 1958).

For some traits, assortative mating can lead indirectly to a change in gene frequencies and consequently a change in the population mean by increasing the effects of selection. A trait like intelligence, which is subject to a high degree of assortative mating in many societies, may also be subject to some degree of selection. In the population as a whole, the

[2] A single generation of random mating would not completely remove the effects of past assortative mating. The association of genes with similar effects (so-called "linkage disequilibrium") decreases at 50% per generation for unlinked genes and at a correspondingly slower rate for linked genes.

degree of selective mating for IQ is slight as compared with the degree of assortative mating, but near the lower end of the IQ distribution there is more or less a selection threshold below which persons either cannot find mates or are discouraged from mating by parents, social workers, and the like. It is known that, at least in the Caucasoid population of the United States, persons with IQs below about 75 are much less likely to marry. The average number of offspring born to all persons with IQs below 75 is less than for the general population mainly because most intellectually retarded persons never marry (Bajema, 1963; Higgins, Reed, & Reed, 1962). A high degree of assortative mating for intelligence, which increases the IQ variance in the population, results in the birth of a larger proportion of persons who fall below the mating threshold, so that their genes for low intelligence are less likely to be passed on to the next generation, with the result that the population gene pool for intelligence is slightly improved. There is evidence that a eugenic effect of this nature has been taking place in the Caucasoid population of the United States (Gottesman, 1968, pp. 42–46).

Inbreeding is the mating of individuals related by common ancestry. Unlike assortative mating, inbreeding is not trait specific. The genotypes of inbred mates are correlated not only for specific genes for particular traits, as in assortative mating, but for *all* gene loci. The degree of correlation, of course, depends on the closeness of relation, quantified by the coefficient of inbreeding, which is taken up in a later section. The effects of inbreeding are only faintly like those of assortative mating. For quantitative traits, inbreeding in human populations is less important than assortative mating in its net effect on the genetic structure of populations because inbreeding is so much less frequent than assortative mating; the average degree of genetic correlation between mates resulting from inbreeding in the population is but a minute fraction of the correlation attributable to assortative mating. However, inbreeding affects *all* segregating gene loci, while assortative mating affects only those related to the trait involved. Inbreeding, like assortative mating, has no direct effect on gene frequencies but changes the frequencies of genotypes, making for more extreme types (homozygotes) and fewer intermediate types (heterozygotes). Inbreeding, therefore, increases the total population variance. However, mates who are related by a common ancestor are less apt to be alike genetically for a given trait than unrelated individuals who are phenotypically matched on the trait, assuming the phenotype has high heritability and is therefore a good indication of the individual's genotype.

Inbreeding has its most conspicuous effect on the frequency of appearance of rare recessive characteristics, which most often are of an undesirable nature. When there is some degree of positive dominance for a polygenic trait, inbreeding will depress the trait among inbred offspring, a phenomenon known as *inbreeding depression.*

Outbreeding, also called crossbreeding, is the mating of individuals from different breeding populations. In humans, it usually refers to interracial matings. It implies no common ancestry between mates, at least for many generations back. But outbreeding is really a relative concept and can arbitrarily be defined as matings among individuals for which the inbreeding coefficient is *lower* than the average inbreeding coefficient within either of the parent populations from which the individuals come. The effects of outbreeding in humans are not at all certain or clear-cut; they are reviewed in a later section on empirical findings.

Another important factor affecting the genetic structure of a population but technically not regarded as a mating system is *selection.* Selection means that some traits or patterns of traits are more generally sought after by prospective mates than are others, for whatever reason. Persons possessing such traits will be more apt to be sought as a mate and will mate sooner in life or are more likely to produce a larger number of offspring, other things being equal. Conversely, persons lacking such traits or possessing their opposites are less likely to find mates, may take longer to do so, or may be left out altogether, with the result that they produce fewer offspring. There is also negative selection for many undesirable characteristics and lower fertility, as, for example, among the mentally retarded and schizophrenics.

Assuming that the trait in question does not have negligible heritability, the net effect of selection genetically is that in each generation the total frequency of the alleles involved in the particular trait will be changed in the population. The number of alleles for the desirable traits will increase, while the alleles for the undesirable traits will decrease. The rate of change in allele frequencies, of course, will depend on the degree of selection and on the heritability of the trait (i.e., the extent to which the observed phenotype reflects the underlying genotype). Since selection changes allele frequencies, it alters the mean and variance of the trait in the population. Animal breeders know this well and have capitalized on it to create breeds that better fulfill certain purposes, as in various specialized breeds of dogs and in animals bred for their agricultural products, such as the egg-laying capacity of chickens, the milk yield of cows, and the lardiness of pigs.

In humans, physical attractiveness as well as various mental and personality characteristics are the basis for selection. But there is far from universal agreement, from one culture to another, as to which characteristics are deemed most desirable. Culturally determined standards of physical beauty and of the behavioral characteristics considered the most valuable or attractive may differ considerably from one population to another and among various subpopulations. And what is viewed as desirable in the one sex may not be sought in the other. Thus, selection does not necessarily imply like mating with like for a particular trait, which is positive assortative mating. If all men like beautiful women, and beautiful women seek "brainy" men (perhaps because they can make more money), both "beauty" and "brains" will be selected. The ugliest women and the stupidest men will be the least likely to mate and leave offspring, and the population's gene pool will be enhanced for beauty and brains, which, of course, will be passed on to both sexes in subsequent generations. Because the population seems so very heterogeneous in many traits, it suggests that selection pressures for such traits are only rather slight or that there are a multitude of different values that affect selection in various groups in the population. Moreover, any strong directionality of genetic change may be precluded by short-term cultural changes in the general standards, values, and ideals that affect mate selection. Some degree of directional selection must constantly take place, however, to offset the constant degenerative effects of mutation. A relaxation of selection results in degeneration. But the amount of selection needed to overcome mutation pressure for IQ or other behavior traits is unknown at present; it depends on the number of genes involved in the trait and on the mutation rate.

GENETIC CONSEQUENCES OF SELECTION AND ASSORTATIVE MATING IN TERMS OF A MENDELIAN MODEL

The best way to understand how and why assortative mating affects the genetic structure of a population is in terms of what is called "Mendelian algebra."[3] By working out particular genetic models in the simple terms of Mendelian algebra, one can gain a direct insight into such general conclusions as, for example, that assortative mating *(a)* increases population variance in the trait in question, *(b)* increases sibling resemblances as measured by the correlation between siblings, *(c)* increases only slightly the amount of homozygosity in the population, *(d)* leaves the mean unchanged unless there is dominance, and *(e)* creates correlations between traits that would be totally uncorrelated in a random-mating population.

The Mendelian algebra can be applied both to single-locus characters and to polygenic traits.[4] Characters controlled by a single locus show

[3] For a more extensive and systematic introduction to quantitative genetics, see Crow (1950) and Li (1955). More advanced treatment are Falconer (1960) and Crow and Kimura (1970).

[4] The term *polygenic* has a loose and a strict meaning in genetic theory. Loosely speaking, polygenic traits are those in which (1) variation is attributable to segregation at a number of loci (and not mostly at one or two loci) and (2) which show fairly continuous variation, without pronounced peaks and ratios. In the strict sense, a polygenic trait is one in which (1) variation is attributable to a large number of genes each having minor effects, (2) the effects are additive, and (3) the genes are interchangeable, i.e., plus (or minus) genes at any locus could be substituted for plus (or minus) genes at other loci.

large and discrete differences among individuals in the population; often the differences are regarded as qualitative rather than quantitative. Most polygenic traits are graded or continuous, and individual differences are quantitative rather than qualitative. Some polygenic traits, however, are expressed in more or less an all-or-none fashion because they are "threshold" traits, that is to say, their phenotypic expression depends on whether some relevant environmental variable is above or below a particular threshold value (Falconer, 1960, Chap. 18). When a single gene accounts for a large part of the population variance in a given trait, it is called a major gene effect. When the population variance is attributable, in whole or in part, to the cumulative effects of a number of genes, each one alone having a small effect relative to the total trait variation in the population, the trait is called polygenic. At the molecular, biochemical level, however, there are *not* two different types of genes, major genes and polygenes. Rather, there are some traits the variance in which is the result of segregation of alleles at a single gene locus, and there are other traits the variance in which is the result of the additive and interactive (i.e., combinations other than additive) effects of alleles of a number of gene loci. Behavioral genetics is more concerned with the latter. Individual differences in most of the characteristics of interest and importance, such as mental abilities and personality traits, are continuously graded. In some cases, variance in a particular trait may be analyzable into both major gene and polygenic effects. A possible consequence of this would be a modal distribution of the trait, the modes of the distribution determined by the major gene and the variation about each mode attributable to the polygenes. A good example is normal variation in height, which is polygenic, and dwarfism, which is a major gene effect superimposed on the polygenic determinants of height. The single gene for dwarfism makes its recipient extremely short, but there is still variation in height among dwarfs, resulting from variation in the normal polygenic determinants of height. The genetic analogue to dwarfism is seen in the psychological realm in certain types of mental deficiency, in which a single gene almost completely overrides the polygenic determinants of the individual's intelligence.

Genes have definite positions, called loci, on the chromosomes, which are the carriers of the genes. Humans have 23 pairs of chromosomes, or 46 in all. Each chromosome is like a long thread banded by a large number (estimated between about 400 and 4000 [see Stern, 1973, pp. 39–41]) of gene loci. One member in each pair of chromosomes is inherited from the mother, and the other member of each pair from the father. Thus, half of the genetic material determining an individual's development comes from the mother and half from the father. In mating, the individual passes on just a sample half of his genetic material to his offspring. Which half of the genes is passed on to the offspring is pure chance. Thus, a parent does not pass on his genotype as a unit to his or her offspring but only a random selection of one from each pair of his or her chromosomes.

The gene loci can be represented by the letters of the alphabet—*A*, *B*, *C*, etc. In order for a gene to produce any differences among individuals in the population, it must have two or more forms, each of which has different effects on the trait controlled by the gene. These two or more forms of the gene are called *allelomorphs,* or simply *alleles.* The alleles that have different effects (or values) at a given locus can be designated by capital and lowercase letters, while the letter itself designates the locus. Thus, **A** and **a** are two alleles at locus *A*. (There may be more than two forms of the gene in the population, but no individual can possess more than two alleles at a given gene locus since each individual possesses only two homologous chromosomes with corresponding gene loci, each of which can harbor one or another form of the gene. There are 23 such pairs of chromosomes in humans, but a given gene locus, say, *A*, occurs only on a single homologous pair. Gene locus *B* may occur on the same chromosome pair as locus *A* or on a different chromosome pair.)

Now, consider the effects at a single locus. There are two alleles, **A** and **a**. The effect of **A** is to *enhance* the trait; the effect of **a** is to *diminish* the trait. We assume the effects of **A** and **a** to be additive, that is, **AA** − **Aa** = **Aa** − **aa**. In order to quantify the effects, we can assign the arbitrary values **A** = $+\frac{1}{2}$ and **a** = $-\frac{1}{2}$. Since every individual has two homologous chromosomes with the *A* locus, any given individual can possess any one of the following possible combinations: **AA, Aa, aA, aa.** Thus, individual differences in this trait would have only three possible genotypes and, assuming no environmental influences, only three possible values, +1, 0, and −1, thus:

$$
\begin{aligned}
\mathbf{AA} &= \tfrac{1}{2} + \tfrac{1}{2} = +1 \\
\mathbf{Aa} &= \tfrac{1}{2} - \tfrac{1}{2} = 0 \\
\mathbf{aA} &= -\tfrac{1}{2} + \tfrac{1}{2} = 0 \\
\mathbf{aa} &= -\tfrac{1}{2} - \tfrac{1}{2} = -1
\end{aligned}
$$

If we assume that the **A** and **a** alleles exist in equal frequencies in the population, the frequencies and proportions of these values would be:

Genotype	*Value*	*Frequency*	*Proportion*
AA	+1	1	.25
Aa and **aA**	0	2	.50
aa	−1	1	.25

Now, let us assume every possible mating combination in this population, occurring with equal frequency, and assume they produce equal numbers of offspring and that each possible resulting genotype occurs with the average frequency expected by chance. Recall that each off-

TABLE 4.1

Genotypes and Values Produced in a Random Mating Population

	Mothers			
Fathers	AA	Aa	aA	aa
AA	**AA** = +1	**AA** = +1	**AA** = +1	Aa = 0
	AA = +1	**AA** = +1	**AA** = +1	**Aa** = 0
	AA = +1	**Aa** = 0	**Aa** = 0	**Aa** = 0
	AA = +1	**Aa** = 0	**Aa** = 0	**Aa** = 0
Aa	**AA** = +1	**AA** = +1	**AA** = +1	**Aa** = 0
	AA = +1	**Aa** = 0	**Aa** = 0	**Aa** = 0
	aA = 0	**aA** = 0	**aA** = 0	**aa** = −1
	aA = 0	**aa** = −1	**aa** = −1	**aa** = −1
aA	**AA** = +1	**AA** = +1	**AA** = +1	**Aa** = 0
	AA = +1	**Aa** = 0	**Aa** = 0	**Aa** = 0
	aA = 0	**aA** = 0	**aA** = 0	**aa** = −1
	aA – 0	**aa** = −1	**aa** = −1	**aa** = −1
aa	**aA** = 0	**aA** = 0	**aa** = −1	**aa** = −1
	aA = 0	**aA** = 0	**aa** = −1	**aa** = −1
	aA = 0	**aa** = −1	**aA** = 0	**aa** = −1
	aA = 0	**aa** = −1	**aA** = 0	**aa** = −1

spring receives, at random, one-half of the parent's genes. Table 4.1 shows all the genotypes and their values produced by the various combinations of the parental alleles.[5] The frequencies and proportions of the various genotypes of all the offspring in this hypothetical population are as follows:

Genotype	*Value*	*Frequency*	*Proportion*
AA	+1	16	.25
Aa and **aA**	0	32	.50
aa	−1	16	.25

This is a simple illustration of the *Hardy–Weinberg* principle, which is fundamental in quantitative genetics. This law states, essentially, that in the absence of mutation, migration, or selection, and with random mating, the gene frequencies in the population remain the same from one generation to the next, and the proportions of the different possible genotypes also remain the same. This "steady state" in the genetic structure of the population is called Hardy–Weinberg equilibrium.

[5] In the notation for genotypes, it is conventional that the first letter in each pair represents the allele received from the father.

In Table 4.1, notice the frequencies of the parental generation (Fathers and Mothers). **AA, Aa** (or **aA**), **aa** are in the ratios of 1, 2, 1. The ratios of the three genotypes are exactly the same in their offspring.

Also, notice that each of the 16 mating combinations[6] occurs with equal frequency (i.e., random mating), and each produces the same number (i.e., 4) of offspring, so there was no selection or favoring of any one mating combination over another. Further, notice that no "new" or mutated alleles (such as **A′** or **a′**) appear in any of the offspring—a necessary condition of the Hardy–Weinberg principle. It is obvious that this is a probability statement, the precision of which depends on the size of the population. It is the basis for predicting the quantitative consequences of factors that can act on a population to change its genetic structure—such factors as migration, selection, differential fertility (which is a special case of selection), and various types of nonrandom mating.

The Hardy–Weinberg principle can be generalized beyond the one locus example illustrated above, so that it will apply to any number of gene loci and to any proportions of the **A** and **a** alleles in the population, assuming the same proportion (i.e., p and q) at all loci. The general formulation is:

$$(p\mathbf{A} + q\mathbf{a})^{2n},$$

where p is the proportion of **A** alleles in the population; q is $1 - p$ or the proportion of **a** alleles in the population, and $p + q = 1$; and n is the number of gene loci involved (with 2 alleles at each locus, thus $2n$).

The expansion of this binomial yields the proportions of all the possible genotypes that exist or could exist and are maintained in such a population from generation to generation under the conditions of the Hardy–Weinberg principle. In the example given above, the proportions of **A** and **a** alleles are .5 and .5, and there is only one locus, so we have

$$(.5\mathbf{A} + .5\mathbf{a})^2$$

which, when expanded, is:

$$.25\mathbf{AA} + .50\mathbf{Aa} + .25\mathbf{aa}.$$

Thus, in general, the probabilities of genotypes **AA, Aa,** and **aa** are p^2, $2pq$, and q^2, respectively. The possible genotype mating combinations are therefore given by the expansion of

$$(p^2\mathbf{AA} + 2pq\mathbf{Aa} + q^2\mathbf{aa})^2.$$

[6] Often in the genetic literature the sexes (i.e., fathers and mothers) are pooled and the **Aa** and **aA** genotypes are pooled, so that there are then only six mating combinations, that is, **AA** × **AA**, **AA** × **Aa**, **AA** × **aa**, **Aa** × **Aa**, **Aa** × **aa**, **aa** × **aa**, with relative frequencies of 1, 4, 2, 4, 4, 1, respectively.

The greater the number of loci (n) contributing to a trait, the larger is the number of grades or genotypic values assumed by the trait. Assuming that the genetic values of the alleles at all loci are equivalent (i.e., **A** = **B**, and **a** = **b**), the number of grades is $2n + 1$, and as noted above, their proportions are given by the binomial distribution $(p + q)^{2n}$. The mean genetic value of the distribution will be (p**A** + q**a**), when **A** and **a** are assigned values of $+\frac{1}{2}$ and $-\frac{1}{2}$, respectively, representing their separate effects on the trait. The variance of the distribution is $2pqn$, and the standard deviation is $(2pqn)^{1/2}$

When $p = q$, the distribution of genotype values will be symmetrical, and as n becomes very large, the values become graded more and more finely, and the distribution approaches the so-called normal, bell-shaped, or Gaussian curve. When $p \neq q$, the distribution of values will be skewed, the amount of skewness depending on the difference between p and q and on the number of loci, n. The larger the n, the less skewed will the distribution be for any given values of p and q.

The effect of genetic selection is to change p and q, that is, the proportions of the two alleles in the population, and this consequently affects the population mean and variance of the trait. In our simple example of (.5**A** + .5**a**)^2, the mean and variance are 0 and .50. (This assumes the

TABLE 4.2

Genotypes and Values Resulting from Random Mating for Two Gene Loci

Male gametes (sperms)	Female gametes (eggs)			
	AB	Ab	aB	ab
AB	**AABB** = +2	**AABb** = +1	**AaBB** = +1	**AaBb** = 0
Ab	**AAbB** = +1	**AAbb** = 0	**aAbB** = 0	**Aabb** = −1
aB	**aABB** = +1	**aABb** = 0	**aaBB** = 0	**aaBb** = −1
ab	**aAbB** = 0	**aAbb** = −1	**aabB** = −1	**aabb** = −2

Frequency distribution[a]

Values	Frequencies	Proportions
−2	1	.0625
−1	4	.2500
0	6	.3750
+1	4	.2500
+2	1	.0625
Total	16	1.000

[a] The mean of the distribution is 0 and the variance is 1.

values **A** = $+\frac{1}{2}$, **a** = $-\frac{1}{2}$.) Now, say that some selective pressure acts on the population to decrease the frequency of **aa** genotypes; for example, **aa** individuals might be less resistant in an epidemic and die off in large numbers, thereby depleting the population gene pool of the **a** type alleles. Then the proportions might be, say, .8**A** + .2**a**. So the population mean now would be +.30 (instead of 0), and the variance would be .32 (instead of .50). If the selection to eliminate **a** alleles continues generation after generation, the mean will increase, and the variance will decrease until eventually the **a** allele disappears completely and all members of the population are homozygous for the trait, that is, everyone is **AA**. Then the mean would be +1 and the variance 0.

To go from the single gene case to the polygenic case is a straightforward matter. One simply represents the two or more loci by different letters, for example **A** and **B**, and works it out in the same manner as the single-locus example. The simplest possible polygenic case involves two loci. In such a population, it is simpler if we represent all the possible parental gametes (sex cells) that can unite to form individuals rather than all the possible parental genotypes. (It should be recalled that the gametes contain only a random half of the genes comprising the parental genotype.) Shown in Table 4.2 are all the resulting genotypes and their values, on the assumption **A** = **B** = +.5 and **a** = **b** = −.5.

Dominance

We have thus far considered only *additive* gene effects; that is to say, the alleles at a given locus do not have any influence on one another, and their separate effects are simply additive in determining the net value of the genotype. Under this condition, the value of the heterozygotes, **Aa** or **aA**, are exactly intermediate between the values of the homozygotes **AA** and **aa**. When there is interaction between the alleles **A** and **a**, their combined effect is not the sum of their separate values but is some nonadditive function of the two. This type of interaction between alleles at the same locus is called *dominance*. If the allele that enhances the trait is dominant, there is *positive* dominance; if the allele that diminishes (or does not enhance) the trait is dominant, it is called *negative dominance*. The nondominant allele is called *recessive*. Dominance is complete when the value of the heterozygote equals the value of one of the homozygotes (e.g., **Aa** or **aA** = **AA**). There is *overdominance* when the value of the heterozygote is more extreme than the value of either of the homozygotes. These conditions are illustrated in Figure 4.1.

Dominance affects the distribution of phenotypes in the population, as can be seen, for example, if we assume complete positive dominance

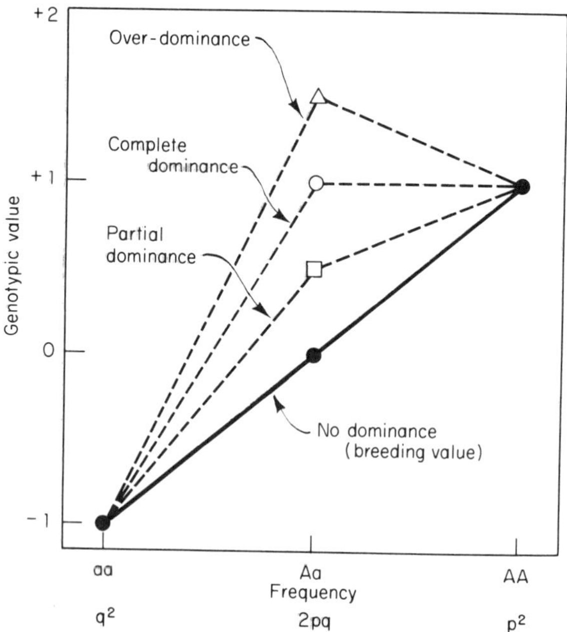

Figure 4.1 Graphic representation of additive genetic effects (no dominance) and various degrees of dominance deviation.

(i.e., **Aa** = **AA** and **Bb** = **BB**) and assign phenotypic values to the various genotypes shown in Table 4.2. The distribution would be as follows:

Values	*Frequencies*	*Proportions*
−2	1	.0625
−1	0	0
0	6	.3750
+1	0	0
+2	9	.5625
Total	16	1.000

The mean is now 1, and the variance is 1.5. Thus, dominance, in this example, has left the variance unchanged but has raised the mean by one standard deviation (σ), from 0 to 1. Dominance need not occur at all loci. If, in the above example, dominance occurred only for **A** but not for **B**, the resulting frequencies would be:

Values	*Frequencies*	*Proportions*
−2	1	.0625
−1	2	.1250
0	4	.2500
+1	6	.3750
+2	3	.1875
Total	16	1.000

The mean is now .5, and the variance is 1.25; note that these values are simply the average of the values (for either the mean or the variance) under complete additivity and complete dominance. (Thus, the total variance V_T can be partitioned into additive and dominance components, V_A and V_D, respectively: In the first example, with complete additivity [lower part of Table 4.2], $V_A = 1$, $V_D = 0$, and $V_T = 1$; in the second example, with complete dominance, $V_A = 1$, $V_D = .5$, and $V_T = 1.50$; and in the third example, with partial dominance [i.e., additivity at locus *A* and dominance at locus *B*], $V_A = 1$, $V_D = .25$, and $V_T = 1.25$. Note that in all cases the additive variance, V_A, remains the same value [= 1 in this example]). Notice, too, that dominance skews the shape of the distribution.

Dominance also affects the resemblance between parents and children and among siblings. Degree of resemblance for a given trait may be indexed by the correlation coefficient.[7] Let us tabulate all possible genotypes based on a single locus A, resulting from all possible mating combinations in the population, and then compute the parent–child and sibling correlations based on the phenotypic values when there is *(a)* no dominance, and *(b)* dominance of **A.** (A similar example could be given for two or more loci, but the tabular presentation of the results would be extremely cumbersome, involving 16^2 possible mating combinations producing 16^3 offspring.) The essential results, however, would be exactly the same as the much simpler single gene example shown in Table 4.3.

[7] When the Pearson *r* is used to determine the correlation between relatives, a double-entry procedure is used. If it is entirely arbitrary which member of any kinship pair is assigned to X or Y in the formula for correlations, each pair is entered twice, each member being entered first as X or as Y and then the reverse. The intraclass correlation is better suited to obtaining kinship correlations, and is essential for sibling correlations when there are more than two siblings per family. For a reasonably large *N* the intraclass *r* approximates Pearson *r* when the means and variances of the X and Y arrays are the same, which is, of course, insured by the double-entry method for computing Pearson *r*. Both the Pearson *r* and the intraclass correlation are estimates of the same correlation ρ in the population. The Pearson *r*, however, has the larger sampling error. (Cf. Fisher, 1950, Chap. 7.) When the intraclass correlation ρ is computed from population values or theoretical values, the formula is

$$\rho = \frac{\sigma_B^2}{\sigma_B^2 + \sigma_W^2}$$

where σ_B^2 and σ_W^2 are the variances between and within classes (e.g. families), respectively. When estimates of these population values, s_B^2 and s_W^2, are obtained from samples of the population, the formula for the intraclass correlation is

$$r_i = \frac{s_B^2 - s_W^2}{s_B^2 + (\bar{n} - 1)s_W^2}$$

where $\bar{n}$ is the arithmetic mean of the number of members in each class (e.g., siblings in each family). Whenever the number of cases differs considerably from one class to another the harmonic mean rather than the arithmetic mean is used for obtaining the value of *n* (See Blalock, 1960, pp. 266–269.)

TABLE 4.3

Genotypes of All Possible Offspring in a Random Mating Population and Genotypic Values with and without Dominance[a]

Father's genotypes	Mother's genotypes: AA = +1 (+1)	Aa = 0 (+1)	aA = 0 (+1)	aa = −1 (−1)
AA = +1 (+1)	AA = +1 (+1) AA = +1 (+1) AA = +1 (+1) AA = +1 (+1)	AA = +1 (+1) AA = +1 (+1) Aa = 0 (+1) Aa = 0 (+1)	AA = +1 (+1) AA = +1 (+1) Aa = 0 (+1) Aa = 0 (+1)	Aa = 0 (+1) Aa = 0 (+1) Aa = 0 (+1) Aa = 0 (+1)
Aa = 0 (+1)	AA = +1 (+1) AA = +1 (+1) aA = 0 (+1) aA = 0 (+1)	AA = +1 (+1) Aa = 0 (+1) aA = 0 (+1) aa = −1 (−1)	AA = +1 (+1) Aa = 0 (+1) aA = 0 (+1) aa = −1 (−1)	Aa = 0 (+1) Aa = 0 (+1) aa = −1 (−1) aa = −1 (−1)
aA = 0 (+1)	AA = +1 (+1) AA = +1 (+1) aA = 0 (+1) aA = 0 (+1)	AA = +1 (+1) Aa = 0 (+1) aA = 0 (+1) aa = −1 (−1)	AA = +1 (+1) Aa = 0 (+1) aA = 0 (+1) aa = −1 (−1)	Aa = 0 (+1) Aa = 0 (+1) aa = −1 (−1) aa = −1 (−1)
aa = −1 (−1)	aA = 0 (+1) aA = 0 (+1) aA = 0 (+1) aA = 0 (+1)	aA = 0 (+1) aA = 0 (+1) aa = −1 (−1) aa = −1 (−1)	aa = −1 (−1) aa = −1 (−1) aA = 0 (+1) aA = 0 (+1)	aa = −1 (−1) aa = −1 (−1) aa = −1 (−1) aa = −1 (−1)

[a] Values not in parentheses = no dominance. Values in parentheses = complete dominance of **A** (i.e., value of **Aa** = **AA**). Separate values of **A** and **a** are +.5 and −.5, respectively. The mating combinations for perfect assortative mating with no dominance are enclosed by the broken line. Assortative mating combinations with dominance are enclosed by a solid line; the frequency of the **aa** × **aa** combination, however, must be increased threefold so that the number of offspring from each genotype in the population will be the same. (See text for explanation.)

Table 4.4 presents the statistics on this hypothetical population when there is no dominance and when there is complete dominance of **A**.

The statistics in Table 4.4 for random mating are based on all the 64 genotypes (in 16 "families") shown in Table 4.3. The 16 genotypes resulting from assortative mating with no dominance are the four "families" in the principal diagonal going from the upper left to the lower right in Table 4.3 (enclosed by a broken line). Note that in order for the statistics calculated on the results of assortative mating to be directly comparable to the statistics for random mating, every parental genotype

TABLE 4.4

Statistics Derived from Genetic Model for Random Mating and Perfect Assortative Mating with and without Complete Dominance

Statistic	Random mating		Assortative mating[a]	
	No dominance	With dominance	No dominance	With dominance
Mean of population	0	.50	0	.333
Total variance (σ_T^2)	.50	.75	.75	.889
Standard deviation (σ)	.707	.866	.866	.943
Variance between families (σ_B^2)	.25	.313	.50	.639
Correlations				
Mother and child	.50	.333	.816	.816
Father and child	.50	.333	.816	.816
Midparent and child[b]	.707	.726	.816	.816
Midparent and midchild[c]	1.000	.730	1.000	.963
Mother and father[d]	0	0	1.000	1.000
Siblings[e]	.50	.417	.667	.719

[a] These are the results after only one generation of assortative mating, that is, random mating prevailed in all generations previous to the parental generation that produced the offspring on whom these statistics are based.

[b] Midparent is the arithmetic average of both parents. Midchild is the average of all children born to the same pair of parents.

[c] The correlations given here are the theoretically expected values based on the assumption that all the parents have had equal numbers of children representing every possible genotype that the parental gametes are capable of producing. Under these conditions, the correlation between midparent and midchild is equal to the narrow heritability h_N^2 of the trait, that is, the proportion of total variance attributable to additive genetic effects. In actual samples, however, the midparent–midchild correlation $r_{\bar{p}\bar{o}}$ underestimates h_N^2 and will be smaller than the theoretical values given in Table 4.4. The theoretically expected midparent midchild correlation when each family produces a random sample of size N of all the possible genotypes that could result from the union of the parental gametes is

$$r_{\bar{p}\bar{o}} = h_N^2 \, \tfrac{1}{2}N/[1 + r_{oo}(N - 1)],$$

where h_N^2 is the narrow heritability; N is the number of offspring in each family; r_{oo} is the correlation among offspring. It can be seen that as N increases, $r_{\bar{p}\bar{o}}$ more closely approaches h_N^2.

[d] This is the coefficient of assortative mating, which in the first two columns is 0 since mating is assumed to be random.

[e] Note that the sibling intraclass correlation is equal to σ_B^2/σ_T^2, that is, the ratio of the variance between families to the total variance. In working out a theoretical model such as this, we are dealing with errorless population values, not sample estimates that contain sampling error. Thus, the formula for the intraclass correlation is different when based on population values than on sample values. (See Footnote 7, p. 65.)

must have equal fertility, that is to say, must enter into an equal number of matings producing equal numbers of children. (For this reason, only the four "families" in the principal diagonal are used in the analysis of assortative mating with no dominance.)

Assortative mating with dominance necessarily involves more mating combinations since assortative mating is based on phenotypes; with complete dominance of **A**, the genotype **Aa** is phenotypically equivalent to the genotype **AA.** The "families" resulting from assortative mating with dominance are enclosed within the solid line in Table 4.3. However, in order to have equal fertility of all the genotypes, we have to increase the **aa** × **aa** combination (in the lower right diagonal) threefold. Note that with dominance and assortative mating each genotype enters into matings with three other genotypes and produces a total of 12 offspring (4 in each family). So we have to add two more families to the **aa** × **aa** mating combination in order that our calculations will be directly comparable to all the other examples, in which all of the genotypes have equal numbers of matings and offspring.

Most of the major effects of assortative mating can be seen in Table 4.4 by comparing the statistics based on the random and assortative mating populations. Note that assortative mating has slightly different effects when there is dominance.

It should be remembered that this example, for the sake of simplicity, is based on *perfect* assortative mating, that is, a mother–father correlation of 1, which, of course, is much greater than the degree of assortative mating known for any human trait (except the assortative mating coefficient for sex, which is, of course, −1!). The correlations between mates for phenotypic intelligence (e.g., IQs) are found to fall mostly between .3 and .6. Furthermore, the hypothetical trait in this example has a broad heritability of 1, that is to say, nongenetic factors are assumed to contribute absolutely nothing to the total variance, which therefore consists entirely of genetic variance. Finally, the generality of this example is further limited by the particular frequencies of the genotypes when the gene pool contains equal proportions of **A** and **a.** Thus, the example is intended only for the didactic purpose of showing the reader how dominance and assortative mating affect means, variances, and kinship correlations, and, while the exact numerical values in Table 4 do not hold for all possible polygenic systems, all the quantitative relationships that can be expressed in terms of "greater than" or "less than" do hold with respect to the indicated effects of dominance and assortative mating.

Note that assortative mating increases the variance, the parent–child correlation, and the correlation among siblings. It is apparent that both the parent–child and sibling correlations are simultaneously increased by assortative mating and decreased by dominance. Thus, any empirically obtained value of such correlations cannot be taken at face value as agreeing or disagreeing with a genetic model without taking into consideration other types of evidence that provide information concerning assortative mating and dominance as well as the possible environmental contribution to the correlation.

The effects of assortative mating on the population variance and the degree of correlation between relatives, as illustrated by the simple two-allele example summarized in Table 4.4, can be generalized to polygenic systems with any number of gene loci and any proportions of the **A** and **a** alleles in the population. One need not work out the results from scratch, so to speak, as we have done for didactic purposes in the above example. Obviously, when a number of gene loci are involved, it would be prohibitively tedious to work out every mating combination and their progeny, as was done in Table 4.3. Fortunately, a bit of quite simple "Mendelian algebra" permits calculation of the theoretical values. It is this type of algebraic formulation, which is directly derived from the kind of Mendelian model we have illustrated, that is most useful in dealing with the analysis of polygenic traits. Since polygenic traits assume a large number of values that for all practical purposes may be regarded as continuous variables that have some kind of distribution in the population, the genetic analysis of such traits is in terms of the analysis of variance. Therefore, the Mendelian model is handled in such a way as to yield *variances,* the components of which are attributable to various genetic effects: those due to the additive genetic effects (called additive genetic variance or genic variance), those due to dominance, and those due to assortative mating. (Epistasis, i.e., variance attributable to interaction or nonadditive interaction effects between genes at different loci, is generally a small source of variance and can be disregarded in the present discussion.) How theoretical means, variances, and kinship correlations are derived from polygenic models by means of Mendelian algebra is explicated in the Appendix.

ASSORTATIVE MATING FOR HUMAN TRAITS: EMPIRICAL FINDINGS

Assortative mating involves many human characteristics, physical, psychological, and social. The variable showing probably the highest degree of similarity between spouses is age. In 1965, the average age difference between spouses was 2.7 years (Rele, 1965), and two studies separated by 45 years both reported a correlation between spouses' ages of +.76 (Hollingshead, 1950; Lutz, 1905). Assortative mating for religion is almost equally high (Hollingshead, 1950), and there is also substantial assortative mating for ethnic background and previous marital status.

Spuhler (1968) has presented a thorough review of the world literature on assortative mating for no fewer than 105 physical characteristics, listing the location, sample, size, and marital correlation for each characteristic. The correlations are generally quite low, nearly all are positive, and there is considerable sampling variation. Spuhler's (1968) summary:

> Correlation coefficients in the range +0.1 to 0.2 are most frequently observed for measurements of body size in Europeans and Americans of European descent, although coefficients smaller than 0.1 and in the range +0.2 to 0.3 are commonly found. Assortative mating coefficients above 0.5 are rare, at least in European peoples. The degree of assortative mating with respect to physical characteristics varies by economic and social class within European national and ethnic groups. . . . The degree of assortative mating experienced in contemporary European peoples probably has a small influence on the population distribution of inherited physical characteristics. Assortative mating leads to increased variability in the population as a whole. [p. 139]

Studies of assortative mating for personality traits also have been summarized by Spuhler (1967, p. 262) and by Vandenberg (1972). The marital correlations for traits such as neuroticism, extroversion–introversion, dominance, and self-sufficiency are all positive but tend to be quite low, ranging from +.02 to +.29, with a mean of about +.15. Similar marital correlations were found on Cattell's 16 Personality Factor Scores; when marriages were grouped as "stable" and "unstable," significant disassortative mating was found for several traits in the "unstable" group, that is, outgoingness (−.50), enthusiasm (−.40), self-sufficiency (−.32), and suspicious, self-opinionated (−.33) (Cattell & Nesselroade, 1967). Studies also have shown significant assortative mating for "social maturity" (Doll, 1937) and for proneness to psychiatric disorders (Gregory, 1959; Kreitman, Collins & Nelson, 1970; Nielsen, 1964; Penrose, 1944).

Assortative mating for mental abilities, especially intelligence, is much higher than for personality traits, as can be seen in Table 4.5, which summarizes all the assortative mating coefficients for mental ability the writer has been able to find in the literature. The median correlation is +.44. The unweighted mean correlation (based on Fisher's Z transformation) is +.45, and the weighted (by N) mean is +.42.

The substantial degree of assortative mating for intelligence is not only a result of men's and women's perspicacity in judging one another's intelligence but of their propinquity in social class and formal education. The sources from which potential mates are derived are generally of similar neighborhood, social class, and occupational background, and there is greater than chance likelihood that persons of the same background will be of similar intelligence. The educational system today is probably the most instrumental in assortative mating. Schools and colleges tend to sort out persons largely according to their intellectual abilities, and the greatest selection takes place in college admissions. It has been said that in America there is a college for almost every level of aptitude. Single persons of marriageable age, mostly between 18 and 22,

TABLE 4.5

Assortative Mating Coefficients (Correlations between Parents) for Various Mental Tests

Test variable	Author	Location	Number of pairs	Correlation
Army Alpha	Jones (1928)	Rural New England	105	+.598
Army Alpha	Conrad and Jones (1940)	"	134	+.52
Army Alpha	Outhit (1933)	United States and Canada	51	+.74
Army Alpha: number series completion	Willoughby (1928)	Palo Alto, California	141	+.20[a]
Army Beta: symbol series completion	"	"	"	+.55[a]
Army Beta: geometric forms	"	"	"	+.33[a]
Army Beta: symbol digit	"	"	"	+.65[a]
National Intelligence Test: opposites	"	"	"	+.37[a]
National Intelligence Test: arithmetic reasoning	"	"	"	+.47[a]
National Intelligence Test: analogies	"	"	"	+.30[a]
National Intelligence Test: checking similarities	"	"	"	+.45[a]
Stanford Achievement: sentence meaning	"	"	"	+.48[a]
Stanford Achievement: science–nature information	"	"	"	+.46[a]
Stanford Achievement: history–literature information	"	"	"	+.58[a]
Sum of five verbal tests	"	"	"	+.44[a]
Sum of six nonverbal tests	"	"	"	+.44[a]
Vocabulary	Carter (1932)	California	108	+.21
Arithmetic	Carter (1932)	California	108	+.03
Scholastic Achievement Test	Burt (1972)	England	95	+.678
Stanford–Binet IQ (foster parents)	Burks (1928)	California	174	+.42
Stanford–Binet IQ (parents raising own children)	Burks (1928)	California	100	+.55
Stanford–Binet Vocabulary (adoptive parents)	Leahy (1935)	Minnesota	174	+.61
Stanford–Binet Vocabulary (parents raising own children)	Leahy (1935)	"	164	+.43
Otis IQ	Freeman *et al.* (1928)	Illinois	150	+.49
Otis IQ (adoptive parents)	Leahy (1935)	Minnesota	177	+.57
Otis IQ (parents raising own children)	Leahy (1935)	"	173	+.41
Group Intelligence Test	Burt and Howard (1956)	England	?	+.453
Group Intelligence Test	Burt (1972)	England	95	+.379

TABLE 4.5 *(Continued)*

Test variable	Author	Location	Number of pairs	Correlation
IQs (various tests) of husbands and wives obtained as schoolchildren	Reed and Reed (1965, p. 57)	Minnesota	1866	+.326
Raven's Standard Progressive Matrices	Halprin (1946)		324	+.76
Raven's Standard Progressive Matrices	Spuhler (1962)	Michigan	180	+.399
Raven's Standard Progressive Matrices	Guttman (1974)	Israel	100	+.26
Primary mental abilities (verbal meaning), total number correct	Spuhler (1962)	Michigan	151	+.305
Primary mental abilities (verbal meaning), proportion of correct answers on items attempted	"	"	151	+.732
Wechsler Adult Intelligence Scale—factor score: verbal comprehension	Williams (1973)	Canada	57	+.65
Wechsler Adult Intelligence Scale—factor score: perceptual organization	"	"	57	+.16
Wechsler Adult Intelligence Scale—factor score: freedom from distractibility	"	"	57	+.20
Wechsler Adult Intelligence Scale—factor score: memory	"	"	57	+.31
Wechsler Adult Intelligence Scale—factor score: *g* (general factor)	"	"	57	+.69
Terman Concept Mastery Test (parents of gifted children)	Stanley (1972)	Baltimore, Maryland	22	+.15
Various tests	Smith (1941)		433	+.19
Various tests	Higgins *et al.* (1962)	Minnesota	1016	+.33
"Mental Grade"	Penrose (1933)	England	100	+.44
Cattell Culture-Free Test of *g*. In "stable" marriages.	Cattell and Nesselroade (1967)	Illinois	102	+.31

[a] Corrected for attenuation.

are thus brought together in terms of aptitudes and interests and also the family background factors that enter into the choice of a college. High school and college dropouts tend to marry other dropouts of comparable education, and graduates marry other graduates. There is a higher degree of assortative mating for sheer number of years of formal schooling than for any other human characteristic yet investigated, with the exception of age, which is of negligible consequence genetically. The correlation between mates in amount of education undoubtedly underestimates the true correlation in educational aptitude and attainments since the reported correlations based on *amount* (meaning years) of education do not take into account the academic standards of the high schools and colleges involved.

The most extensive data on assortative marriage for amount of education were collected in 1962 by the U.S. Bureau of the Census (Warren, 1966). More than 33,000 households were sampled in such a way that the results could be generalized to the noninstitutional population of the United States in the 1960 Census with regard to age, color, and sex. The husband–wife correlations for education, socioeconomic origin status, and number of siblings are shown in Table 4.6. Note that the correlations for education are about the same for Caucasoids and non-Caucasoids. When education is partialed out of the husband–wife correlations for socioeconomic status, the latter correlations are reduced by approximately one-half. The low but significant correlations for number of siblings shows that there is also assortative mating for family size.

The educational, socioeconomic, and ability factors involved in assortative mating are highly interrelated, as shown in Table 4.7, from a recent study by Williams (1973). Each variable is based on a number of assess-

TABLE 4.6

Correlations for Education, Socioeconomic Origin Status, and Number of Siblings of Married Couples by Age of Wife and Color

Color and wife's age	N^a	Education	Socioeconomic origin status	Number of siblings
White				
22–26	3726	.63	.31	.21
27–31	4077	.61	.27	.17
32–36	4685	.59	.25	.19
37–41	4907	.55	.31	.17
42–46	4311	.62	.30	.20
47–51	3785	.60	.29	.16
52–56	2810	.60	.32	.15
57–61	1846	.63	.37	.18
Nonwhite				
22–26	393	.52	.30	−.02
27–31	417	.62	.21	.07
32–36	473	.70	.37	.20
37–41	458	.39	.16	.13
42–61	984	.62	.11	.01

Source: From Warren, 1966.

[a] Estimated population in thousands.

TABLE 4.7

Intercorrelations[a] among Family Variables

Variable	2	3	4	5	6	7
Father's intelligence[b]	.61	.64	.75	.74	.48	.47
Mother's intelligence[b]		.56	.59	.60	.35	.39
Father's occupational prestige			.82	.68	.61	.41
Father's education				.71	.55	.42
Mother's education					.58	.40
Family income						.21
Child's intelligence[c]						

Source: From Williams; 1973.

[a] Correlations based on number of families varying from 55 to 100.

[b] Full Scale IQ on the Wechsler Adult Intelligence Scale.

[c] Full Scale IQ on the Wechsler Intelligence Scale for Children.

ments. The intelligence measure is the Full Scale IQ based on the 11 subtests of the Wechsler Intelligence Scales.

Some traits are subject to assortative mating only through their association with other traits that are more directly related to the variables involved in mate selection. An interesting example is provided by Guttman (1970) in a study of the judgment of visual number. Subjects were shown sets of objects in transparent plastic bags for 3–4 seconds and asked to judge the number of objects. The subject's score was simply the number of objects he judged there to be in each of five bags containing different objects. The correlations between the scores of 595 sets of mothers and fathers and between parents and offspring are shown in Table 4.8. The heritability is estimated by the formula $h^2 = 2r_{\text{po}}/(1 + r_{\text{pp}})$. (This formula for h^2 ignores a possible environmental correlation between parent and child.) Note that there is assortative mating for only two of the tasks and that only three of the tasks show any heritability. It is likely that the first two tasks involve some factor in common with general intelligence, for which there is substantial assortative mating. Mates surely are not attracted to each other in any direct way on the basis of their ability to estimate the number of Ping-Pong balls and marbles. One may wonder in how many other behavioral characteristics mates show unsuspected resemblance.

Since the factors most instrumental in assortative mating for mental abilities—educational and occupational selection—are probably more highly differentiating in the upper and lower extremes than in the middle part of the normal distribution of abilities, we can suspect that there may be a higher degree of assortative mating for intelligence at both the higher and lower levels than at the average level of intelligence. However, there is presently little evidence in the literature that bears directly

TABLE 4.8

Mother–Father, Parent–Child, and Sibling Correlations on Five Measures of Number Estimation

Objects estimated	Father–mother	Parent–offspring	Mother–daughter	Mother–son	Father–daughter	Father–son	First sibling–second sibling	h^2
5 Ping-Pong balls	.62	.46	.42	.41	.45	.52	.42	.57
6 large marbles	.46	.48	.46	.50	.50	.46	.36	.66
34 small marbles	.03	.21	.16	.19	.33	.10	.29	.42
15 large marbles	.04	.04	−.01	−.13	.14	.01	−.14	.00
15 small marbles	.14	−.01	−.07	.02	.04	−.01	.25	.00

Source: From Guttman, 1970, pp. 60–61.

on this hypothesis. (See Eckland, 1972b.) But there are two predictable genetic consequences of a higher degree of assortative mating in the upper than in the lower half of the IQ distribution, and evidence for both of these has been reported in the literature: *(a)* a larger proportion of very high IQs (over 130) than would be expected in a normal or Gaussian distribution (Burt, 1943) and *(b)* a higher sibling correlation for scholastic achievement in the upper than in the lower half of the IQ distribution (Burt, 1943, p. 91). For scholastic achievement, Burt found a correlation of .61 between sibs with IQs over 100 and of .47 between sibs with IQs below 100. Both of these findings are consistent with the hypothesis of a higher degree of assortative mating among the more intelligent parents, but they are also consistent with another hypothesis, namely, a greater degree of covariance between genetic endowment and environmental influences on IQ and scholastic performance among intellectually superior children. On the other hand, there are factors that may work against higher assortative mating at the upper end of the IQ distribution: *(a)* females, on the average, achieve less than males; hence, high-IQ females are not as easy to identify as high-IQ males; *(b)* due to the lower probability density of very high IQs in the population, assortative mating for high IQs would be more difficult, that is, the probability of a very high-IQ person's meeting a person of similar IQ is relatively small. The spouses of Terman's gifted group (whose mean IQ was 152) averaged some 25 points lower than their gifted mates (Terman & Oden, 1959).

Relationship of Assortative Mating to Fertility

If fertility is related to assortative mating, then assortative mating might alter gene frequencies as well as genotype frequencies in the population, and this can have long-term evolutionary consequences for those traits involved in assortative mating. Therefore, it is of interest to geneticists to find out if and how the degree of resemblance between mates for various traits is related to their fertility, that is, their total number of progeny.

Spuhler (1962) reported the correlations, based on 183 married couples in Michigan, between an index of mate similarity in 25 anthropometric variables and the couple's fertility. Only 1 of the 25 *r*s was significant at the .05 level, and it was only +.18; more significant *r*s would be expected by chance alone. All the other correlations were nonsignificant and negligible. In reviewing this study elsewhere, Spuhler (1968) concluded with respect to physical measurements that "while assortative mating acts to change the distribution of genotypes concerned with body size, this selective mating is not an important mode of evolutionary change in this population because there is little or no differential association between fertility and assortative mating in the population (p. 138)."

The same may not hold true for behavioral traits, although the available evidence is too scant for any strong conclusions on this point. There is, however, quite good but not completely definitive evidence for a relationship between fertility and mate resemblance in educational level. Based on extensive data from the 1960 U.S. Census, Kiser (1968) found the following relationships: *(a)* Assortative mating is strongest at the extremes of educational level; those with the most or the least education are most apt to marry someone of highly similar educational level. *(b)* This is a stronger tendency for Caucasoid wives than for Caucasoid husbands or non-Caucasoid wives. *(c)* At the college level, mate similarity in education was greater for non-Caucasoid husbands than for either Caucasoid wives or Caucasoid husbands (who showed the least assortative mating). *(d)* For persons with less than an eighth grade education, assortative mating was strongest for Caucasoid husbands and non-Caucasoid wives and weakest for non-Caucasoid husbands. *(e)* The degree of mate resemblance in educational level is positively related to fertility in Caucasoid and negatively related in non-Caucasoids except that *(f)* for both Caucasoids and non-Caucasoids, assortative mating increased the fertility of those of lowest educational attainment, that is, when *both* mates had very little education, they produced more children than couples of which one mate had more education. *(g)* The overall effect of assortative mating for education, taking into account the different proportions of the population at different educational levels, is an increase in fertility of about 10% above the general average for Caucasoids and a negligible amount for non-Caucasoids (1–2%). Kiser (1968) states: "Although assortative mating appears to have a depressing effect on the fertility of nonwhites at all educational levels except the lowest, the proportion of nonwhites in the lowest educational levels apparently was sufficient to yield a small positive impact of assortative mating on the fertility of the total nonwhite groups considered" (p. 112).

What all this may mean, of course, is that a larger proportion of the next generation comes from parents who are more alike in educational attainment (and probably also in intelligence) than from parents who are less alike. This magnifies all the effects of assortative mating in the population, such as increasing the genetic part of the variance in the traits involved in educational level, especially increasing the between-families portion of the total variance. At the genetic level, the educational haves and have-nots, so to speak, are pulled even farther apart by assortative mating as a result of the fact that the greater the degree of mate resemblance in education, the greater the fertility. The net effect—a 10% increase in fertility over what would exist under random mating—is not negligible, at least in the Caucasoid population. It may also help to account for the greater variance in IQ in the Caucasoid than in the Negroid population of the United States. (See Jensen, 1973b, pp. 211–216.) That assortative mating for education by non-Caucasoids (of which, in the 1960 U.S. Census, Negroids comprised about 98%) depresses fertility at every educational level except the lowest (less than eighth grade) suggests a dysgenic trend for Negroids for traits related to educability. That is to say, a disproportionate number of the next generation of Negroids will come from the least educated. To the extent that educational level reflects genetically conditioned ability factors, this could mean a genetic downgrading of the Negroid population in absolute terms and especially relative to the Caucasoid population. In the Caucasoid population, assortative mating for educational level has the opposite effect, that is, it enhances fertility. These differential trends revealed in Kiser's analysis of the 1960 Census data surely are of sufficient social import to merit much further investigation.

EFFECTS OF INBREEDING

The effects of inbreeding on a trait afford one of the most compelling lines of evidence that genetic factors are involved and particularly that there are genes evincing *directional dominance*. Therefore, evidence of inbreeding effects predictable from genetic theory is of major significance to students of the inheritance of mental ability.

Inbreeding is mating between persons who share genes derived from a common ancestor. The degree of inbreeding is a function of the number of common ancestors and the number of generations they are removed. Brother–sister and parent–offspring matings are the highest degree of inbreeding found in humans. In sibling matings, the ancestry of both mates is shared in common. In parent–offspring matings, the parent and offspring have only half their ancestors in common, but they share as many genes through common descent as a brother and sister.[8] Since, in any finite population, if one could trace pedigrees enough generations back it would be found that nearly everyone has ancestors in common, there is naturally some very low average base rate of inbreeding in any existing population. In theoretical discussions, this average amount of inbreeding in the population is arbitrarily regarded as zero, and matings

[8] Another distinction between brother–sister matings and parent–offspring matings is that a brother and sister share half their genes only on the average, in a probabilistic sense, while parent and offspring share half their genes in an absolute, causal sense. Though exceedingly improbable, theoretically two siblings could have *no* genes in common, while an individual *must* have received half of his genes from each of his parents. Because of this difference, sociologist Robert A. Gordon (personal communication) has suggested the hypothesis that the offspring of incestuous unions that escape severe effects of inbreeding should be more often of brother–sister matings than of parent–child.

that are positive deviations from this baseline are considered instances of inbreeding.

The quantification of inbreeding becomes a highly involved and technical subject that can only be touched on here. It is one of the major topics in population genetics and is treated in detail in textbooks in this field (e.g., Bodmer & Cavalli-Sforza, 1971, Chap. 7; Crow & Kimura, 1970, Chap. 3; Wright, 1969). The method for determining the coefficient of inbreeding from pedigrees was originally proposed by Sewall Wright (1922). Wright signified the inbreeding coefficient as f (sometimes given as F) because of the tendency of inbreeding to *fix* homozygotes, that is, to produce and maintain gene combinations such as **AA** and **aa** at the expense of the heterozygote **Aa.** All the effects of inbreeding follow from its tendency to increase homozygosity at all gene loci.

A few simple examples will serve to illustrate how the value of the inbreeding coefficient, f, is determined from pedigrees. (The working out of much more complicated pedigrees is well explicated by Crow & Kimura, 1970, Chap. 3.) A father–daughter mating can be represented as follows:

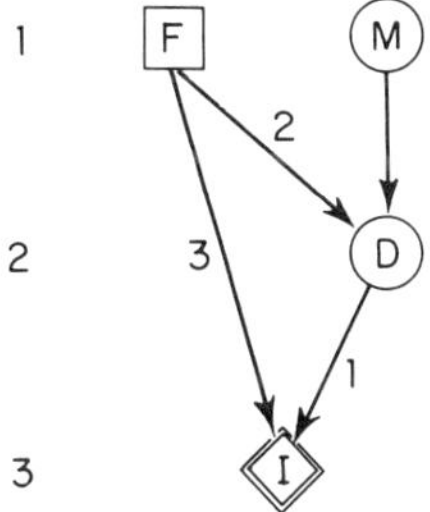

Squares stand for males, circles for females, and diamonds for either sex. Father, mother, and daughter are indicated by their initials, and I stands for the inbred individual resulting from inbreeding, in this case between father and daughter. Each generation is indicated by the numbers on the left. Now, the coefficient of inbreeding, f, is the probability that for any given gene locus, I will receive the same allele, say **A,** on both homologous chromosomes from the common ancestor, in this case, the father. Notice the lines in the diagram. They connect only individuals between whom genes are directly transmitted, and the arrows indicate the direction of transmission. The probability that allele **A** or **a** will be passed from F to D, of course, is 1; the probability that the same allele will be passed from D to I is $\frac{1}{2}$; and the probability that the same allele will be passed from F to I is also $\frac{1}{2}$. The joint probability that I will receive two of the same alleles from the father, therefore, is $1 \times \frac{1}{2} \times \frac{1}{2} = \frac{1}{4}$, which is f, the coefficient of inbreeding for this case. There is a simple rubric for getting f: count the number (n) of individuals through whom the arrows pass in completing the loop (I)-D-F-(I) from I back to I (but not counting I). The inbreeding coefficient, then, is simply $\frac{1}{2}$ raised to the power n, that is, $f = (\frac{1}{2})^n$. If one or more of the common ancestors is himself an inbred individual, the formula becomes $f = (\frac{1}{2})^n (1 + f_A)$, where f_A is the inbreeding coefficient for the inbred ancestor. This formula is more generally written

$$f = \Sigma(\tfrac{1}{2})^n(1 + f_A)$$

because in more complicated pedigrees there is more than one loop from I back to I and the joint probabilities of each of the loops must be summed.

A brother–sister mating is the simplest example of this:

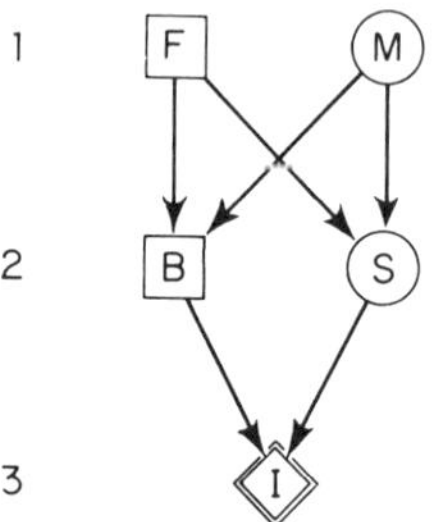

Here there are two common ancestors of I; that is, F and M, and two loops to be traversed: (I)-S-M-B-(I) and (I)-B-F-S-(I). The number of individuals through whom the arrows pass (not including I) is 3 in each case. So $f = (\frac{1}{2})^3 + (\frac{1}{2})^3 = \frac{1}{4}$, which is the same coefficient of inbreeding that resulted from the father–daughter mating.

For a slightly more complex pedigree, there is the mating of first cousins:

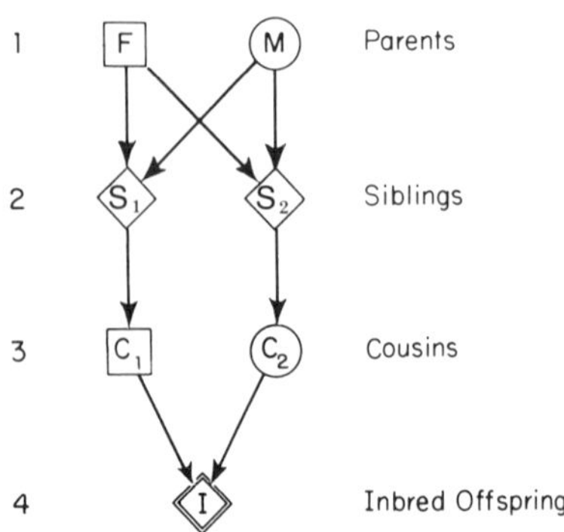

The loops are: (I)-C_1-S_1-F-S_2-C_2-(I) and (I)-C_2-S_2-M-S_1-C_1-(I). (Note that only two complete loops are needed to traverse all the paths; one never completes more loops than are needed to traverse all the paths and to pass through each of the remotest common ancestors [in this case, F and M] no more than once.) So here $f = (\frac{1}{2})^5 + (\frac{1}{2})^5 = \frac{1}{16}$.

An uncle–niece mating:

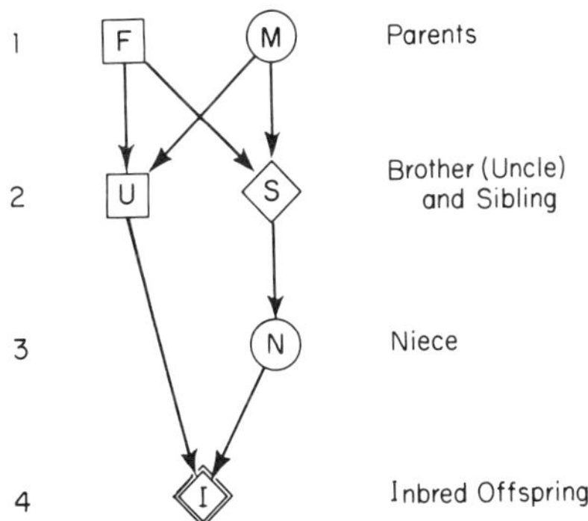

The loops are: (I)-U-F-S-N-(I) and (I)-N-S-M-U-(I), and $f = (\frac{1}{2})^4 + (\frac{1}{2})^4 = \frac{1}{8}$.

In the same manner, one can work out the inbreeding coefficients for many other pedigrees. Table 4.9 gives some examples of the values of f for various types of mating.

TABLE 4.9

Inbreeding Coefficient f for Various Matings

Relationship of parents	f
Self-mating	$\frac{1}{2}$
Parent–offspring	$\frac{1}{4}$
Full siblings	$\frac{1}{4}$
Half siblings	$\frac{1}{8}$
Grandparent–grandchild	$\frac{1}{8}$
Uncle–niece (or aunt–nephew)	$\frac{1}{8}$
Double first cousins	$\frac{1}{8}$
First cousins	$\frac{1}{16}$
Double half cousins	$\frac{1}{16}$
First cousins once removed	$\frac{1}{32}$
Second cousins	$\frac{1}{64}$
Second cousins once removed	$\frac{1}{128}$
Third cousins	$\frac{1}{256}$
Random mating	0[a]

[a] Arbitrarily set at zero, though it actually has some very small value, varying from one population to another.

The value of f in every case is just half the genetic correlation between the mates, assuming that the mates themselves are not inbred. Self-mating does not occur in humans, but in species in which it may (as in corn and wheat), the f value of the inbred offspring is $\frac{1}{2}$.

The inbreeding coefficient f has two other meanings besides the one already mentioned, namely, the probability that the offspring will receive two of the same alleles from the same ancestor.

The second meaning of f is this: f is the proportion by which heterozygosity is decreased, that is, the reduction in frequency of **Aa** combinations relative to all possible combinations (**AA**, **Aa**, and **aa**). Note that inbreeding per se, like assortative mating, does not change gene frequencies in the population but only genotype frequencies. It converts heterozygosity to homozygosity at all gene loci at a rate directly proportional to the average inbreeding coefficient.

The third meaning of f is expressed as a correlation coefficient: If the different alleles carried at each locus by each parent were numbered (e.g., **A** = 1, **a** = 0), the coefficient of correlation r between the arrays of numbers from each of the parental gametes uniting to form the offspring is equal to the inbreeding coefficient f.

From a practical standpoint, inbreeding is especially important for two main reasons:

First, it greatly increases the probability of producing homozygous recessives, and since most deleterious characteristics are recessive, inbreeding increases the risk of genetic defects in the progeny; this is especially marked in the case of single-gene characteristics such as albinism, alkaptonuria, limb–girdle muscular dystrophy, congenital deafness, and causes of severe mental defect such as phenylketonuria, galactosemia, microcephaly, and Tay Sachs disease. The latter, for example, occurs with a frequency of about 1 in 40,000 births in the general population, but something between 11 and 40% of these cases result from first-cousin matings, varying in different populations (Penrose & Haldane, 1969, p. 163).

Second, for polygenic traits, inbreeding alters the mean of the inbred progeny, depending on the degree and direction of dominance variance in the trait in question. This effect is wholly dependent on the existence of some directional dominance. Since genes themselves are a product of the evolution of the species, the more advantageous genes tend to be dominant, and generations of selection pressures acting in one direction increase the number of dominant genes (and the proportion of dominance variance). Genes for deleterious or nonadaptive characteristics are gradually eliminated from the population to the point that their frequency approximates the rate of spontaneous mutations in the population. Many harmful genes, almost always recessive, are maintained at some low and more or less constant frequency in the population by mutations not induced by man. In some cases, however, the heterozygous condition confers some benefit to the individual that maintains the recessives at a high frequency in the population even though the recessive homozygous condition may often be lethal. A good example is the recessive gene for the sickling trait, which, in the homozygote, produces severe, often fatal anemia, while the heterozygous condition confers a high resistance to malaria. This obvious benefit maintains the sickling gene at a fairly high frequency in tropical areas in which malaria is common. Another advantage of heterozygosity is related to what Lerner (1954) has termed genetic homeostasis, which consists, in part, of the fact that for many characteristics the heterozygous individuals are more buffered against harmful environmental influences.

The effect of inbreeding on the mean can be expressed as follows. As shown in the Appendix, the mean $(\bar{X})$ of the population for a given trait under random mating is

$$\bar{X}_O = \Sigma a(p - q) + 2\Sigma pqd,$$

where

$\bar{X}_o$ = the mean under random mating;
a = the difference between the values of the homozygotes (e.g., **AA** − **aa**);
p = the proportion of the trait-enhancing allele (e.g., **A**) in the population;
q = 1-p, that is, the proportion of the other allele (**a**) in the population;
d = the dominance deviation, which is equal to the difference between the value of the heterozygote and the average value of the two homozygotes, that is, **Aa** − (**AA** + **aa**)/2.

The summation sign Σ indicates that for a polygenic trait these values are summed over all contributing loci. The effect of inbreeding is to depress the mean by an amount directly related to the inbreeding coefficient, that is, $2f\Sigma pqd$, so that the mean under a given degree of inbreeding $(\bar{X})$ is

$$\bar{X}_F = \bar{X}_O - 2f\Sigma pqd$$

This effect is known as *inbreeding depression.* From this formulation, it can be seen that without directional dominance deviation (i.e., when $\Sigma d = 0$), inbreeding has no effect on the mean.

Like assortative mating, inbreeding increases the genetic variance, but not so quickly and not so much, especially at the low levels of inbreeding, which are all that occur in humans. If there is no dominance, the total genetic variance, V_G, is increased to $(1 + f)V_G$, the variance *between* inbred families (or lines) is increased to $2f V_{GB}$, and the variance *within* inbred families is decreased to $(1 - f)V_{GW}$. When there is dominance, the effect of inbreeding on these variance components involves a much more complicated formulation, for which readers are referred elsewhere (Crow & Kimura, 1970, pp. 343–344). Suffice it to say here that the total genetic variance, V_G, and the genetic variance *between* families are monotonically increasing functions of f, but the genetic variance *within* families has a curvilinear (quadratic) relationship to f, increasing up to $f = \frac{1}{2}$ and decreasing thereafter. Obviously, this is relevant only to very highly inbred lines, which do not occur in human populations.

Inbreeding in Human Populations

Before taking up the subject of the specific effects of inbreeding on human characteristics, we should gain some impression of the "baseline" or average amount of inbreeding in modern human populations. Doubtless, the average amount of inbreeding has decreased markedly in modern times. There was a much higher average degree of inbreeding during the first 99% of human evolutionary history. From ethnographic data, Spuhler (1967, p. 251) claims that in primitive populations human bands did not exceed a total of 200–300 persons of all ages, or 50–75 breeding pairs, throughout the history of the human species before agriculture. This is still true today for food-gathering peoples without agriculture or domestic animals other than the dog. Some inbreeding occurs in any small society that remains isolated for a few generations simply due to a restriction of the breeding population. With the advent of agriculture, there began a marked increase in the size of human communities. This provided a wider range of mate selection and consequently decreased the amount of inbreeding. The increased mobility of modern man has worked in the same direction.

The average inbreeding coefficients estimated in various modern populations are very low by comparison with small and isolated primitive societies. Spuhler (1967, p. 251) and Morton (1961, p. 264) reported some representative figures for the mean f of various populations, as shown in Table 4.10. The percentage of cousin marriages in these populations varies from close to zero to more than 10, and these percentages are highly related to the mean f of the population. Spuhler (1967, p. 251) notes that in various smaller subpopulations such as some of the rural populations of Europe, North America, and black Africa, numbering some 1500 to 4000 persons, with about 80% endogamy, most of the members are related to one another as third cousins, with $f = .0039$. The highest inbreeding reported for any human group occurs among the Samaritans, with $f = .043$ (Bonné, 1963).

While the overall effects of inbreeding in any given generation are deleterious because of inbreeding depression and the increased frequency of homozygosity of recessive genes with major harmful effects, in a long-term evolutionary sense, inbreeding has a beneficial effect on the

TABLE 4.10

Mean Coefficient of Inbreeding in Various Modern Populations

Population	Mean f
France (1926–1945)	.00066
England and Wales (1924–1929)	.00028
England (hospitals)	.0004
Japan (three cities)	.0035
Brazil (Sáo Paulo)	.0005
Germany (Munster)	.0002
India (Marathas)	.0064
United States (rural)	.0006

population by exposing deleterious genes so that they can be eliminated from the gene pool by selection. As the degree of inbreeding is reduced, however, more and more undesirable recessives remain hidden, carried and transmitted by heterozygous individuals. Animal breeders at times strongly inbreed their lines in order to rid them of undesirable characteristics, which are systematically culled, and then the lines are outcrossed again to counteract inbreeding depression.

Effects of Inbreeding: Empirical Evidence

Inbreeding depression is manifested in many quantitative human characteristics: birth weight, height, head circumference, chest girth, muscular strength, fetal and infant viability, resistance to infectious disease and dental caries, rate of physical maturation, and age of walking, to name a few that are well documented in the genetics literature. Mental ability is similarly subject to inbreeding depression to as great or even a greater degree than most of the physical features mentioned above.

A demonstration of inbreeding depression in human populations depends on comparing an inbred group with an outbred or control group. The choice of a proper control group can be especially problematic in the study of inbreeding depression in behavioral traits, such as mental abilities, since these traits themselves may be correlated with inbreeding. In Japan, for example, in which there is a relatively high percentage (about 5%) of cousin marriages, it has been found that in urban communities inbreeding is negatively correlated with socioeconomic status, while in rural communities there is a positive correlation (Schull & Neel, 1972). Therefore, a control group must be carefully matched with the inbred groups on relevant characteristics other than the trait under study.

The most striking feature of inbreeding is the greater risk it incurs for the appearance of rare defects due to single recessive genes in the homozygous condition. Morton (1961, p. 277) has pointed out, for example, that a lowering of the inbreeding coefficient from .006 to .0006 should result in 50% reduction in the frequency of rare recessive defects. Thus, such genetic risks are amplified considerably by consanguineous matings such as first cousins, where $f = \frac{1}{16}$, to say nothing of incestuous matings (sibs and parent-child), with $f = \frac{1}{4}$. The average risk of genetic defects is approximately doubled in the offspring of first-cousin matings as compared with the offspring of unrelated mates (Bodmer & Covalli-Sforza, 1971, p. 369), and since the risk is a linear function of f, it would be increased threefold in uncle–niece matings, and eightfold for incestuous matings (parent–child and brother–sister). For individuals born with quite rare genetic defects, particularly various forms of severe mental deficiency, the percentage of their parents who are first cousins is some 20–40 times as great as the percentage of first–cousin matings in the general population (Bodmer & Cavalli-Sforza, 1971, Table 7.14, p. 370).

Inbreeding Effects on Mental Abilities

Some of the most impressive evidence for the polygenic inheritance of mental abilities is afforded by studies of the effect of inbreeding on measured intelligence and scholastic performance. That these characteristics generally show somewhat greater inbreeding depression than most quantitative physical characteristics studied suggests that they involve more genetic dominance for genes that enhance mental ability. It also suggests but does not prove the inference, in accord with genetic principles, that mental ability has evolved in humans under somewhat greater selection pressure than has been the case for most anthropometric traits.

Cohen, Block, Flum, Kadar, and Goldschmidt (1963) compared 38 offspring of cousin marriages in Israel with 48 offspring of unrelated couples who were matched with the cousin parentages in ethnicity and other variables related to intelligence. The offspring of the cousin marriages averaged about .2–.3 standard deviation below the noninbred offspring on every one of the seven subtests of the Wechsler Intelligence Scale used in this study.

A much larger and methodologically more sophisticated study was conducted in Japan by Schull and Neel (1965). Approximately half of the subjects were the offspring of various degrees of cousin marriage, and half of them (the "controls") were the offspring of unrelated parents. The mean f of the inbred children was about 0.05, or slightly less than the degree of inbreeding resulting from first-cousin parentage with $f = .0625$. The two parental groups were, in effect, statistically equated by multiple regression analysis for a number of relevant background factors: age, education, and occupational level (all on both parents), number of persons in the household, and various indices of socioeconomic status. The inbred and control offspring themselves were statistically equated for birth rank, age at examination, and month of examination. A multitude of physical and behavioral measurements were obtained on all the children, including the Japanese version of the Wechsler Intelligence Scale for Children (WISC). The amount of inbreeding depression on the 11 subscales of the WISC can be expressed in various ways, based on the elaborate statistical analyses of Schull and Neel. The most readily interpretable indices are those suggested by Vandenberg (1971, pp. 201–202) in terms of *(a)* the amount of depression of inbred children's scores expressed as a percentage of the noninbred mean, *(b)* the *expected* decrease in score for offspring of incestuous (i.e., parent–child or brother–sister) matings, as estimated from the observed effects of cousin matings, and *(c)* the amount of lowering in subjects' age that would decrease the test score by an amount equivalent to the inbreeding depression resulting from first-cousin matings. These figures, based on approximately 1850 children given the WISC in Hiroshima, are presented in Table 4.11.

In school marks for various subjects, the amount of inbreeding depression was only slightly less than for the WISC subscales, as can be seen in Table 4.12, based on approximately 4650 children.

The catastrophic inbreeding effects resulting from incestuous mating, it has been convincingly argued, is the basis for the strong universal taboo found in all known societies, including primitive peoples (Lindzey, 1967). Three studies illustrate the severity of these effects. The offspring of seven brother–sister and six father–daughter unions were investigated at 4–6 years of age by Carter (1967). Only 5 of the 13 were in the normal range of IQ; 1 child was too severely retarded to be tested, and 4 were retarded, with IQs between 59 and 76; 3 had died of recessive genetic diseases that are extremely rare in the general population.

A similar study of 18 incestuous matings (12 brother–sister and 6 father–daughter) included a noninbred control group of children in which each mother was carefully matched with another of an inbred child for age, height, race, socioeconomic status, and intelligence (Adams & Neel, 1967; Adams, Davidson, & Cornell, 1967). The authenticity of consanguinity was checked by extensive blood-group testing. The IQs of the 18 mother-matched pairs are shown in Table 4.13. Some of the incestuous parents were from middle-class families and included college graduates. Note that several of the inbred progeny have fairly high IQs, which would be expected to result from the great increase in homozygosity when $f = \frac{1}{4}$ and when there is partial dominance, that is, the dominant homozygote is superior to the heterozygote. Recall that although inbreeding lowers the mean, that is, inbreeding depression, it also increases the variance.

TABLE 4.11

Effects of Inbreeding on Subtest Scores of the Wechsler Intelligence Scale for Children (WISC)

WISC Subtest	Depression as percentage of noninbred mean[a]	Expected depression (in σ units[b]) of offspring of incestuous matings	Lowering of age producing a score decrease equivalent to inbreeding depression from first-cousin mating[c]	
			years	months
Information	8.1– 8.5	.97	1	2
Comprehension	6.0– 6.1	.61	1	5
Arithmetic	5.0– 5.1	.50		11
Similarities	9.7–10.2	.96	1	9
Vocabulary	11.2–11.7	1.00	1	3
Picture completion	5.6– 6.2	.54	2	6
Picture arrangement	9.3– 9.5	.90	2	1
Block design	5.3– 5.4	.48	1	4
Object assembly	5.8– 6.3	.52	10	11[d]
Coding	4.3– 4.6	.44	1	1
Mazes	5.3– 5.4	.54	4	3
Verbal score	8.0– 8.0	.96	1	3
Performance score	5.1– 5.1	.78	1	0
Total IQ	7.0– 7.1	1.01	1	4

Source: After Schull and Neel, 1965.

[a] These figures merely give the range defined by the means of the sexes and do not compare male versus female, in that order. The values given are merely the range values ordered by size.

[b] σ based on Japanese norms for the WISC (Kodama & Shinagawa, 1953).

[c] Mean age of the children in the Schull and Neel study was 8 years 7 months.

[d] The scores of this subtest have such a low correlation with age that it requires a very large lowering of age to be equivalent to first-cousin inbreeding depression.

The largest study of children of incestuous matings involved 161 such births, but in this study the controls were the inbred children's half siblings, that is, they were born to the same mothers when impregnated by men not genetically related to the mother (Seemanova, 1971). Among the inbred offspring, there were 40 cases of moderate and severe mental retardation; there were no retardates in the control group. As in previous studies, the inbred children also showed much higher rates of mortality and physical malformations.

The results of these studies of the effects of inbreeding on mental ability are virtually impossible to explain in environmental terms, yet

TABLE 4.12

Inbreeding Depression in Offspring of Cousin Matings for Marks in Various School Subjects

Subject	Depression as percentage of noninbred mean[a]	
	Hiroshima	Nagasaki
Language	5.0–5.3	5.0–5.2
Social studies	4.4–4.5	4.3–4.3
Mathematics	6.4–6.5	6.3–6.3
Science	6.6–6.8	6.4–6.7
Music	6.1–6.9	6.1–6.6
Fine arts	4.7–5.2	4.7–4.9
Physical education	6.2–6.3	6.1–6.3

Source: From Schull & Neel, 1965, p. 308.

[a] These figures merely give the range defined by the means of the sexes and do not compare male versus female, in that order. The values given are merely the range values ordered by size.

TABLE 4.13

IQs of Offspring of Incestuous Matings and of Unrelated Parents (Controls) with Mothers Matched for IQ and SES

Case number	IQ	
	Incest group	Control group
1	Died at 2 months	101
2	Died after 15 hours	100
3	Died after 6 hours	104
4	Untestable: severely retarded	107
5	Untestable: severely retarded	93
6	64	100
7	64	133
8	64	109
9	85	103
10	68	81
11	92	108
12	98	108
13	110	91
14	112	105
15	113	91
16	114	85
17	118	121
18	119	95

Source: From Adams *et al.*, 1967.

they are predictable from genetic principles. Such studies indicate that genetic factors, including directional dominance, are involved in individual differences in human mental abilities.

OUTBREEDING AND INTERRACIAL CROSSES

Outbreeding, or exogamy, is the mating of persons from different, genetically isolated populations. Interracial crosses are an example of outbreeding. The degree of genetic isolation is a relative matter. In modern times, with improvements of transportation and with the breaking down of many social, religious, and economic barriers, the degree of genetic isolation of population groups has greatly diminished, with a consequent decrease in the average inbreeding coefficient in all populations, with small and rare exceptions. Outbreeding involves matings with a markedly lower inbreeding coefficient than exists for the average of the parent populations. Since virtually all human groups are already hybrids, technically speaking, the outcrossing of purebred lines, as seen in some plant and animal breeding, has no true counterpart in human genetics.

Unfortunately, from the standpoint of this review, the effects of outbreeding have been much less adequately investigated, especially with respect to behavioral traits, than the other topics discussed in this chapter. In terms of genetic theory, the effects of outbreeding should be the opposite of inbreeding, for just as inbreeding increases homozygosity, outbreeding increases heterozygosity. And just as inbreeding depresses traits that involve genetic dominance, so outbreeding enhances the same traits, an effect known as hybrid vigor or *heterosis*. In the outbred or hybrid generation, and particularly in the second filial generation (F_2), the total genetic variance of polygenic traits is redistributed; genetic variance between population means is converted to genetic variance among individuals. Skin color of racial hybrids is a good example. The pigmentation of African Negroids and of North Europeans, for example, are each homogeneous, but the mean difference is great. The first generation of offspring from interracial matings will be uniformly café au lait in color, but subsequent crosses among the hybrids will yield offspring with great variance in skin color, as the mean difference is distributed among individuals. (The observable variation of skin color among Negroid–Caucasoid hybrids can be explained on the basis of three or four loci with alleles of equal effect at the different loci and no dominance; Stern, 1970.)

One of the main difficulties of investigating empirically the outbreeding effects expected from genetic considerations is that, unlike inbreeding, extreme deviations from the mean f in the population do not occur in the outbreeding direction. If the opposite effects of inbreeding and outbreeding were perfectly symmetrical, even interracial crosses would not approach the extreme deviations from the average f in the population as represented by closely consanguineous and incestuous matings. Any observed effects of outbreeding or racial crossing in existing human populations, therefore, are found to be relatively small and hard to detect except for traits that show an extreme mean difference between the parent populations, such as skin color and hair form in African Negroids, say, as contrasted with European Caucasoids. Moreover, there is little chance of detecting heterosis in traits with large nonadditive genetic variance. The marked heterosis seen in agricultural plants and animals, such as hybrid corn, is made possible by a severe degree of selection to produce highly inbred lines emphasizing specific characteristics. It has no parallel in human populations.

The closest examples in the research literature consist of small, isolated communities that, because of their geographical separation and their small breeding population, have a markedly higher average degree of inbreeding, by at least an order of magnitude, than larger, nonisolated human groups. When individuals from two such isolate populations mate, their offspring seem to manifest some of the genetically expected effects of outbreeding, such as heterosis for some traits. Hulse (1957), for example, found an increase of 2 centimeters in the adult height of the offspring of parents who came from different, relatively isolated villages in Switzerland as compared with offspring whose parents were born in the same village. This effect has been noted as a partial explanation for the general increase in body size in Europe and America during the past century, with the increased mobility of people increasing heterozygosity. Some geneticists examining this evidence, however, have expressed doubts along highly technical lines as to its genetic interpretation and suggest other factors of an environmental nature that might explain the observed phenomena (e.g., Morton, 1961; see also Chung & Morton, 1966, which presents a concise review of the research on the genetics of interracial crosses). In the absence of sufficient evidence, there has been considerable and varied speculation (e.g., Kuttner, 1967, see Index on "hybridization, human," p. ix) about the possible and probable consequences of interracial crossing, based on theoretical genetic considerations. But such discussion is at best only a source of yet untested hypotheses, often of doubtful testability, considering the methodological limitations of human genetic research. Provine (1973) provides an interesting historical account of the shifting opinions and attitudes in the study of the genetic effects of race crossing.

The most adequate study of interracial crosses, in terms of genetic methodology, is that of Morton, Chung, and Mi (1967) in Hawaii. The study involved 180,000 births between 1948 and 1958 in Hawaii, in which one-third of the matings were interracial, among native Hawaiians, Chinese, Japanese, Puerto Ricans, Koreans, Filipinos, and

Caucasoids. Unfortunately from our standpoint, no behavioral measures were included. Only physical characteristics and morbidity rates were studied, and these showed few if any consistent outbreeding effects. It was concluded that the total effect of outbreeding relative to the opposite effects of inbreeding is less than that due to an inbreeding coefficient of .003 and might even be zero. In other words, most of the measured effects are simply additive, that is, the characteristics of the progeny are an arithmetic average of the parent populations. In the authors' words, "We find no evidence of heterosis or recombination effects in man. It may be that a sample much larger than our 180,000 observations would reveal such effects, but the burden of proof is clearly on him who would claim to detect such effects in human populations (Chung & Morton, 1966, p. 679)." (*Recombination effects* are characteristics of the hybrids that are not found in either parent population but result from new combinations of genes producing novel genotypes. Many of these may be less adaptive in the existing environment because they have not been subjected to the sieve of selection that produced the parent populations.) One characteristic that displayed significant heterosis in these interracial crosses was the frequency of dizygotic (DZ) twins born to hybrid mothers. The frequency of DZ twins in Mongoloids is less that half that in Caucasoids, but the racial hybrid mothers showed only a slightly higher frequency than the Mongoloids, which suggests that DZ twinning in these human populations may be attributable to a recessive gene. This is consistent with other findings. African Negroids, whose DZ twinning rate is almost double that of Caucasoids, when crossed with Caucasoids produce fewer DZ twins than the average of the twin rates in the two races (Bulmer, 1960). (A detailed account of the genetics of twinning is presented by Bulmer, 1970, Chap. 6).

Probably the first important study of racial hybrids in which behavioral measures were included is that of Davenport and Steggerda (1929) on race crossing in Jamaica. Though Davenport was a noted geneticist of his day, the methodology of population genetics was too little developed at that time for anyone to appreciate the technical requirements such a study must meet if any valid conclusions of a genetic nature are to be drawn. For these reasons, the Jamaican study is quite inadequate. The racial hybrids were various unknown mixtures of the original African Negroid slave stock and Caucasoids. The Jamaican population consisted mainly of Caucasoids, Negroids, and "colored," who were the racial hybrids from over many generations. Without knowledge of pedigrees, of specific degrees of interracial crossing, or of the number of filial generations represented by the hybrid progeny under study, the investigation really amounts to no more than a comparison of Caucasoids, hybrids, and Negroids (100 in each group) on 63 physical measurements and a number of psychological tests. Except that the number of different measurements is much more extensive than in other studies, therefore, the study is not unique. There have been numerous other studies comparing the intelligence and other characteristics of groups of Negroids differing in skin color and presumed amount of Caucasoid ancestry. (For a review and main references, see Jensen, 1973b, Chap. 9.)

For the numerous physical measurements, Davenport and Steggerda generally found, consistent with genetic expectation, that on those traits in which Negroids and Caucasoids genetically differ the most, the browns tended to be considerably more variable than either the Negroids or Caucasoids. The authors claim there was no evidence of heterosis in the browns, which is consistent with the results of the much more satisfactory study of racial crosses in Hawaii.

Davenport and Steggerda used 12 psychological tests covering a wide variety of abilities: The Seashore Musical Aptitude Test (pitch, intensity, time, harmony, rhythm, and auditory memory), tests of spatial visualization ability (visual form discrimination, figure copying, form board, folding and cutting paper [from the Binet test]), verbal and numerical reasoning (criticism of absurd sentences, Army Alpha), and short-term memory (memory span for digits, imitation of tapping blocks). The authors presented the results in terms of the tests on which the browns performed better or worse than the Negroids and Caucasoids. Browns scored below the other two groups in four of the Army Alpha subtests, two of the spatial visualization tests, and the harmony discrimination test of the Seashore. Browns scored above the other groups in rhythm discrimination, visual form discrimination, digit span memory, criticizing absurd sentences, and one of the Army Alpha subtests ("common sense"). These results are exceedingly difficult to interpret, if they are interpretable at all. Two points seem consistent with many subsequent studies (reviewed by Shuey, 1965), namely, the largest racial differences (Negroids and browns versus Caucasoids) are observed in tests involving reasoning or complex mental manipulation, as in some of the spatial visualization tests, and the smallest racial differences are observed in the tests involving short-term memory and basic auditory abilities such as involved in the Seashore tests. Also, the study included Gesell's tests of motor development on 25 Negroid infants, 12–13 months of age, whose performance was, if anything, slightly more advanced than the norms for New Haven Caucasoid children at this age. This accords with the findings of subsequent large-scale studies (e.g., Bayley, 1965). Davenport and Steggerda concluded their 477-page monograph with the following: "While, on the average, the Browns are intermediate in proportions of mental capacities between Whites and Blacks, and although some of the Browns are equal to the best of the Blacks in one or more traits, still among the Browns, there appear to be an excessive percent over random expectation who seem not to be able to utilize their native endowment (p.

477)." Evaluation of such statements is well nigh impossible with the insufficient information about the relative social and environmental conditions of the three groups.

But going from 1929 to the most recently published studies of the mental abilities of the offspring of Caucasoid–Negroid matings unfortunately shows but little advance in methodology. A study by Willerman, Naylor, and Myrianthopoulos (1970) found that the interracial (Negroid × Caucasoid) offspring of Caucasoid mothers had higher Stanford–Binet IQs at age 4 than the interracial offspring of Negroid mothers, and they interpreted this effect as attributable to nongenetic factors. But since there were no data on the parental intelligence, the results are quite uninterpretable and cannot reduce uncertainty with respect to any hypothesis of interest. (For more extensive discussion of this and related studies, see Jensen, 1973b, pp. 228–230.)

Although skin color is correlated with the amount of Caucasoid admixture in the hybridized Negroid population, the low positive correlation (.12–.30 in various studies) between lightness of pigmentation and IQ is, by itself, not crucially informative concerning racial genetic differences in intelligence. The correlation could be attributable to either one of two main possible causes or to some combination of these. It could result from the lighter colored Negroids also having inherited more intelligence-enhancing genes along with the other genetic characteristics received from their Caucasoid ancestry, which assumes that the two parent populations differ in mean genetic potential for the development of intelligence; or the correlation could be attributable to selection in which both intelligence and lightness of skin color are socially favored, so that the genes underlying both characteristics are segregated together. The causal interpretation of such correlations depends on much other information and the development of more powerful analytic methods, as I have indicated elsewhere (Jensen, 1973b, Chap. 9). At present, there exists no evidence that would support any scientifically worthy conclusions concerning the effects of outbreeding or interracial crossing on mental abilities or other human behavior.

SUMMARY

The aim of this chapter was to give the reader some understanding of how and why the culturally determined mating practices of a society can have profound effects on the genetic structure of the population. Much of the character of any society is importantly conditioned by the genetic structure of its population in a host of polygenic traits having obvious physical and behavioral consequences. The means and variances and the forms of the distribution of many traits in the population, as well as the degree of resemblance among relatives and the frequency of appearance of severe physical and mental defects due to single mutated and recessive genes, are all affected, for good or ill, by various types of departure from completely random mating.

Assortative mating, as we have seen, has a number of genetic effects that can cause a characteristic to become a more salient feature in a population, thereby exposing it to possibly greater chances for selection. Assortative mating, which exists for many human traits, perhaps most importantly for intelligence, does not by itself affect the population mean unless there is dominance, but it directly increases the genetic variance (hence the phenotypic variance as well) of the trait and increases the genetic correlation among relatives, at the same time increasing differences *between* families. If there is a selection threshold for a trait below which mating does not occur or occurs with much lower probability (e.g., mental retardation), then the increase in variance due to assortative mating results in a larger proportion of the population falling below the threshold and consequently a more rapid elimination of the relevant genes from the population's gene pool. The consequences of assortative mating are magnified when assortative mating for a particular trait is correlated with fertility.

Assortative mating may also affect the structure of abilities in a population by bringing about genetic correlations among two or more abilities or other traits that have independent genetic determinants and would therefore be uncorrelated in a random mating population.

Inbreeding, a special case of assortative mating, has some similar effects but also important differences. Unlike assortative mating, which is character specific, inbreeding affects all gene loci and increases genetic variance in all characteristics simultaneously. It is a less important factor than assortative mating in the genetic structure of most present-day human populations because it occurs at such a low frequency. Inbreeding has its most conspicuous consequences in increasing the relative frequency of appearance of physical and mental defects due to rare recessive genes, an effect that is extremely accentuated in consanguineous, and especially in incestuous, matings. When a part of the trait variance is attributable to genetic dominance, polygenic traits show some degree of *inbreeding depression* in the offspring of consanguineous matings, with a greater diminution of the trait value the closer the relatedness of the mates. Such inbreeding depression has been demonstrated for IQ (and many other traits) in the offspring of cousin marriages. Inbreeding depression is seen in a much more exaggerated form in the offspring of incestuous (i.e., father–daughter and brother–sister) matings.

Outbreeding, or the mating of individuals from different breeding populations, apparently has only slight or indiscernible genetic effects in humans. Interracial crosses with few exceptions seem to result only in additive effects in the progeny, that is, the quantitative characteristics of the progeny are merely the arithmetic average of the parental characteristics. Research so far has revealed no other effects of outbreeding on important polygenic traits in humans.

Finally, an important subject for study, barely touched in the present chapter, is how cultural and social forces influence the mating patterns, selection, and fertility, which, in turn, shape the genetic determinants of human variation. Besides leading to a better understanding of mankind's present situation, such a study would seem to be most worthwhile from the standpoint of potential benefit to future generations.

APPENDIX: MENDELIAN ALGEBRA

Mendelian algebra is a method for obtaining theoretical values of means, variances, and kinship correlations and regressions from genetic models. For polygenic systems, the procedure would be too cumbersome if these parameters had to be worked out in the simple and direct manner as was illustrated in the text in connection with Table 4.3. A more economic method of Mendelian algebra is used in quantitative genetics, which can be most easily explained in terms of a single-locus character but which can be generalized to any number of loci since in polygenic models the variances at each locus are assumed to be additive.

To obtain the mean and total variance, we proceed as follows. First, the values a, d, and $-a$ are assigned to the genotypes:

Genotypes	**AA**	**Aa** or **aA**	**aa**
Values	a	d	$-a$

a = half the difference between the two homozygotes. Note that this value will be different for different numbers of loci if all **A** alleles are arbitrarily given a value of +.5 and all **a** alleles are given a value of −0.5. (Do not confuse the value a with the allele **a**.) In our one locus example, $a = 1$, that is, $\frac{1}{2}$(**AA** − **aa**) or $\frac{1}{2}[+1 - (-1)] = 1$. If there were two loci, then the homozygotes would be **AABB** = 2 and **aabb** = −2 and the value of a would be $\frac{1}{2}[+2 - (-2)] = 2$.

d = the value of the heterozygote, **Aa,** which is positive if **A** is dominant and negative if **a** is dominant. If there is no dominance, $d = 0$. That is, the independent effects of **A** and **a** in the heterozygote *Aa* (or *aA*) combine additively, so that $(+0.5) + (-0.5) = 0$. If there is complete dominance, **Aa** = **AA** = 1, and $d = 1$. If there is partial dominance, then $0 < d < 1$.

Given these values, the mean of the population will be:

$$m = (p - q)a + 2pqd.$$

(Recall that p and q are the proportions of **A** and **a** alleles in the population, and $p + q = 1$.)

From this formula for the mean, we can clearly see the effect of genetic selection on the population mean. If the proportions of **A** and **a** alleles are equal, that is, $p = q = .5$, and there is no dominance, then the mean = 0. With complete dominance, the mean would equal .50. If the genotype **AA** were reproductively favored over **aa,** the proportion of **A** alleles in the population would increase, and the mean would be correspondingly raised. For example, if the proportions of **A** and **a** were .8 and .2, respectively, the population mean (without dominance) would become .6, and with complete dominance, .92.

The total genetic variance, V_G, is comprised of two parts, the variance due to additive gene effects, V_A, and the variance due to dominance, V_D. Thus, $V_G = V_A + V_D$. In terms of the above values, the additive genetic variance (also called genic variance) is

$$V_A = \Sigma 2pq[a + d(q - p)]^2$$

The dominance variance is:

$$V_D = \Sigma(2pqd)^2$$

The summation sign (Σ) in each case indicates that the values are summed over all loci, that is, the total variance is the sum of all the variances produced by each gene locus. Notice that the variance is a linear function of the number of loci involved. Also note that the dominance variance is maximal when $p = q$, and if there is no dominance, the additive variance is maximal when $p = q$.

These formulas may be checked against the example shown in Tables 4.3 and 4.4 for random mating. All the formulas presented so far give the means and variances obtained with random mating. They are a necessary step for getting to the values obtained under assortative mating.

Assortative mating, among other effects, increases the additive genetic variance, V_A. After one generation of completely assortative mating (i.e., assuming previous generations have mated at random), the additive variance is:

$$V'_A = V_A(1 + r_G/2)$$

where

V'_A = the additive genetic variance of the progeny after one generation of assortive mating;
V_A = the variance of the randomly mating parent population;
r_G = the additive genetic correlation between mates, that is, the correlation between mates' breeding values. (The *breeding* value of a genotype is the sum of the independent additive genetic values of each of the alleles.)

The variance attributable to assortative mating, then, is simply $V'_A - V_A = V_{AM}$, that is, the additional additive genetic variance due to assortative mating.

When dominance is involved, the formulation for the effect of assortative mating is more complicated. (The formulas for the effects of assortative mating with dominance in the single locus [two-allele] case have been derived by Reeve [1961]). If the number of loci is large, geneticists usually make the simplifying assumption that the dominance variance is unaffected by assortative mating and that the total variance is simply $V_A + V_{AM} + V_D$ (as these have been defined above). This makes for little error unless V_D is large, the degree of assortative mating is great, and there are few loci. For this reason, the precise values in Table 4.4 under assortative mating with dominance are discrepant from those that would be obtained with the formula based on the simplifying assumption that assortative mating does not affect the dominance variance. Usually, assortative mating decreases V_D. (The exceptions to this are when there is only one or a very few loci involved in the trait variation and particularly when inbreeding [a special case of assortative mating] is involved.)

With each generation of assortative mating, the additive variance is further increased, up to a limit. If there is perfect assortative mating, and all genetic variance is additive, the frequency of heterozygotes will diminish over generations until eventually they are completely eliminated from the population. (With 10 loci, heterozygosis would be decreased about one-third in 15 generations; with dominance, heterozygosis would decrease more slowly.) The variance would then have reached the maximal limit. When assortative mating is less than perfect, as is always the case in nature, the variance is increased at a negatively accelerated rate from generation to generation and finally becomes stabilized at a level called its equilibrium value. Most human traits that have been subject to assortative mating for several generations, like height and intelligence, are probably close to equilibrium. The equilibrium value will, of course, be raised if there is an increase in the degree of assortative mating.

The additive variance at equilibrium, assuming a constant degree of assortative mating, is:

$$\hat{V}_A = \frac{V_A}{1 - r_G[1 - (2n)^{-1}]}$$

where n is the number of loci, and the other symbols are as previously defined. The total genetic variance at equilibrium will be simply $V_G = \hat{V}_A + V_D$.

In actual practice, all we can generally determine directly is the phenotypic correlation between mates, r_{pp}. Assuming the correlation between marriage partners is entirely a consequence of their phenotypes, then the genetic correlation, r_g, between mates is estimated by

$$r_g = h_N^2 r_{pp}$$

where h_N^2 is the narrow heritability,

$$h_N^2 = \frac{V_A}{V_A + V_D + V_E}$$

where V_E is environmental variance. (The rationale of this formulation of r_g is given by Crow and Kimura, 1970, pp. 156–158. (Also see Footnote 9, pp. 100–101.)

We can now proceed to consider the effects of assortative mating on the correlations between relatives.

The theoretical genetic correlations between various kinships can be derived directly from our simple Mendelian model. The procedure can be generalized as follows: The values of the genotypes to be correlated are expressed in terms of a and d, and their proportions in the population (for the case of random mating) in terms of the expansion of $(p + q)^{2n}$. From this information, one can calculate the covariance between any two sets of relatives in the Mendelian population. For example, let us compute the covariance between single parents (i.e., either fathers or mothers) and their offspring (taken singly). First, what are the population frequencies of the various possible pairs of parent and offspring genotypes; second, what are the values of the genotypes? These are listed in Table A.1. The total covariance of parent and offspring is obtained in the usual way, simply by obtaining the cross-products of parent × offspring genotypic values, multiplying each cross-product by the frequency, summing over all the cross-products and frequencies, and dividing by the total number of parent–offspring pairs. This is the familiar formula for the covariance: $\text{cov}(xy) = \Sigma xy/N$, where x and y represent the individual parent and offspring values, measured as deviations from the population mean, and N is the total number of pairs. Then, the correlation, which is simply the standardized covariance, is

$$r_{xy} = \text{cov}(xy)/(V_x \times V_y)^{1/2}.$$

If one wishes to go through the rather tedious algebraic exercise of working out the covariance, using the algebraic expressions for the genotypic values as given in Table A.1, the covariance turns out to be $pq[a + d(q - p)]^2$, and the square root of the product of the variances of parents and offspring is $2pq[a + d(q - p)]^2 + [2pqd]^2$, so the correlation between parent and offspring is

$$r_{po} = \frac{pq\,[a + d(q - p)]^2}{2\,pq\,[a + d(q - p)]^2 + [2\,pqd]^2}$$

(With complete dominance, the parent–offspring genetic correlation is simply $q/(1 + q)$, where q is the frequency of the recessive gene.)

If $a = 1$, and there is no dominance (i.e., $d = 0$), then $r_{po} = .25/.50 = .50$ or $\frac{1}{2}$. If there is complete dominance, $d = 1$, and $p = q = .5$, then $r_{po} = .25/.75 = .333$ or $\frac{1}{3}$. If $p \neq q$, and there is dominance, then the parent–child correlation depends on the relative frequencies of the dom-

TABLE A.1

Genotypic Frequencies and Genetic Values (Measured as Deviations from the Population Mean) of Parents and Offspring in a Randomly Mated Mendelian Population

Genotypes[a]		Genotypic values		Frequency of parent–offspring combinations
Parent	Offspring	Parent	Offspring	
AA	AA	$2q(a - pd)$	$2q(a - pd)$	p^4
AA	Aa	”	$a(q - p) + d(1 - 2pq)$	$2p^3q$
Aa	AA	$a(q - p) + d(1 - 2pq)$	$2q(a - pd)$	$2p^3q$
Aa	Aa	”	$a(q - p) + d(1 - 2pq)$	$4p^2q^2$
Aa	aa	”	$-2p(a - qd)$	$2pq^3$
aa	Aa	$-2p(a + qd)$	$(q - p)a + d(1 - 2pq)$	$2pq^3$
aa	aa	”	$-2p(a + qd)$	q^4

Note: Population mean: $M = a(p - q) + 2dpq$.

[a] The parent-offspring combinations **AA-aa** and **aa-AA** cannot occur and are therefore omitted in this table.

inant and recessive alleles, p and q, respectively. Note that the parent–child covariance turns out to be $\frac{1}{2}$ V_A, that is, one-half of the additive genetic variance. The grandparent–grandchild covariance is $\frac{1}{4}$ V_A. The covariance for more distant direct-line ancestors is $(\frac{1}{2})^n V_A$, where n is the number of generations apart of ancestors and progeny.

All other kinship covariances and correlations may be worked out in the same fashion. The correlations for various degrees of kinship can be expressed most conveniently in terms of the variance components that are common to the two sets of kins, that is,

$$r = \frac{\text{Variance components in common}}{\text{Total variance}}$$

Thus, the parent–offspring correlation can be represented as

$$r_{po} = \frac{\frac{1}{2}V_A}{V_A + V_D} \quad \text{or} \quad \frac{\frac{1}{2}V_A}{V_T},$$

where V_T is the total variance of the population. We can simplify our expression of kinship correlations by expressing the fraction V_A/V_T as the narrow heritability, H (sometimes labeled h_n^2 or just h^2), and the fraction V_D/V_T as the proportion of dominance variance, D. Then the correlations between various kinships under random mating and under assortative mating[9] at equilibrium are as shown in Table A.2.

[9] It is assumed that the assortative mating is based on mate selection for the phenotypic trait itself and not on propinquity. An example of assortative mating for IQ due to propinquity would be that of marriage among college students even if there were no correlation

TABLE A.2

Correlations between Relatives under Random Mating and at Equilibrium under Assortative Mating Where r is the Phenotypic Correlation between Mates[a]

Relatives	Correlations	
	Random mating	Assortative mating[b]
Identical twins	H	H
Parent–offspring	$\frac{1}{2}H$	$\frac{1}{2}\hat{H}(1 + r)$
Grandparent–offspring	$\frac{1}{4}H$	$\frac{1}{4}\hat{H}(1 + r)(1 + r\hat{H})$
Greatgrandparent–offspring	$\frac{1}{8}H$	$\frac{1}{8}\hat{H}(1 + r)(1 + r\hat{H})^2$
Full Siblings	$\frac{1}{2}H + \frac{1}{4}D$	$\frac{1}{2}\hat{H}(1 + r\hat{H}) + \frac{1}{4}\hat{D}$
Half Siblings	$\frac{1}{4}H$	$\frac{1}{4}\hat{H}(1 + r)(1 + r\hat{H})$
Double first cousins	$\frac{1}{4}H + \frac{1}{16}D$	$\frac{1}{4}\hat{H}(1 + 3r\hat{H}) + \frac{1}{16}\hat{D}$
Uncle–niece	$\frac{1}{4}H$	$\frac{1}{4}\hat{H}(1 + r\hat{H})^2 + \frac{1}{8}\hat{D}r\hat{H}$
First cousins	$\frac{1}{8}H$	$\frac{1}{8}\hat{H}(1 + r\hat{H})^3 + \frac{1}{16}\hat{D}r\hat{H}^2$

Source: After Crow and Felsenstein, 1968.
[a] A simplifying assumption is that the number of effective loci is large.
[b] The caret (⋀) indicates the value at equilibrium.

between mates' IQs, but all were above the population average because they came from families belonging to social classes having a genetic potential for IQ above the mean of the whole population. The genetic correlation r_G between mates on a given trait is equal to rH (i.e., the phenotypic correlation between mates × the heritability of the trait) only if assortative mating is based on the phenotype itself. If propinquity as well plays a part in assortative mating and propinquity is based in part on genetic factors, the genetic correlation between mates could be higher than rH.

The genetic correlation r_g between mates can perhaps best be understood in terms of the following expressions:

1. If we match individuals only on their genotypes, and environments are random, then $r_g = r_{pp}/H$, where r_{pp} is the phenotypic correlation between mates and H is the heritability.
2. If we match individuals only on phenotypes, and genotypes and environments are random, then $r_g = r_{pp}H$.
3. If we match individuals only on their environments, and phenotypes and genotypes are random within environments, then $r_g = [r_{pp} - r_e(1 - H)]/H$, where r_e is the correlation between mates' environments.

Mating by propinquity involves some weighted average of Expressions 1 and 3, that is, mates may resemble one another genetically because of their origins and they may resemble one another in environmental background. But mate selection in terms of the phenotype itself, and not just propinquity, is usually involved in assortative mating, and so in reality r_g should be thought of as some weighted average of the values defined by expressions 1, 2, and 3.

REFERENCES

Adams, M. S., Davidson, R. T., & Cornell, P. Adoptive risks of the children of incest—a preliminary report. *Child Welfare*, 1967, *46*, 137–142.

Adams, M. S., & Neel, J. V. Children of incest. *Pediatrics*, 1967, *40*, 55–62.

Allen, G. Within group and between group variation expected in human behavioral characters. *Behavior Genetics*, 1970, *1*, 175–194.

Bajema, C. Estimation of the direction and intensity of natural selection in relation to human intelligence by means of the intrinsic rate of natural increase. *Eugenics Quarterly*, 1963, *10*, 175–187.

Bayley, N. Comparisons of mental and motor test scores for ages 1–15 months by sex, birth order, race, geographical location, and education of parents. *Child Development*, 1965, *36*, 379–411.

Blalock, H. M., Jr. *Social statistics*. New York: McGraw-Hill, 1960.

Bodmer, W. F., & Cavalli-Sforza, L. L. *The genetics of human populations*. San Francisco: W. H. Freeman, 1971.

Bonné, B. The Samaritans: A demographic study. *Human Biology*, 1963, *35*, 61–89.

Bulmer, M. G. The twinning rate in Europe and Africa. *Annals of Human Genetics*, 1960, *24*, 121–125.

Bulmer, M. G. *The biology of twinning*. Oxford: Clarendon Press, 1970.

Burks, B. S. The relative influence of nature and nurture upon mental development: A comparative study of foster parent-foster child resemblance and true parent-true child resemblance. *Yearbook of the National Society for the Study of Education*, 1928, *27* (I), 219–316.

Burt, C. Ability and income. *British Journal of Educational Psychology*, 1943, *13*, 83–98.

Burt, C. The inheritance of mental ability. *American Psychologist*, 1958, *13*, 1–15.

Burt, C. Inheritance of general intelligence. *American Psychologist*, 1972, *27*, 175–190.

Burt, C., & Howard, M. The multifactorial theory of inheritance and its application to intelligence. *British Journal of Statistical Psychology* 1956, *9*, 95–131.

Carter, C. O. Risk to offspring of incest. *Lancet*, 1967, *1*, 436.

Carter, H. D. Family resemblances in verbal and numerical abilities. *Genetic Psychology Monographs*, 1932, *12*, 1–104.

Cattell, R. B., & Nesselroade, J. R. Likeness and completeness theories examined by 16 P.F. measures on stably and unstably married couples. *Journal of Personality and Social Psychology*, 1967, *7*, 351–361.

Chung, C. S., & Morton, N. E. Genetics of interracial crosses. *Annals of the New York Academy of Sciences*, 1966, *134*, 666–687.

Cohen, T., Block, N., Flum, Y., Kadar, M., & Goldschmidt, E. School attainments in an immigrant village. In E. Goldschmidt (Ed.), *The genetics of migrant and isolate populations*. Baltimore: Williams & Wilkins, 1963.

Conrad, H. S., & Jones, H. E. A second study of familial resemblance in intelligence: environmental and genetic implications of parent-child and sibling correlations in the total sample. *Yearbook of the National Society for the Study of Education*, 1940, *39* (2), 97–141.

Crow, J. F. *Genetics notes*. Minneapolis: Burgess, 1950.

Crow, J. F., & Felsenstein, J. The effect of assortative mating on the genetic composition of a population. *Eugenics Quarterly*, 1968, *15*, 85–97.

Crow, J. F., & Kimura, M. *An introduction to population genetics theory*. New York: Harper & Row, 1970.

Davenport, C. B., & Steggerda, M. *Race crossing in Jamaica*. Washington, D.C.: Carnegie Institution, 1929.

Doll, E. A. The inheritance of social competence. *Journal of Heredity*, 1937, *28*, 153.

Eckland, B. K. Trends in the intensity and direction of natural selection. *Social Biology,* 1972, *19,* 215–223. (a)

Eckland, B. K. Evolutionary consequences of differential fertility and assortative mating in man. In Th. Dobzhansky, M. K. Hecht, & W. C. Steere (Eds.), *Evolutionary biology,* Vol. 5. New York: Appleton, 1972. (b)

Falconer, D. S. *An introduction to quantitative genetics.* New York: Ronald Press, 1960.

Fisher, R. A. *Statistical methods for research workers.* 11th ed. Edinburgh: Oliver & Boyd, 1950.

Freeman, F. N., Holzinger, J., & Mitchell, B. C. The influence of the environment on the intelligence, school achievement and conduct of foster children. *Yearbook of the National Society for the Study of Education,* 1928, *27*(1), 103–217.

Gottesman, I. I. Biogenetics of race and class. In M. Deutsch, I. Katz, & A. R. Jensen (Eds.), *Social class, race, and psychological development?* New York: Holt, 1968.

Gregory, I. Husbands and wives admitted to mental hospitals. *Journal of Mental Science,* 1959, *105,* 457–462.

Guttman, R. Parent-offspring correlations in the judgment of visual number. *Human Heredity,* 1970, *20,* 57–65.

Guttman, R. Genetic analysis of analytical spatial ability: Raven's Progressive Matrices. *Behavior Genetics,* 1974, *4,* 273–284.

Halperin, S. L. Human heredity and mental deficiency. *American Journal of Mental Deficiency,* 1946, *51,* 153–163.

Higgins, J., Reed, S., & Reed, E. Intelligence and family size: A paradox resolved. *Eugenics Quarterly,* 1962, *9,* 84–90.

Hollingshead, A. B. Cultural factors in the selection of marriage mates. *American Sociological Review,* 1950, *15,* 619–627.

Hulse, F. S. Exogamie et heterosis. *Archives Suisses de Anthropologie Général,* 1957, *22,* 103–125.

Jensen, A. R. *Genetics and education.* New York: Harper & Row, 1973. (a)

Jensen, A. R. *Educability and group differences.* New York: Harper & Row, 1973. (b)

Jinks, J. L., & Fulker, D. W. Comparison of the biometrical genetical, MAVA, and classical approaches to the analysis of human behavior. *Psychological Bulletin,* 1970, *73,* 311–349.

Jones, H. E. A first study of parent-child resemblance in intelligence. *Yearbook of the National Society for the Study of Education,* 1928, *27* (1), 61–72.

Kiser, C. V. Assortative mating by educational attainment in relation to fertility. *Eugenics Quarterly,* 1968, *15,* 98–112.

Kodama, H., & Shinagawa, F. *WISC Chinō Shindan Kenshanō* [The WISC Intelligence Test]. Tokyo: Nihon Bunka Kagakusha, 1953.

Kreitman, N., Collins, J., Nelson, B., *et al.* Neurosis and marital interaction: I. Personality and symptoms. *British Journal of Psychiatry,* 1970, *117,* 33–46.

Kuttner, R. E. (Ed.) *Race and modern science.* New York: Social Science Press, 1967.

Leahy, A. M., Nature-nurture and intelligence. *Genetic Psychology Monographs,* 1935, *17,* 241–305.

Lerner, I. M. *Genetic homeostasis.* New York: Wiley, 1954.

Li, C. C. *Population genetics.* Chicago: Univ. of Chicago Press, 1955.

Lindzey, G. Some remarks concerning incest, the incest taboo, and psychoanalytic theory. *American Psychologist,* 1967, *22,* 1051–1059.

Lutz, F. Assortative mating in man. *Science,* 1905, *16,* 249–250.

Morton, N. E. Morbidity of children from consanguineous marriages. In A. G. Steinberg (Ed.), *Progress in medical genetics.* New York: Grune & Stratton, 1961.

Morton, N. E., Chung, C. S., & Mi, M. P. *Genetics of interracial crosses in Hawaii.* New York: S. Karger, 1967.

Nielsen, J. Mental disorder in married couples (assortative mating). *British Journal of Psychiatry,* 1964, *110,* 682–697.

Outhit, M. C. A study of the resemblance of parents and children in general intelligence. *Archives of Psychology,* 1933, *149,* 1–60.

Penrose, L. S. A study in the inheritance of intelligence. The analysis of 100 families containing subcultural mental defectives. *British Journal of Psychology* (General Section), 1933, *24,* 1–19.

Penrose, L. S. Mental illness in husband and wife: A contribution to the study of assortative mating in man. *Psychiatric Quarterly Supplement,* 1944, *18,* 161–166.

Penrose, L. S., & Haldane, J. B. S. *The biology of mental defect.* London: English Language Book Society, 1969.

Provine, W. B. Geneticists and the biology of race crossing. *Science,* 1973, *182,* 790–797.

Reed, W. E., & Reed, S. C. *Mental retardation: A family study.* Philadelphia: Saunders, 1965.

Reeve, E. C. R. A note on nonrandom mating in progeny tests. *Genetic Research,* 1961, *2,* 195–203.

Rele, J. R. Trends and differentials in the American age at marriage. *Milbank Memorial Fund Quarterly,* 1965, *43,* 224.

Schull, W. J., & Neel, J. V. *The effects of inbreeding on Japanese children.* New York: Harper & Row, 1965.

Schull, W. J., & Neel, J. V. The effects of parental consanguinity and inbreeding in Hirado, Japan. V. Summary and interpretation. *American Journal of Human Genetics,* 1972, *24,* 425–453.

Seemanova, E. A study of children of incestuous matings. *Human Heredity,* 1971, *21,* 108–128.

Shuey, Audrey M. *The testing of Negro intelligence.* (2nd ed.) New York: Social Science Press, 1965.

Smith, M. Similarities of marriage partners in intelligence. *American Sociological Review,* 1941, *6,* 697–701.

Spuhler, J. N. Empirical studies on quantitative human genetics. In *The use of vital and health statistics for genetic and radiation studies.* New York: United Nations, 1962.

Spuhler, J. N. Behavior and mating patterns in human populations. In J. N. Spuhler (Ed.), *Genetic diversity and human behavior.* Chicago: Aldine, 1967.

Spuhler, J. N. Assortative mating with respect to physical characteristics. *Eugenics Quarterly,* 1968, *15,* 128–140.

Stanley, J. C. Study of mathematically and scientifically precocious youth. *Annual Report to the Spencer Foundation on its Five-year Grant to the Johns Hopkins University Covering the First Year of the Grant,* 1 September 1971 through 31 August 1972, October, 1972.

Stern, C. Model estimates of the number of gene pairs involved in pigmentation variability of the Negro American. *Human Heredity,* 1970, *20,* 165–168.

Stern, C. *Principles of human genetics* (3rd ed.). San Francisco: W. H. Freeman, 1973.

Tenopyr, M. L. Race and socioeconomic status as moderators in predicting machine-shop training success. Paper presented at the annual meeting of the American Psychological Association, Washington, D.C., September 4, 1967.

Terman, L. M., & Oden, M. *The gifted group at mid-life.* Stanford, Cal.: Univ. Press, 1959.

Tharp, R. G. Psychological patterning in marriage. *Psychological Bulletin,* 1963, *60,* 97–117.

Vandenberg, S. G. What do we know today about the inheritance of intelligence? In R. Cancro (Ed.) *Intelligence: Genetic and environmental influences.* New York: Grune & Stratton, 1971, 182–218.

Vandenberg, S. G. Assortative mating, or who marries whom. *Behavior Genetics*, 1972, *2*, 127–157.

Warren, B. L. A multiple variable approach to the assortative mating phenomenon. *Eugenics Quarterly*, 1966, *13*, 285–290.

Willerman, L., Naylor, A. F., & Myrianthopoulos, N. C. Intellectual development of children from interracial matings. *Science*, 1970, *170*, 1329–1331.

Williams, T. Cultural deprivation and intelligence: Extensions of the basic model. Paper presented at the Annual Meeting of the American Educational Research Association, New Orleans, 1973.

Willoughby, R. R. Family similarities in mental test abilities. *Yearbook of the National Society for the Study of Education*, 1928, *27* (1), 55–59.

Wright, S. Coefficients of inbreeding and relationship. *American Naturalist*, 1922, *56*, 330–338.

Wright, S. *Evolution and the genetics of populations.* Vol. 2. *The theory of gene frequencies.* Chicago: Univ. of Chicago Press, 1969.

5

The Nature and Development of Intellectual Abilities

JOHN L. HORN
University of Denver

The principal purpose of this chapter is to describe behavior that is believed to be indicative of important processes and capacities of human intelligence. The behavior to be described is primarily that displayed in response to tasks (tests) that can be given under standardized conditions. Our principal systematized scientific understanding of intellectual abilities has come from efforts to construct problems, puzzles, questions—that is, tasks in general—that can be administered in much the same way to a number of people in order to elicit behavior that can be systematically analyzed, studied, and interpreted.

In accordance with the dominant theme of this book (human variation), the emphasis in this chapter is on individual differences. The study of individual differences in abilities is one, but not the only, reasonable way to gain insight into the nature of intellectual functioning. The information derived in this way does not exclude knowledge derived from other forms of study, but the reader should be careful not to try to interpret individual-difference results in ways that eliminate individual differences. He should see clearly that most of the results of this chapter derive

The work on this chapter was supported by grants from the Army Research Institute, Grant No. DAHC19-74-G-0012, and the National Science Foundation, Grant Number GB-41452. A somewhat more detailed, earlier draft of this chapter is available from the author on request.

from efforts to maximize human differences and to describe qualities in respect to which humans differ, not qualities in respect to which most humans are highly similar.

OVERVIEW OF THEORETICAL ORIENTATIONS

There is a tremendous variety of behavior that is said to be indicative of important aspects of human abilities. To help explain this behavior, many theories, concepts, hypotheses, and measurement procedures have been advanced. Only a few of the major theories and concepts can be mentioned here.

Theories of Intelligence: General Features

The most widely studied and generally accepted theories about human abilities involve a concept of intelligence. But there are many theories of intelligence. No one of them has yet earned acceptance by a majority of the scientists who have studied in this area. There are widely used tests of intelligence, and there is evidence of considerable convergence in measurement from different tests; thus, in this sense, there is some agreement in measurement definitions of intelligence. Yet exactly the same measurement definition will be treated quite differently in one as compared to another theory. Also, rather different measurement definitions will be treated similarly in different accounts of the nature of intelligence. It is in these senses that there is no single agreed-upon concept of intelligence.

Most commonly (but by no means always) the concept of intelligence is used to designate *(ex hypothesi)* an innately determined quality, a potential that will become manifest as a function of normal maturation. One implication of this concept is that intelligence should be measured so that individual differences in the measures will reflect primarily differences in genetic determiners. The exceptions should be attributable to postconception loss or injury of structures and processes essential to intellectual functioning, to environmental circumstances that exceed those for which the species is adapted, and to errors of measurement.

The most widely used measures of intelligence—tests such as the Stanford–Binet, Wechsler, Lorge–Thorndike, Otis, and so on—are usually interpreted in accordance with this concept of intelligence. That is, the IQ score derived from these tests is frequently regarded as a reasonably accurate indicant of one's innate potential. There are problems with this interpretation, however.

For one thing, the items in the most widely used tests measure achievements that would seem to be influenced mainly by differences in child-rearing environments (homes, neighborhoods, and so on) and differences in schooling circumstances. There is evidence (Humphreys, 1972) that the different intelligence tests intercorrelate among themselves about to the same extent that they correlate with scholastic achievement tests. Thus, it can be argued that because achievement tests measure achievement and different intelligence tests correlate as much with these other tests as they correlate with each other, the proper inference is that intelligence tests measure primarily achievement rather than intelligence. This, however, overlooks the possibility that scholastic achievement differences in an open society, such as that of the United States, may be largely a consequence of genetic variations; it ignores the fact that differences in child-rearing and schooling environments are not independent of genetic factors and may not be large enough to account fully for obtained results.

In considering controversies about the inheritance of intelligence, it is important to recognize that the fundamental issues do *not* pertain to whether or not it is reasonable to expect that there are inherited individual differences in the biological substrate—the physiological structures and process capacities—upon which development of human abilities is based. Even casual study of genetics and pedigrees leads the objective investigator inexorably to the conclusion that there are inherited individual differences in structure that are likely to relate to important individual differences in the ease with which various skills are acquired and the quality of abilities ultimately attained. It is hard to doubt that if human breeding were controlled in the manner of mouse breeding in the laboratory, or even as is done by dog fanciers and cattle raisers, then almost certainly one could demonstrate the equivalent of "maze bright" and "maze dull" humans. In this sense, there is little doubt that the capacities of intelligence are inherited. This is *not* the issue. Rather, the issues pertain to whether or not inherited differences in the most important capacities of intelligence are properly, or even at all, represented in the measures of intelligence used by researchers.

As soon as the issues are framed in this way, one sees that the questions about the inheritance of intelligence depend fundamentally on adequate answers to questions about the nature of intelligence. There must be a well worked out, empirically supported scientific theory indicating the essential processes of intelligence—what it is not as well as what it is—and how these processes are affected and not affected by the myriad of influences that appear to be important. In its major features, such theory would have the endorsement of most scientists in this domain of study, just as most chemists endorse the major ideas of the ion theory of reactivity. As noted before, such a scientific theory of human intelligence has not yet been worked out (although some believe we may be at the brink of such a development, cf. Horn, 1976a). It is precisely because there is no such theory that there can be so much controversy among scientists about such issues as the inheritance of intelligence.

The different tasks with which one is asked to cope in intelligence tests are intended to tap different capacities and indicate different basic processes. Different tests involve rather different collections of tasks and thus seemingly measure rather different things. This is true not only of tests having different titles but also of the same tests used at different age levels. Research into the nature of intellectual functions has not produced definitive indications of the capacities and processes that are both necessary and sufficient to describe what can be agreed upon as the sine qua non of intelligence. Thus, we do not know quite what to measure or how to combine the measurements of relevant functions we do know how to measure. The domain of behavior that is believed to be indicative of intelligence includes (among many other things) short-term memories of several kinds, recall from the distant past, different kinds of reasoning, ability to learn and retain, ability to specify the reverse of a course of events (decentration), a variety of ways of coping with novel situations, capacity for detaching oneself from concrete situations (abstraction), ability to recognize common and unique characteristics and form concepts on this basis, verbal production, and a great variety of other such abilities, capacities, and styles of thinking.

It is almost certain that manifest individual differences in some of the qualities of intelligence reflect primarily innate differences. But it is certain, too, that observed individual differences in some of the qualities largely reflect experiential differences. There are tremendous difficulties in controlling for potentially distorting influences in demonstrating that a quality is or is not innately determined. Individual differences in information-processing capacity might be thought to be innately determined, for example, but substantial individual differences in sensory detection of information can influence results. And this is but one example.

It is extremely difficult to specify behavioral functions that correspond to innately determined capacities for intellectual development. There is no compelling evidence to indicate precisely what these functions are. No test, or combination of tests, provides thoroughly valid and comprehensive measurement of the necessary and sufficient functions of innate intelligence.

One should not interpret this as revealing chaos. In any scientific field, there is imprecision about its most important concepts. One can see this historically in the development of the energy concept in physics and chemistry. There is still controversy and confusion in these disciplines about the nature of the elementary particles of matter.

At a broad level, many scientists accept a definition of intelligence similar to that stated by Humphreys (1972), that intelligence is the entire repertoire of acquired skills, knowledges, learning sets, and generalization tendencies considered to be intellectual and available to an individual at a given time. Thus, the concept represents a broad domain of abilities that people accept as indicating intelligence, and the task of measuring intelligence is one of properly sampling from this universe of abilities. Many investigators agree that tests labeled "intelligence tests" provide at least rough samplings (not fully representative) of important abilities of intelligence. Research is converging toward increasingly clearer understandings of these abilities.

Multiability Theories

While historically theories about intelligence (usually containing the innateness feature) have predominated in the study of human abilities, there have always been multiple-concept theories, encompassing both genetically determined and other influences. It has been recognized for some time that widely used tests of intelligence measure a mélange of abilities that relate in different ways to life achievements, to the course of development in childhood and adulthood, and to other variables. There have been efforts either to replace the concept of intelligence with a concept of multiple abilities or to differentiate the concept into multiple components.

One kind of effort in this regard has been to identify a particular ability and define it as a form of intelligence, an aspect of intelligence, or a quality quite separate from intelligence. For example, it has been observed:

1. There are reliable individual differences in paired-associates learning and memory.
2. These differences are not highly correlated with individual differences in vocabulary, verbal problem solving, spatial reasoning, and other highly regarded subtest measures of intelligence.

On the basis of these two sets of observations, investigators have postulated two forms of intelligence. For example, Jensen (1971) has suggested that the first of these sets of abilities represents an "intelligence I" in which there are little or no social-class and ethnic-group differences, and the second set of abilities represents an "intelligence II" in which there are notable social-class and ethnic-group differences. Other investigators have noted that the verbal production of connotations of a word, of uses for a common object, or of possible ways of construing a given event all correlate relatively lowly with the indicants of intelligence mentioned previously, and have suggested that the former kinds of behaviors represent a form of creativity that is quite distinct from intelligence (Cropley, 1972; Kogan, 1971; Torrance, 1972; Wallach & Kogan, 1965).

Primary Mental Abilities

These and similar efforts to develop multiple concepts to describe the phenomena that are referred to under the heading of intelligence have generated confusion as to the nature of intelligence, but they have also helped to make investigators aware of the diversity of behaviors that can be accepted as indicative of intelligence. This has led to efforts to identify the essential features of the diversity in intellectual behavior. Many of these efforts have involved use of factor analyses of batteries of tests designed to measure different fundamental abilities.

Over 100 factor analytic studies of human abilities have been conducted. The results provide a wealth of information, more than can be summarized within the scope of this chapter. The flavor of this information can best be savored by considering the descriptive listings of the abilities that have been identified through factor analytic research. These are readily available in such sources as French (1951), French, Ekstrom, and Price (1963), Guilford (1967), Guilford and Hoepfner (1971), Horn (1972, 1976b), and Pawlik (1966).

Facet Theories

L. L. Thurstone identified nine primary mental abilities in the pioneering studies in this area. In the years following Thurstone's early efforts, the paradigm he introduced was applied repeatedly. Particularly with the advent of the modern computer (making possible the analyses of very large batteries of tests), a welter of ability factors emerged. In accordance with Thurstone's labeling, these are often referred to as primary mental abilities.

Approximately 50 primary abilities have been identified at this time. The thought of replacing a theory of a single attribute of intelligence with theories about this number of primary abilities is, if not overwhelming, then at least inhibiting. Hence, those interested in developing multiability theories have been pushed back, as it were, toward simplifying the system of primaries or finding some integrative principle that will make the system more easily comprehended. Perhaps the best known and most influential effort of this kind is the structure of intellect (SI) model, developed by Guilford and his co-workers.

The SI theory is one of several facet theories of human abilities (Carroll, 1975; El Koussy, 1955; Guttman, 1965, 1970; Harris & Harris, 1971; Humphreys, 1962; and reviewed in Horn, 1972, 1976b). There are two fundamentally different ideas about a facet in such theories. One idea is that ability tasks can be cross-classified in terms of distinct functions, capacities, abilities, or processes involved in the task performances. The other notion is that it is useful to cross-classify in terms of the features to which test constructors can attend when they construct tests. These two different conceptions of the facets of ability tests have quite different implications, of course; yet it is by no means always clear which conception is stressed in a particular facet theory. This is true of SI theory, for example.

SI theory suggests that it is useful to organize thinking about intellectual tasks and abilities in terms of five kinds of Operations (Evaluation, Convergent Production, Divergent Production, Memory, Cognition), four kinds of Contents (Semantic, Symbolic, Figural, Behavioral), and six kinds of Products (Units, Classes, Relations, Systems, Transformations, Implications). But are these categories supposed to indicate different human functions (in respect to which there are individual differences), are they intended to serve merely as a guide for test construction, or are both interpretations intended? For example, is production of semantic units an ability reliably different from production of semantic classes, or does this distinction merely represent the fact that one can construct tests that require one response or a class of responses? SI theory is not very explicit in respect to these kinds of questions.

In much of the work advanced in support of SI theory, however, the analyses have implied that the facets represent reliably distinct abilities. To support a claim of this kind, it is necessary that the factors be psychometrically independent. This means that the set of component variables (items, subtests, tests) that define a given factor (representing an ability, function, process) must have internal-consistency reliability in excess of the correlation of the factor with a linear combination of component variables with respect to which the factor is expected to be independent. In the present context of an ability factor defined by several tests, internal-consistency reliability is the correlation of a factor with itself, that is, the pooled correlations for the tests, combined linearly, of which that factor is comprised. This internal-consistency estimate (r^2_{xt}, the fallible factor correlation [squared] with its own true part) should indicate a better estimate of the factor than is produced by a similar linear combination of tests (of other factors) that are, by hypothesis, independent of the factor ($r^2_{x.12\ldots m}$, the fallible factor correlation [squared] with all that it has in common with other factors).

Independence in this sense has not been demonstrated in most of the research designed to indicate support for SI theory. The theory, therefore, cannot be regarded as well supported (see Carroll, 1972, 1975; Eysenck, 1973; Hammond, 1976; Horn, 1970b; Horn & Knapp, 1973, 1974; Humphreys, 1968; Knapp & Horn, 1977; Matarazzo, 1972; Undheim & Horn, 1977, for evaluations of this evidence). Nevertheless, the theory has a number of interesting implications and continues to be influential.

Hierarchical Theories: Outgrowths of General Intelligence Theories

The earliest efforts by Spearman and Binet (in the 1890s and early 1900s) to develop a theory of intelligence involved recognition that diverse kinds of abilities are involved and an assumption that all are manifestations of a single underlying function. The basic idea is that the single function may be manifested in different ways in consequence of different environmental influences but that one and only one basic capacity of intelligence produces the diversity of observed abilities.

Giving support to this view is the now well-established finding that almost all ability measures are positively correlated. Even abilities for which there is little claim that they are of the essence of intelligence (simple reaction time, for example) correlate positively with abilities for which there is strong claim that they represent important aspects of intelligence. (A possible exception to this generalization is the existence of zero or low negative correlations between very simple abilities measured under highly speeded conditions and abilities involving high degrees of carefulness and persistence.) If almost all the abilities accepted as indicating intellectual processes are positively correlated, there is suggestion that they are all manifestations of a single function. This kind of evidence, and common belief, provide perhaps the most compelling argument for a unitary concept of intelligence.

Those who state unitary intelligence theories suggest that the separate abilities are subfunctions or different manifestations of intelligence. When subfunctions are proposed, there must be indications of how they are linked together. The facet theory of Guttman (1970) and the developmental theory of Piaget (e.g., 1946) illustrate ways in which subfunctions are said to be interrelated in theories involving a singular concept of intelligence. In Piaget's theory, for example, intelligence is defined as a mixture of awarenesses and understandings, referred to as schemata. These are realized through development in which the person strives to obtain equilibrium between comprehensions of the environment in accordance with the way it is portrayed by sensory–perceptual processes (assimilation) and his predilections to impose structure on the environment in accordance with conceptions already developed.

But perhaps the most influential theories about subfunctions of intelligence are those in which it is proposed that the subfunctions are organized in the form of a hierarchy (of which there are several possibilities; see Cattell, 1965; Horn, 1972). One of the earliest and most important of these theories is that of Burt (e.g., 1949). Burt suggested that the mind is organized in terms of a subdivided hierarchy. Processes of the lowest level consist of simple sensations or simple movements, such as can be measured with tests of sensory threshold and reaction time. The next level includes the more complex processes of perception and coordinated movement, as demonstrated in experiments on the apprehension of form and pattern. The third level involves associative memory and habit formation. The fourth, and highest, level involves the perception of relations and the eduction of correlates, as specified in Spearman's work (e.g., Spearman, 1927). Intelligence is defined as the "integrative capacity of the mind." It is manifested at *every* level, but these manifestations differ in degree and quality. Although individual differences appear and are noteworthy at every level in the hierarchy, those that mainly represent intelligence are at the fourth level. Progressively less of the variance at each successively lower level is attributable to intelligence (i.e., more of the variance is attributable to other factors, such as sensory acuity or motivation).

Several currently influential hierarchical theories are extensions and refinements of Burt's conception of intelligence. One such theory is that of fluid (Gf) and crystallized (Gc) intelligence, here referred to simply as Gf–Gc theory, largely developed by Cattell (one of Burt's students) and Horn (one of Cattell's students; see Cattell, 1957, 1971; Horn, 1967, 1968, 1970b, 1972, 1976a). This theory can be used to illustrate the general nature of hierarchical theories and to indicate some of the findings that support this kind of view of the organization of human intellectual processes.

Gf–Gc theory derives in large measure from factor analytic evidence of structure among primary mental abilities. It specifies that at a level just below a general integrative function of the mind, there are two broad functions, Gf and Gc, comprised of primary abilities. The primary abilities that define the two broad functions represent much of what is typically referred to as being indicative of intelligence. Because this is true of both Gf and Gc, each can be referred to as a kind of intelligence. Yet the two are operationally independent (i.e., different operations—tests—are used to measure each) and psychometrically independent in the sense described previously. (I.e., the internal consistency of each exceeds the multiple estimation of one from the components of the other.) Moreover, the two appear to be independent also in terms of their development throughout the life span, in terms of predictions of life achievements, in terms of their relationships to pathology, and in a number of other ways.

In particular, the theory can be viewed as a broad-gauge effort to provide an integrative framework for comprehending major findings pertaining to:

1. the structural interrelationships in performances believed to be indicative of important aspects of intelligence (factor analytic evidence, as indicated above, but other evidence as well);
2. the effects of brain damage on abilities (early in life and in maturity);

3. the relations between test performances and opportunities to acquire the abilities most valued, or most tutored, in a culture;
4. the feasibility of constructing intellectual tests that are little affected by cultural and subcultural differences;
5. the life-span development of abilities.

The findings in all of these areas are at least mildly supportive of the view that, indeed, at a broad level there are two distinct and important forms of intelligence, each influenced by rather different developmental factors and each, therefore, having a different course of development over the life span.

Gf and Gc represent fundamental processes of perceiving relationships among stimulus patterns, inferring meanings for these relationships, stepping up and down the abstraction ladder in interpreting relationships, forming concepts, bringing learned concepts to bear in interpretations and problem solving. These processes are commonly referred to as essential in reasoning, abstracting, language formation, and cognitive functioning generally. The quality of functioning of these processes is commonly thought to be characteristic of the intellectual behavior that most clearly distinguishes the mature human from other animals and the young human. It is in this sense that Gf and Gc best represent the concept of intelligence as this term has been used traditionally. It is in this sense that both are referred to as forms of intelligence.

But other broad functions are indicated in factor analytic study of the structure among primary abilities. Two such prominent functions are Gv, Broad Visualization, and Ga, Broad Audition. Gv and Ga are exemplars of perceptual–sensory functions that provide organized input for the Gf–Gc processes outlined above. Similar functions must be associated with touch, kinesthesis, smell, and the other senses, but these have not been isolated in structural analyses of performances on intellectual tasks.

Broad Speediness (Gs), Short-term Acquisition and Retrieval (SAR), and Verbal Productive Thinking (VPT) also contribute noteworthy variance in performances on intellectual tasks. These functions can be seen to condition the processes of Gf and Gc in somewhat the same manner as Gv and Ga condition, or give different qualitative character to, manifestations of intelligence. SAR and VPT, particularly, indicate the quantities of information that can be available at a given instant for the processing of Gf and Gc. In this sense, they limit intellectual functioning (as through Gf and Gc) because ultimately this function can be only as good as the information that is processed. Processes that moderate the expression of intelligence are referred to as *anlage* functions in Gf–Gc theory (Horn, 1967, 1968, 1970b).

The SAR dimension appears to be composed of two elementary *anlage* functions that have been described rather fully in the general experimental literature on short-term memory and retrieval. These subfunctions are referred to as primary memory (PM) and secondary memory (SM).

Primary memory represents retention over periods of less than roughly) 30 seconds in which such organization as exists among the retrieved elements is peripheral, as in relation to the sensory organ at which the stimulus is received (Broadbent, 1954). Organization in terms of logical categories is minimal for this kind of immediate recall.

The amount that a person remembers for longer than about 30 seconds is a function of the organization he imposes on the material. If one can organize elements into meaningful categories—as, for example, odd numbers and even numbers—then if one can recall the category, he can recall most of the elements in the category (up to a limit of about four elements). Memory over seconds and minutes of this kind is the secondary memory subfunction of SAR. It seems that this subfunction is prominently implicated in expressions of Gf and Gc (Horn, 1976b, c; Hundall & Horn, 1977).

PM and SM together make up the memory span primary ability (Ms). The mean for this span is about six or seven; the sigma is about two. These statistics define what Miller (1956) has referred to as the magical number seven, plus or minus two. This represents a kind of limit in the information-processing capacity of the human.

The breakdown of Ms into PM and SM suggests that the information-processing capacity of the human involves a subfunction for keeping near-meaningless information in awareness and a subfunction for retaining information that is organized meaningfully. Unpublished work of the present writer (Horn, 1976b, c) indicates that individual differences in these PM and SM subfunctions can be reliably ($r_{xx} > .70$) and independently ($r_{xy} < .40$) measured. The mean for PM appears to be about 3 ± 1 and that for SM about 4 ± 1. Thus, the magical number seven, plus or minus two, appears to be composed of two magical numbers, three and four, each having a plus and minus sigma range of about one.

One implication of this evidence of noteworthy individual differences in PM and SM is that persons may fail in intellectual tasks either because of failure in retention of unorganized elements (PM) or because of failure in the organizational processes of SM. These possibilities need to be considered carefully in evaluations of evidence pertaining to relationships between aging in adulthood and intellectual functioning. Some of this evidence will be considered later.

There is suggestive evidence from a number of sources that the capacities of memory reach asymptote at relatively early stages of phylogenetic and ontogenetic development. That is, at least primary memory, and perhaps secondary memory as well, seems to be about the same for 8-year-old children as for young adults. Similarly, certain apes (chimpanzees, gibbons) and dolphins may have primary memory spans comparable to those of humans. The ability to retain low association material in immediate awareness is necessary in solving many intellectual problems. Yet this awareness is not a sufficient condition for solution. It is in this sense that PM moderates the expression of intelligence but is not indicative of intelligence, as such. SM also may be such an *anlage* function, for this capacity, too, appears to reach maximum relatively early in phylogenetic and ontogenetic development. Thus, it may also represent a function that is a necessary but not sufficient condition for defining adult human intelligence that is distinct from the emerging forms of intelligence observed in children and the great apes (Horn, 1976b, 1976c).

The verbal productive thinking (VPT) function represents facility in retrieving verbally tagged elements (concepts, ideas, and so on) from what can be referred to as quasi-permanent storage (QPS)—"quasi" because the probability of retrieval of all information acquired by an individual appears to decrease with the time elapsed between initial acquisition (or rehearsal) and retrieval (Riegel, 1973). Terms such as tertiary memory (TM) or semantic memory are used to identify retrieval from QPS. The behavior evinced in this retrieval can be seen to reflect:

1. the size of store of QPS, for example, the number of relevant concepts available, and or
2. facility (rate, flexibility, ease) in accessing that which is available in QPS.

There is some evidence that these two possible subfunctions can be separated. Work now under way is designed to indicate the extent to which each of the two subfunctions is involved in changes that may occur in intellectual functioning in adulthood aging (Horn, 1976b, c). Some of the ideas guiding this work will be outlined in subsequent sections dealing with the development of intelligence.

DEVELOPMENT: INFANCY

For the behavioral scientist interested in measuring intellectual capacities, there is not much to observe and measure in the very young human. Until recently, efforts to measure infant intelligence have been dominated by what might be called a medical model or a model of physical education. That is, the measures have focused on whether or not the child is healthy and whether or not its motor and sensory capabilities are intact. Knowing (as we do) that athletic ability has only a very low (but positive) relationship to intellectual ability and that good health certainly is not a sufficient condition—and may not even be a necessary condition—for advanced intellectual development, it should not be surprising to learn that infant tests measuring sensorimotor alertness do not provide very good measures of intelligence (Hofstaetter, 1954; Lewis & McGurk, 1972; McCall, Hogarty, & Hurlburt, 1972; Pease, Walins, & Stockdale, 1973). Indeed, there are now several bits of evidence to suggest that precocity in the sensorimotor development of infancy is *negatively* related to intellectual development in childhood and thereafter (Kagan & Klein, 1973; McCall *et al.*, 1972; Werner & Bayley, 1966). In any case, measures of the age (in days or weeks) at which a child rolls from back to stomach, smiles in recognition of the mother, grasps a block, follows a moving object with his eyes, or evinces a variety of other such awarenesses and skills are at best only very lowly predictive of behaviors that are accepted as indicative of intelligence in, say, 5-year-olds. This failure of infant tests to predict intellectual development at later ages cannot be attributed to unreliability of the measures. The infant scales have sufficient internal consistency and short-period (over weeks and months) test–retest reliability to indicate that they measure something. They do not predict measures of intelligence because, quite simply, health and sensorimotor alertness are not of the essence of intelligence.

But the concerned citizen may plead: Are there not some reliable indicants of intelligence to be observed in the first weeks and months of life? Is it not very important to diagnose intelligence early in life so that plans to ameliorate undesirable conditions and to maximize the potential of desirable conditions can be formulated early, before it is too late? Indeed, has not the government of the United States recently ordained that such diagnosis shall take place (Hobbs, 1974)? While the answer to these last two questions may well be yes, the answer to the first one must be that at present very little of what is known about individual differences in infancy is at all predictive of intellectual status in childhood and adulthood. There are some very low predictive correlations, however.

As noted before, the essence of human intelligence is seen in responses to problems requiring the perception of complex relationships, the imposition of order on these relationships, and the drawing of inferences based on these awarenesses. These processes depend in part on elementary *anlage* functions such as primary and secondary memory. Although *anlage* functions are not the most characteristically human aspects of intelligence, they seem to be easier to measure in infancy than functions that are of the essence of intelligence. Some prediction of level of intelligence can be obtained from careful measures of *anlage* functions in infancy.

Recent work of researchers such as Bayley, Piaget, Hunt, Elkind, McCall, Campos, and Kagan holds promise of improving the measurement of intelligence as this can be manifested in infancy. Kagan's (1972) work, for example, suggests that by distinguishing between passive responsiveness and involved responsiveness to stimuli, one might be able to identify "mental work" of the kind that indicates concept awareness (see also Kagan & Klein, 1973). From birth onward, the proportion of passive to involved responsiveness to stimuli appears to decrease. This is suggested by such things as duration of fixation in contemplating stimuli and heart rate change to stimuli. For example, at the earliest ages, cardiac deceleration is the most usual response to stimuli that later excite cardiac acceleration (Campos, 1975). Similarly, there is a decrease through the first year of life in duration of a kind of bland fixation in contemplating objects. Duration of fixation reaches a nadir near the end of the first year and then increases in the months of the second year; in this case, however, the fixation appears to represent a kind of alert involvement with the stimulus. Thus, it is possible that measures of the age (in days or weeks) of onset of cardiac acceleration and "alert fixation" will be indicative of the kind of thinking that is predictive of later (in childhood) manifestations of intelligence.

It has been suggested that awareness of certain kinds of concepts appears universally in normal humans at roughly the same age, plus or minus the months that indicate the normal range of human variation. That is, perhaps children of all cultures inevitably learn conservation, as indicated by awareness of the idea that squeezing a ball of mud into a patty need not change the amount of mud, and that one can recreate the ball from the patty by appropriate squeezing. To the extent that there are such universals, to the extent that infant tests measure precocity versus retardation in this development, and to the extent that such development is a sine qua non of intelligence, such tests can predict intellectual development in childhood and beyond. Research along these lines promises yet another possibility for measuring important intellectual abilities in infant development.

These are examples only of the many approaches now being taken in efforts to describe the first indications of intelligence in infants. There is much work of this kind. As yet, however, it represents only promises.

To briefly characterize infant intelligence in terms of what is now known with some assurance we can say only:

1. There is prominent development of a sensorimotor alertness that relates primarily only to health and physical skill and is not at all, or only very weakly, possibly even negatively, related to development of the basic functions of intelligence within the normal range; severe retardation of sensorimotor alertness tends to be associated with injuries and birth anomalies that are also indicative of mental retardation, but these represent abnormal development.

2. There is development of *anlage* functions (classical conditionability as well as PM and SM) that are essential for intellectual functioning but represent only a small part of this functioning and, moreover, are not the

functions that are most characteristic of mature human expressions of intelligence (Watson, 1975).

DEVELOPMENT: CHILDHOOD

As the infant passes into the preschool stage of development, more and more of his behavior evinces awareness of concepts. The processes of intelligence are largely directed at producing such awareness. Thus, to measure awareness of concepts is to measure important aspects of intelligence. Although concept awareness is indicated in behavior other than language behavior, it is most readily seen in the child's use of the language of his culture. Intelligence, as measured in childhood, is defined largely in terms of ability to comprehend and express language (e.g., as in the item "Show me the doggie").

But intelligence is manifested also in the child's facility in using generalized problem solution techniques, called aids. Algebra is an example of an aid. Knowing algebra enables one to solve more complex problems than otherwise would be the case. This is an example of a conventional aid, the development of which is linked closely to the accultural process. Other aids are somewhat idiosyncratic, just as awareness of a concept need not be linked to any particular cultural (semantic) representation. But it is easier to understand and measure a child's use of conventional aids than to comprehend and measure his development of idiosyncratic aids. For this reason, to the extent that intelligence tests measure facility in use of aids (and this seemingly is less involved than measuring awareness of concepts), they measure mainly only the child's use of conventional aids.

If there is high correlation between the development of awareness of conventional concepts and ability to use conventional aids on the one hand, and similar development in respect to idiosyncratic concepts and aids, then tests that measure the former should be highly indicative of intelligence as such. In some conditions, as when all children develop under very similar conditions, obtained measures in which there is emphasis on conventional aids and concepts can be highly indicative of the general intellective capacity also manifested in development of idiosyncratic aids and concepts. However, it seems that the conditions for intellectual development vary rather considerably from one person to another, beginning perhaps even in infancy and surely in childhood. Indeed, systematic variation in these conditions over the course of childhood development appears to produce the distinction between fluid and crystallized intelligence. What are some of the features of this development?

The development of intelligence is fundamentally linked to learning. But it is important to recognize that not all learning is indicative of intelligence. Classical conditioning, for example, can be induced readily in organisms as simple as the paramecium, and it occurs almost as readily in persons of low IQ as in those of high IQ. The kind of learning that is most indicative of intelligence can be referred to as meaningful learning, in which multiple associations are comprehended and held in mind (mediated) as one considers consequences (i.e., the perception of relations and eduction of correlates). In this respect, childhood development is characterized by change from use of only one association for a given stimulus to use of multiple associations. It is characterized also by transition from learning that is mainly reactive to learning that is guided by inquiry and hypothesis testing. These general principles of development interact with a variety of other important influences, some of which may be briefly listed as follows.

Acculturational Influences

Education, even in the most formal sense of this term, accounts for much development and differentiation of intellectual abilities, particularly in childhood. But education does not encompass all of the shaping of intellectual abilities that occurs through the influence of the dominant culture. For example, a family that is supportive of the achievement values of a culture will produce a different environment for absorption of education than a family in which there is rejection of the society's achievement values. Some of the means for bringing about major accultural effects are described briefly in the following sections.

Positive Transfer

In general, learning one concept or aid makes it easier to learn related concepts and aids (although there are a number of qualifiers to this assertion). Learning Spanish makes it easier to learn Italian. In concept learning, one also learns aids that facilitate other concept learning even when the concepts, as such, are not very similar. The "learning sets" of Harlow's well-known experiments, for example, represent aids in paired associates learning that may be applied to the learning of quite different sets of associations.

As Ferguson (1954, 1956) insightfully pointed out in pioneering work on this theme, the effect of positive transfer over the long course of development of abilities in childhood (and thereafter) must be to group together expressions of similar abilities in individual differences. The child who learns one ability of a group of similar abilities finds it relatively easy to, and thus tends to, learn the other abilities of the group, while the opposite is also true. Such positive transfer must account in part for the emergence in childhood of at least some of the primary ability factors.

Selection and Deselection

Some individuals are exposed to rather extended acculturation of a particular kind (e.g., that associated with a form of religion that is different in major respects from the major religions of a society), while other individuals encounter virtually none of this kind of acculturation. The abilities that are formed under this kind of influence constitute a similarity group that may look as if it resulted from positive transfer but is due instead to the fact that those who learn one ability of the group tend to learn the others, while those who do not receive this form of learning do not acquire any of the abilities of the group. This kind of development produces part of a sentiment (Cattell, 1957, 1971; Horn, 1966).

Labeling

When one is selected for some programs of acculturation, he also is labeled, and such labeling can be an influence in further development. A child placed in the program for "exceptionals," for example, may take on a concept of self associated with the label and subsequently learn in conformance with, or in opposition to, this characterization. Others with whom he is in frequent contact may react to his label and thus tend to "call out" particular forms of behavior from him. (See Finn, 1971, on the so-called Pygmalion effect.) These factors can notably influence achievement, perhaps across a wide spectrum of abilities (Clark, 1955).

Avoidance Learning

In one sense, this is the other side of positive transfer. More generally, however, it represents a rather broad category of influences associated not only with learning to avoid circumstances in which certain kinds of learning can occur but also learning not to learn even when placed in those circumstances. Girls in our culture may tend to learn not to learn mathematics, for example. When certain areas are avoided, and others are not, the result over the course of development is to produce similarity groupings of abilities.

Interpersonal Configurations

Zajonc (1976) has brought together evidence from several sources to suggest that a child's ability development, particularly in respect to primary abilities such as verbal comprehension, is systematically influenced by the number of, and intellectual maturity of, the persons with whom the child has principal contact during formative years. Zajonc used this idea to help explain data showing: *(1)* decrease in children's ability scores with increase in birth order and size of family; *(2)* lower scores for twins than nontwins; *(3)* higher scores for children in home settings in which adults in addition to the parents (e.g., the grandparents) also reside; *(4)* change in standardized high school and college achievement scores (SAT, ACT) associated with change in birth rates. With an increase in the birth rate between 1946 and 1950, there was a decrease in SAT scores obtained in the 1964–1968 period, for example. In general, the notion is that ability groupings reflect (in part) the amount and kind of interaction a child has with exemplars of the culture during the early (e.g., first 10) years of life.

Values of Significant Others

A general expression of a theory of interpersonal influences on the development of abilities should contain references to configurations representing the dominant kind of learning the child encounters in interactions with parents, with peers, in neighborhoods, in schools, and so on. For example, Lesser, Fifer, and Clark (1965) have presented evidence indicating that children raised in Jewish homes tend to score high on verbal comprehension relative to spatial abilities but that children brought up in Chinese–American homes score high in spatial abilities relative to verbal comprehension. One suggestion is that in the homes of one subcultural group one kind of learning is emphasized, perhaps at the expense of underemphasis on another type of learning. (See also Hill, 1967; Horn, 1970a; Levinson, 1961; Werts, 1967 for more of this kind of evidence for home and subcultural influences on the formation of primary abilities.)

Incidental Learning Influences

Not all the learning that becomes manifest in the abilities of intelligence is indicative of individual differences in acculturation. Acculturation represents only that learning that is sponsored, as it were, by the culture. But the child can learn much that is not so sponsored, some children learn more in this way than others, and such learning can affect performance on tasks that are accepted as indicating intelligence. For example, parents and teachers typically do not conduct much tutoring aimed at teaching a child the street lore of his neighborhood (e.g., how to use the alleys and lots and yards, as well as the streets, to get around quickly in a neighborhood), and some children learn this much better than others. This is referred to as incidental learning, to contrast it with accultural learning. Although most tests constructed to assess aspects of intelligence are not explicitly designed to measure outcomes of this kind of learning, nevertheless, that which is learned in this manner can facilitate performance on tests, and on some tests more than others. For example, one who acquires exceptional ability in getting through the mazes of a city or a forest can find this ability of use in working with paper and pencil tests involving spatial concepts. In some respects, creative expressions of intelligence are encouraged more by incidental learning than by acculturation.

Physiological Influences

Physiological, particularly neurological, structures can be seen to affect intellectual development and to be affected by it. How well one can comprehend, learn, and so on is determined by the adequacy of functioning of the physiological substratum. But how well one learns then determines (in part) how well one maintains the physiological substratum. There is thus a dynamic interaction between physiological and acquisition processes in the development of intelligence. Such interactive influences determine, in part, the development realized through acculturation, but there are many factors in the latter that are not preordained by the former, and vice versa. A brain injury can occur largely independently of the influences associated with prior acculturation, for example. Thus, over the course of development, the intellectual outcomes of physiological influences can be quite distinct from the outcomes produced mainly by acculturation.

It seems likely, too, that some important genetic influences have the quality of time capsules and for this reason produce effects through the physiological processes that are relatively independent of acculturational influences. A genetically determined capacity or limitation need not be displayed in early stages of development but can emerge relatively late (perhaps, but not necessarily, as a consequence of catalytic environmental events). There are several examples of physical ailments that have this quality (e.g., Huntington's chorea), although little is known about such influences on intellectual development. Even less is known about "time capsule" influences that enhance development of intelligence, although it seems that the concept of the "late bloomer" may represent an action of this kind.

No doubt a substantial proportion of the direct influences associated with the effective physiological base (EPB) for intellectual development derives from genetically determined structure. Some of these genetic factors are directly related to acculturation but not all. The genetic difference between children conceived by the same parents can be quite notable—a purely genetic theory predicts about an 11 IQ point difference on the average, for example (Loehlin, Lindzey & Spuhler, 1975)—yet such children can be exposed to many highly similar acculturational influences, influences that differ notably from those to which children of other families are exposed.

Emergence of Major Dimensions of Intelligence

The influences outlined in previous sections operate in complex interaction over extended periods of time to produce the primary mental abilities and the organization among these, Gf, Gc, Gv, SAR, VPT, and Gs. Genetic, physiological, and incidental learning influences are more prominent in the development of some abilities than in others, and the same is true for acculturational influences as well. Some of the primary abilities are most closely linked to genetic indicants. (See Loehlin *et al.*, 1975, and Vandenberg, 1962, for example.) Similarly some of the primaries relative to others are more immediately, and less reversibly, affected by injuries to the central nervous system (e.g., Payne, 1961; Sterne, 1969). At a broader level, these kinds of distinctions are manifested in the separation of Gf from Gc.

In the earliest years of childhood, the distinction between Gf and Gc is not clearly drawn. This appears to be true partly because good measures of idiosyncratic concept awareness and aid development have not yet been constructed to provide indications of Gf independently of Gc. It seems to be true, also, because not much acculturation can occur in the early few years, and what does occur is not grossly dissimilar for different children of a given culture. Throughout development, too, there are many factors that produce systematic, positive correlation between the abilities of Gf and Gc. But a host of acculturational influences operate largely independently of genetic–physiological and incidental learning influences. Consider, for example, the child's area of residence, mother's interests, father's encouragement, number and ages of siblings, surrounding morale, philosophy of school, characteristics of particular teachers, attitudes and actions of peers and playmates. As these determiners impinge over the course of childhood development, the correlations between earlier and later measures of aspects of intelligence decrease. Each step in acculturation is a kind of prerequisite for further steps. The child who makes these steps finds the whole universe of the collective intelligence of the culture gradually opening up for inspection and use, while the child who misses steps tends to be shut away from this view. In this way, acculturation builds on acculturation to produce Gc. But if one so shut away is fortunate in having good genetic potential, favorable conditions for growth and maintenance of neural structures, and an environment that is conducive to incidental learning and idiosyncratic (creative) development of concepts and aids, then there can be notable development of Gf even in the absence of many favorable features of acculturation. Thus, in individual differences, expressions of Gf can be largely independent of (although positively correlated with) expressions of Gc.

DEVELOPMENT: ADULTHOOD

Over the last 40 years, there has been considerable debate as to whether or not intelligence declines in adulthood. (There has been less concern with questions about if and how intelligence improves in this period of development.) In the 1930s, the debate was premised on an implicit assumption that the really important issues pertained to a single dimension of intelligence. The controversy then revolved around two principal kinds of questions: (1) Does speediness in either the expression or measurement of intelligence produce what appears to be aging decline but really is not? (2) Do results from cross-sectional sampling of subjects indicate what appears to be decline but really is not? More recently, beginning with the seminal work of Cattell and Hebb in the 1940s, investigators have more frequently questioned the assumption that the major issues pertain to a single dimension of intelligence. The focus has shifted toward questions about which (if any) important intellectual abilities decline. Still, in the 1970s, the debate continues to pertain to subject-sampling issues and concerns about the nature of speediness effects (e.g., Baltes & Schaie, 1974, 1976; Cunningham, 1976; Horn & Donaldson, 1976; Schaie & Gribbin, 1975).

Subject Sampling Issues

In cross-sectional gathering of a sample, individuals of different ages are measured and compared on one particular occasion. If the means for performances of older cohorts (i.e., groupings of older individuals) are found to be systematically lower than the means for younger cohorts, then the results may be interpreted as indicating aging decline. There have been many studies to suggest this conclusion for measures of general intelligence. However, since older individuals have gone through their most formative years of intellectual development at different periods of history than younger persons, the differences between the means for the cohorts may mainly reflect historical–cultural change. To deal with this problem, it was reasoned that samples should be gathered longitudinally. Individuals tested on a first occasion should be followed up and tested again (and, if possible, again and again). Some results from single follow-up sampling of this kind suggested little or no noteworthy decline in intelligence. These results were often assumed to be more trustworthy than those derived from cross-sectional studies. However, longitudinal sampling also presents noteworthy problems for inference because there can be bias in resampling and bias in repeated measurements.

To provide a better data base, a few investigators (notably Schaie) first gathered a cross-sectional sample, then obtained retest measurements on the part of this sample that could be found at later times, and also obtained new cross-sectional samples on each occasion of retesting. Data gathered in this way were treated by analysis of variance (ANOVA), and the results were presented as indicating separate age, cohort, and time of measurement effects. For a given age, there could be as many cohorts as there were times of measurement; similarly, for a given cohort, there could be several age groupings. From analyses of such data Schaie and his co-workers argued that most of the intellectual differences between adults of different ages reflect generational differences associated with cohort, not processes inherent in aging, as such.

This conclusion was questioned on grounds that the analyses on which it was based were inappropriate and misinterpreted (Botwinick & Arenberg, 1976; Horn, 1976a; Horn & Donaldson, 1976). Age, cohort, and time of measurement are thoroughly confounded in such analyses. There is much arbitrariness in decisions to regard results as indicating mainly cohort or mainly age differences. The very results that were interpreted by Schaie and Baltes as indicating little aging decline in intelligence were interpreted by Horn and Donaldson as indicating noteworthy decline.

Variable Sampling Issues

Review of longitudinal and cross-sectional studies suggested that the results were not as contradictory as sometimes seemed to be assumed if one considered the ages at which measurements were obtained, and that the studies differed not only in terms of sampling of subjects but also in terms of sampling of variables to represent intelligence (Horn, 1968, 1970a). In the cross-sectional studies, the variables that best represented crystallized intelligence were often found to indicate little or no aging decline, perhaps even aging increments. The variables that best represented fluid intelligence, on the other hand, usually indicated aging decline. In the longitudinal studies that had indicated little decline, it was found that the measures of intelligence tended to be dominated by tests representing Gc. When a distinction could be made between Gf and Gc tests, the decline indicated by Gf tests could be seen to be canceled in the overall measurements by the increments indicated by Gc tests. Also, in some of these studies, the first measurements were taken in adolescence before the peak in intellectual development would have been obtained, and the follow-up measurements were obtained in midadulthood, before notable aging decline would be expected.

Most of the longitudinal and cross-sectional studies seem to be telling much the same story, namely:

1. Some of the important abilities of intelligence, principally those of Gc but also those of VPT, decline very little or not at all, perhaps even improve with increasing age in adulthood, up to about 60. Beyond 60, there may be average decline in Gc abilities, although this decline would appear to be less than for Gf abilities.

2. Other important abilities of intelligence, notably Gf and SAR, show declines beginning perhaps as early as young adulthood (the twenties) but almost certainly in evidence by age 40. The decline is gradual until about 60 but becomes pronounced (in the averages) in the years following this age (Cattell, 1971; Hooper, Fitzgerald & Papalia, 1971; Horn, 1970b, 1972, 1975, 1976a; Matarazzo, 1972; Nesselroade, Schaie, & Baltes, 1972).

It is important in evaluating the results from both longitudinal and cross-sectional studies to remember that the findings pertain to averages for groups of individuals, not to any particular individual. A relatively few individuals who score exceptionally low or high can very materially affect such results.

Speediness Issues

Many studies show that in a wide variety of tasks older adults tend to work more slowly than younger adults. The kinds of variables for which there is evidence of speediness-decline with age include a number of perceptual comparison measures, in which the subject must find a particular symbol or check to see if one symbol is the same as another, a number of reaction time measures, in which a person must respond as quickly as possible to a point stimulus or pattern of stimuli (auditory and tactile as well as visual), and a number of measures of motor speed, as in performing a small muscle or large muscle task. There is considerable evidence that older individuals hear less well and see less well than younger individuals, and there are suggestions that slowness in perceptual and motor tasks may relate mainly to defects in hearing and/or seeing. Since many intellectual tasks are speeded, some of the aging decrements in performance on these tasks are due to decreases in perceptual–motor speediness brought on by creeping defects in peripheral sensory processes. Existing evidence suggests, however, that at least some of the aging decrement in intellectual functions is due to central intellective functions, that not all of it reflects only sensory defect, changes in attitude toward intellectual tasks, or similar nonintellective factors.

Three kinds of evidence have cast doubt on assertions that aging decrement in intellectual abilities is due only to peripheral sensorimotor speediness (SPS) and/or simple reaction time (SRT) functions. One set of evidence derives from studies in which decrement is found for tests given under unspeeded conditions. For example, Horn and Cattell (1966) found that when Matrices and Letter Series measures of fluid intelligence were administered under conditions in which all subjects attempted all items, the aging decrements recorded with these measures were virtually the same as for other Gf tests administered with time limits.

A second kind of evidence derives from studies in which both speediness and accuracy are measured separately in intellectual tasks and the two separate measures are related to aging. Welford (1958) has brought together a wealth of evidence of this kind, and more recent reviews have been provided by Botwinick (1976), Arenberg (1973), Horn (1970a, 1975, 1976a), and others. While speediness often declines with age, so, too, do other nonspeeded aspects of performance. In particular, as the difficulty of problems increases, there is increase in working time and increase in the number of errors of older as compared to younger subjects.

In the third kind of study, measures designed to indicate SPS and SRT are obtained from separate sets of operations, and analyses are directed at determining if the differences measured in this way will account for the age-related differences indicated for Gf or other abilities. When the effects associated with SRT and SPS are controlled in this manner by partial correlational or covariance analyses, there is some reduction in the intellectual decline indicated for Gf and SAR, but the decrements are not eliminated. The reliably measured individual differences in SRT and SPS do not account for the measured age differences in abilities of the kind represented in Gf and SAR (Horn, 1976b, c).

Complex reaction time (CRT) seems to account for some of the aging decrement in ability tests. A CRT task is one in which a subject must anticipate several contingencies and/or do one of several different things. As task complexity increases, age decrements become more pronounced. These decrements are related to those recorded in Gf tasks. Several findings suggest that the decrement in this regard is shown mainly in the processes of initiating a response—that is, digesting, as it were, the stimulus in relation to the contingencies and utilizing this food for thought to energize a response (Horn, 1976a; see also Chapter 10).

A second kind of speediness test that seems to account for at least some of the aging decrement seen in Gf tests is that which requires the subject to maintain close attention to distinct elements among many irrelevant elements. This is the perceptual speed factor mentioned previously. The age decrements found here are not notably reduced by controlling for writing speed or for the broad visualization factor, Gv.

Memory Functions

The tasks that define the verbal productive thinking (VPT) dimension are measures of recall from quasi-permanent storage (QPS) and thus are indicants of tertiary memory (TM). Existing evidence suggests that there is little or no aging decline in TM. It seems that the size of QPS increases with age, and this does not occasion slower access rate. Hence, if anything, older persons tend to have better tertiary memory than younger persons.

But span memory (Ms) and its two more elementary components, primary memory (PM) and secondary memory (SM), do appear to decline with age in adulthood (Horn, 1976b, c). It is not clear just how this change should be interpreted, but it seems that it may be a part of the same process of maintaining close awareness that accounts for some (but not all) of the decline in Gf. Both PM and SM account for part of the age decrements of Gf, and they do so somewhat independently. That is, partialling PM alone in a Gf-age relationship reduces the correlation, but partialling SM further reduces it. Both also account for some of the aging decrement in perceptual speediness measures, and vice versa. Thus, it seems that the facility involved in maintaining keen awareness of the elements of a problem are indicated also in PM and SM measures.

Some recent findings of Botwinick and Storandt (1974) show that the ability to write *slowly* declines markedly with age in adulthood and is notably related to measures of SAR. The slow writing task can be seen to require an ability to sustain close attention.

SUMMARY

1. There is a great variety of ideas about just what intelligence is and how it is manifested. The definitions of psychology point to behaviors that are indicative of distinct and essential processes, or functions, or abilities of intelligence.

2. In the best-known theories about intelligence, individual differences in the attribute are specified as mainly due to inherited predispositions. It is difficult to accumulate evidence in support of, or opposition to, this theory partly because the aspects of intelligence that are due mainly to genetic factors have not been identified, even in theory, and partly because it is so difficult to control for possible confounding influences in assessing the effect of genetic factors.

3. There have been movements away from theories about intelligence toward theories about multiple abilities (multiple intelligences). The principal outcome of these efforts has been operational definitions of many abilities that have at least some claim to representing important aspects of intelligence or distinct intelligences.

4. The empirical data reduction methods of factor analysis have indicated that the many, many reliable tests that have been invented to provide operational definitions of intellectual abilities can be largely explained in terms of some 50 or so primary mental abilities (PMAs).

5. Several different kinds of efforts have been directed at providing valid simplification of the system of 50 PMAs. Some of these efforts have

focused on classification in terms of facets of either test construction or function. The best known of these theories is the structure of intellect model. This has been found to be deficient largely because sufficient attention has not been given to establishing independence among the abilities that are specified as indicating support for the theory.

6. In hierarchical theories, there is effort to describe accurately the interrelationships among PMAs in terms of a system of abilities in which very broad abilities encompass and are defined by narrower abilities. Existing hierarchical theories specify intelligence as the broadest function subdivided into several second-order functions, which in turn are subdivided into the PMAs. One of the better known hierarchical theories can be represented as follows:

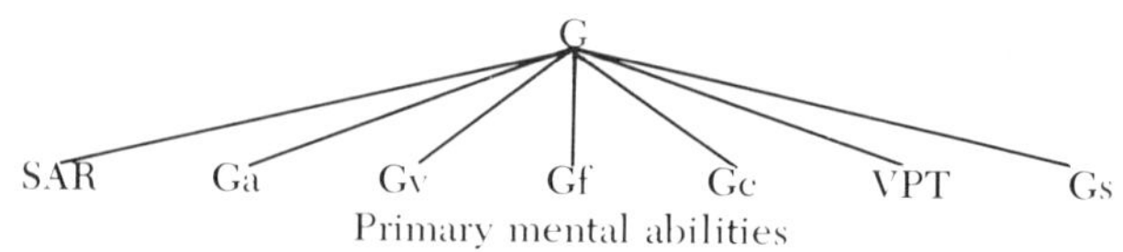

in which G stands for general intelligence; Gf, fluid intelligence; Gc, crystallized intelligence; Gv, broad visualization; Ga, broad auditory function; Gs, broad speediness; SAR, short-term acquisition and retrieval function; VPT, verbal productive thinking.

7. Dimensions of broad visualization and broad auditory ability appear to represent important influences of sensory–perceptual functions on the development and expressions of intelligence. Short-term acquisition and verbal productive thinking dimensions appear to represent the independent influences of short-term and long-term memory in intelligence. Broad speediness may represent motor or perhaps central intellective quickness, most likely the former. These functions, while essential aspects of general intelligence, do not well represent the relation-perceiving and correlate-educing functions that are the sine qua non of adult human intelligence, as seen in Gf and Gc.

8. Observations and measurements of infancy do not provide a basis for estimating intelligence, as this is measured at later stages of development. The essence of intelligence is indicated by awareness of concepts and formation of aids, little of which is manifested in easily measured form in the first 2 years of life. New techniques for assessing a child's activation and display of possibly universal modes of thinking hold promise for yielding valid measures of intelligence in infancy.

9. Acculturational influences accumulate over the course of childhood development to produce primary mental abilities and the broad collection of these abilities that has been identified as crystallized intelligence, Gc. As measured and as a concept, this is very close to vernacular conceptions of intelligence.

10. Physiological and incidental learning influences operate somewhat independently of accultural influences to produce PMAs that tend to coalesce in a broad collection of abilities indicating fluid intelligence, Gf.

11. The abilities of intelligence improve throughout most of childhood, up through adolescence into young adulthood. Exceptions to this may be the most elementary aspects of memory and perception, development of which may reach asymptote in midchildhood. Although neurological damage all along in infant and child development can be expected to occur and to set limits on further intellectual development, such effects are masked in childhood by the rapid growth of physiological structures supportive of intelligence and the rapid expansion of the learning that produces the intelligence that is measured.

12. In adulthood, the growth and learning that sustain intelligence are considerably slowed, with the result that decline in some of the important functioning of intelligence can begin to be manifested. Of these declines, the most general and noteworthy is that of fluid intelligence, although the decline in SAR no doubt is also important. This decline in Gf and SAR does not seem to be due to loss of sensorimotor function, at least as it affects speediness of performance. In part, the decline seems to be due to loss of capacity for maintaining close awareness of different aspects of a problem and thus sustaining the capacity for perceiving complex relationships.

13. Some important abilities of intelligence decline very little, or not at all, or else improve with age in adulthood. Gc and VPT well represent these abilities. Since Gc and VPT derive from the functions represented in Gf and SAR as well as from functions inherent to Gc and VPT, as such, there can be eventual decline in these latter if the processes of Gf and SAR are gradually eroded. The major portion of Gc that is available to an individual at a given time consists of material that was learned or rehearsed in the immediate, in contrast to the remote, past. Thus, if there is slowing of the build-up of Gc due to decline in the basic functions represented by Gf and SAR, then Gc, too, will ultimately decline. There is some suggestion that, in the averages, some such decline for Gc may begin to appear around age 60.

REFERENCES

Arenberg, D. Cognition and aging: Verbal learning, memory, and problem solving. In C. Eisdorfer & M. P. Lawton (Eds.), *The psychology of adult development and aging.* Washington, D.C.: The American Psychological Association, 1973. Pp. 74–97.

Baltes, P. B., & Schaie, K. W. The myth of the twilight years. *Psychology Today,* 1974, 8, 35–40.

Baltes, P. B., & Schaie, K. W. On the plasticity of intelligence in adulthood and old age: Where Horn & Donaldson fail. *American Psychologist*, 1976, *31*, 720–723.

Botwinick, J. Aging and intelligence. In J. E. Birren & K. W. Schaie (Eds.), *Handbook of the psychology of aging*. Princeton, N.J.: Van Nostrand-Reinhold, 1976.

Botwinick, J., & Arenberg, D. Disparate time spans in sequential studies of aging. *Experimental Aging Research*, 1976, *2*, 55–66.

Botwinick, J., & Storandt, M. *Memory, related functions and age*. Springfield, Ill.: Thomas, 1974.

Broadbent, D. E. The role of auditory localization in attention and memory span. *Journal of Experimental Psychology*, 1954, *47*, 191–196.

Burt, C. Subdivided factors. *British Journal of Statistical Psychology*, 1949, *2*, 41–63.

Campos, J. Heart rate. In L. Lipsett (Ed.), *Psychobiology: The significance of infancy*. Washington, D.C.: Erlbaum, 1975.

Carroll, J. B. Stalking the wayward factors. *Contemporary Psychology*, 1972, *17*, 321–324.

Carroll, J. B. Psychometric tests as cognitive tasks: A new "structure of intellect." In L. Resnick (Ed.), *The nature of intelligence*. Washington, D.C.: Erlbaum, 1975.

Cattell, R. B. *Personality and motivation structure and measurement*. New York: World Book, 1957.

Cattell, R. B. Higher order factor structures and reticular-vs-heirarchical formulae for their interpretation. In C. Banks & P. L. Broadhurst (Eds.), *Studies in psychology*. London: Univ. of London Press, 1965.

Cattell, R. B. *Abilities: Their structure, growth and action*. Boston: Houghton-Mifflin, 1971.

Clark, K. B. *Prejudice and your child*. Boston: Beacon, 1955.

Cropley, A. J. A five-year longitudinal study of the validity of creativity tests. *Developmental Psychology*, 1972, *6*, 119–124.

Cunningham, W. R. Field intelligence vs. the intellectual speed hypothesis. Paper presented at the American Psychological Association Annual Meetings, 1976.

El Koussy, A. H. Trends of research in space abilities. Paper presented at the International Colloquium on Factor Analysis, Paris, 1955.

Eysenck, H. J. (Ed.) *The measurement of intelligence*. Baltimore: Williams & Wilkins, 1973.

Ferguson, G. A. On learning and human ability. *Canadian Journal of Psychology*, 1954, *8*, 95–112.

Ferguson, G. A. On transfer and the abilities of man. *Canadian Journal of Psychology*, 1956, *10*, 121–131.

Finn, J. D. Expectations and the educational environment. *Review of Educational Research*, 1971, *42*, 387–399.

French, J. W. The description of aptitude and achievement tests in terms of rotated factors. *Psychometric Monographs*, 1951, No. 5.

French, J. W., Ekstrom, R. B., & Price, L. A. *Manual for kit of reference tests for cognitive factors*. Princeton, New Jersey: Educational Testing Service, 1963.

Guilford, J. P. *The nature of human intelligence*. New York: McGraw-Hill, 1967.

Guilford, J. P., & Hoepfner, R. *The analysis of intelligence*. New York: McGraw-Hill, 1971.

Guttman, L. A faceted definition of intelligence. *Scripta Hierosolymitana*, 1965, *14*, 166–181.

Guttman, L. Integration of test design and analysis. *Proceedings of the 1969 Invitational Conference on Testing Problems*. Princeton, New Jersey: Educational Testing Service, 1970.

Hammond, S. B. The nature of intelligence. Unpublished manuscript, Psychology Department, University of Melbourne, Australia, 1977.

Harris, M. L., & Harris, C. W. Three systems of classifying cognitive abilities as bases for reference tests. Theory Paper 33, *A structure of concept attainment abilities*, Univ. of Wisconsin, 1971.

Hill, J. P. Similarity and accordance between parents and sons in attitudes toward mathematics. *Child Development,* 1967, *38,* 777–791.

Hobbs, N. (Ed.) *Issues in the classification of children.* San Francisco: Josey-Bass, 1974.

Hofstaetter, P. R. The changing composition of intelligence: A study in F technique. *Journal of Genetic Psychology,* 1954, *85,* 159–164.

Hooper, F. H., Fitzgerald, J., & Papalia, D. Piagetian theory and the aging process: Extensions and speculations. *Aging and Human Development,* 1971, *2,* 3–30.

Horn, J. L. Motivation and dynamic calculus concepts from multivariate experiment. In R. B. Cattell (Ed.), *Handbook of multivariate experimental psychology.* Chicago: Rand McNally, 1966.

Horn, J. L. Intelligence—why it grows, why it declines. *Trans-Action,* 1967, *5,* 23–31.

Horn, J. L. Organization of abilities and the development of intelligence. *Psychological Review,* 1968, *75,* 242–259.

Horn, J. L. Organization of data on life-span development of human abilities. In L. R. Goulet & P. B. Baltes (Eds.), *Life-span development psychology.* New York: Academic Press, 1970. Pp. 423–466. (a)

Horn, J. L. Review of J. P. Guilford's "The nature of human intelligence." *Psychometrika,* 1970, *35,* 273–277. (b)

Horn, J. L. The structure of intellect: Primary abilities. In R. H. Greger (Ed.), *Multivariate personality research.* Baton Rouge, Louisiana: Claitor, 1972. Pp. 451–511.

Horn, J. L. Psychometric studies of aging and intelligence. In Samuel Gershon & Allen Raskin (Eds.), *Genesis and treatment of psychologic disorders in the elderly.* New York: Raven, 1975. Pp. 19–45.

Horn, J. L. Human abilities: A review of research and theories in the early 1970's. *Annual Review of Psychology,* 1976, *27,* 437–485. (a)

Horn, J. L. *Progress report on a study of speed, power, carefulness and short-term learning components of intelligence and changes in these components in adulthood.* Army Research Institute, Grant No. DAHC19-74-5-0012, 1976. (b)

Horn, J. L. *Progress report on learning components of intelligence and adulthood development.* National Science Foundation, Grant No. GB-41452, 1976. (c)

Horn, J. L., & Cattell, R. B. Refinement and test of the theory of fluid and crystallized intelligence. *Journal of Educational Psychology,* 1966, *57,* 253–270.

Horn, J. L., & Donaldson, G. On the myth of intellectual decline in adulthood. *American Psychologist,* 1976, *31,* 701–719.

Horn, J. L., & Knapp, J. R. On the subjective character of the empirical base of Guilford's structure-of-intellect model. *Psychological Bulletin,* 1973, *80,* 33–43.

Horn, J. L., & Knapp, J. R. Thirty wrongs do not make a right: A reply to Guilford. *Psychological Bulletin,* 1974, *81,* 502–504.

Humphreys, L. G. Hierarchical factors in course grades in an aviation high school. *Technical Documentary Report PRL-TDR-62-63.* 6570th Personnel Research Laboratory, Lackland Air Force Base, Texas, 1962.

Humphreys, L. G. The fleeting nature of the prediction of college academic success. *Journal of Educational Psychology,* 1968, *59,* 375–380.

Humphreys, L. G. Theory of intelligence. Unpublished manuscript, Psychology Department, University of Illinois, 1972.

Hundall, P. S., & Horn, J. L. On the relationships between short-term learning and fluid and crystallized intelligence. *Applied Psychological Measurement,* 1977, *1,* 11–22.

Jensen, A. R. A two-factor theory of familial mental retardation. *Proceedings of the 4th International Congress on Human Genetics,* 1971, 263–271. Amsterdam: *Excerpta Medica.*

Kagan, J. Do infants think? *Scientific American,* 1972, *226,* 74–82.

Kagan, J., & Klein, R. E. Cross-cultural perspectives on early development. *American Psychologist,* 1973, *28*(11), 947–961.

Knapp, J. R., & Horn, J. L. Capitalization on chance in research on Guilford's structure-of-intellect model. *Journal of Educational Measurement,* 1977, (in press).
Kogan, N. A clarification of Cropley and Maslany's analysis of the Wallach-Kogan creativity tests. *British Journal of Psychology,* 1971, *62,* 113–117.
Lesser, G. S., Fifer, G., & Clark, D. H., Mental abilities of children from different social-class and cultural groups. *Monographs of the Society for Research in Child Development,* 1965, *30,* No. 4.
Levinson, B. M. Subcultural values and IQ stability. *Journal of Gerontological Psychology,* 1961, *98,* 69–82.
Lewis, M., & McGurk, H. Evaluation of infant intelligence: Infant intelligence scores—true or false? *Science,* 1972, *178,* 1174–1177.
Loehlin, J. C., Lindzey, G., & Spuhler, J. *Race differences in intelligence.* San Francisco: Freeman, 1975.
Matarazzo, J. D. *Wechsler's measurement and appraisal of adult intelligence.* (5th ed.) Baltimore: Williams & Wilkins, 1972.
McCall, R. B., Hogarty, P. S., & Hurlburt, N. Transitions in infant sensorimotor development and the prediction of childhood IQ. *American Psychologist,* 1972, *27,* 728–748.
Miller, G. A. The magical number seven, plus or minus two: Some limits on our capacity for processing information. *Psychological Review,* 1956, *63,* 81–97.
Nesselroade, J. R., Schaie, K. W., & Baltes, P. B. Ontogenetic and generational components of structural and quantitative change in adult behavior. *Journal of Gerontology,* 1972, *27,* 222–228.
Pawlik, K. Concepts in human cognitions and aptitudes. In R. B. Cattell (Ed.), *Handbook of multivariate experimental psychology.* Chicago: Rand McNally, 1966.
Payne, R. W. Cognitive abnormalities. In H. J. Eysenck (Ed.), *Handbook of abnormal psychology.* New York: Basic Books, 1961. Pp. 193–261.
Pease, D., Walins, L., & Stockdale, D. F. Relationship and predictions of infant tests. *Journal of Genetic Psychology,* 1973, *122,* 31–35.
Piaget, J. *The psychology of intelligence.* London: Routledge & Kegan Paul, 1946.
Riegel, K. F. The recall of historical events. *Behavioral Science,* 1973, *18,* 354–363.
Schaie, K. W., & Gribbin, K. Adult development and aging. *Annual Review of Psychology,* 1975, *26,* 65–96.
Spearman, C. *The abilities of man.* New York: Macmillan, 1927.
Sterne, D. M. The Benton, Porteus & WAIS Digit Span Tests with normal and brain-injured subjects. *Journal of Clinical Psychology,* 1969, *25,* 173–175.
Torrance, E. P. Predictive validity of the Torrance tests of creative thinking. *Journal of Creative Behavior,* 1972, *6,* 236–252.
Undheim, J. O., & Horn, J. L. Critical evaluation of Guilford's structure-of-intellect theory. *Intelligence,* 1977, (in press).
Vandenberg, S. G. The hereditary abilities study: Heredity components in a psychological test battery. *American Journal of Human Genetics,* 1962, *14,* 220–237.
Wallach, M. A., & Kogan, M. *Modes of thinking in young children: A study of the creativity-intelligence distinction.* New York: Holt, 1965.
Watson, J. S. Early learning and intelligence. In Lewis, M. (Ed.), *Origins of intelligence.* New York: Plenum, 1975.
Welford, A. T. *Aging and human skill.* New York: Oxford Univ. Press, 1958.
Werner, E. E., & Bayley, N. The reliability of Bayley's revised scale of mental and motor development during the first year of life. *Child Development,* 1966, *37,* 39–50.
Werts, C. E. Paternal influence on career choice. *National Merit Scholarship Corporation Research Report,* 1967, *3,* 19.
Zajonc, R. B. Family configuration and intelligence. *Science,* 1976, *197,* 227–236.

6

Race and Sex Differences in Heritability of Mental Test Performance: A Study of Negroid and Caucasoid Twins

R. T. OSBORNE
University of Georgia, Athens

PREVIOUS STUDIES OF NEGROID AND CAUCASOID TWINS

The paucity of comparative heritability studies of Negroid and Caucasoid twins would tend to support those who claim that the academic community is averse to encouraging research on the racial aspects of human heritability. In addition to the information comparative heritability studies might yield on the nature of the differences between the two groups, comparisons would be of interest for other reasons. For example, heritability estimates might not be stable for both groups over the entire age range; that is, heritability estimates of mental test factors might fluctuate by age, race, and sex as do height and weight. Comparative heritability studies might also reveal that broad heritability varies with IQ range or socioeconomic status, as some investigators have claimed.

Despite the fact that the rate of twinning among Negroids is greater than among Caucasoids, only four Negroid–Caucasoid comparative twin studies have been located. Of these, two were designed especially to compare heritability estimates of Negroid and Caucasoid subjects. The two other studies, part of the Cooperative Twin Studies by Vandenberg and Osborne, were planned not to make racial comparisons of heritability but to investigate heritability of a wide range of mental test factors, including personality, social awareness, school achievement, and the primary mental test factors, space, verbal, and number.

In the independent papers based on the Cooperative Twin Studies data, Vandenberg (1970) and Osborne and Gregor (1967) arrived at diametrically opposite conclusions. In a paper presented to the Instituto Internacional de Sociologia, XXII Congreso, Madrid, October 1967, Osborne and Gregor concluded, "On the basis of data presented in Table 1, the hypothesis of the differential rate of genetic or biological contributions for whites and Negroes on spatial test performance must be rejected. That is, environment does not play a more significant role in the mental development of spatial ability of the disadvantaged (Negro) than of the culturally advantaged." In a later paper (1968), the same writers reported, "The h^2 differences are not remarkable but on seven of the eight spatial tests h^2 was higher for Negroes than for whites suggesting more rather than less genetic or biological contributions for Negro children than for white children on spatial test performance (p. 738)."

Using the Cooperative Twin Studies data, Osborne and Miele (1969) examined the racial differences in heritability estimates for numerical ability and found:

> The agreement among the four heritability ratios suggests that numerical ability is independently inherited, with as much as 59% of the within-family variance accounted for by hereditary factors. Heredity and environment produce greater differences in DZ twins than environmental influences alone produce in MZ twins. The findings cannot support the hypothesis of differential heritability ratios for white and Negro children on tests of numerical ability. The heritability estimates are not significantly higher for white than for Negro children [p. 538].

Vandenberg (1970) analyzed all (20) tests used in the Cooperative Twin Studies and concluded, "It is clear from this tabulation that there is good evidence for the thesis that the ratio between hereditary potential and realized ability was generally lower for Negroes than for whites (p. 283)."

Discrepancies between the Vandenberg and Osborne papers did not go unnoticed by Jensen, who faults both writers for the same very good reason. He writes,

> Vandenberg's data, therefore, provide no statistically reliable support for his conclusion that "there is good evidence for the thesis that the ratio between hereditary potential and realized ability was generally lower for Negroes than for Whites. . . [Vandenberg, 1970, p. 283]." Clearly, the trouble with this study is the small number

> ($N = 14$) of DZ twin pairs in the Negro group. With so few cases, the sampling error of the variance estimates is simply too large to permit any statistically reliable inferences [Jensen, 1973, p. 181].

The Osborne and Gregor paper (1968) gets the same treatment.

> The fact that a statistically significant genetic component of variance shows up on only two of nine tests for Negroes and on all of the tests for whites certainly provides no support for the authors' conclusion that "environment does not play a more significant role in the development of spatial ability of Negro children than of white children." But neither does this study provide any support for the opposite conclusion. Because of the very few cases in the Negro sample, the study throws no light whatever on Negro–white differences in the heritability of mental abilities [Jensen, 1973, p. 182].

In addition, Vandenberg and Osborne used different scoring formulas for several tests. The corrections for guessing were not applied consistently to the multiple-choice tests.

Of the other two comparative studies of Negroid and Caucasoid twins, only that of Scarr-Salapatek (1971) tested a large enough sample to keep away from the wide confidence limits of the Vandenberg and Osborne studies. This investigation, which involved 506 pairs of Negroid and 282 Caucasoid twins from the Philadelphia area, was, however, faulted on other counts. Although the sample size of the Philadelphia study is quite impressive, the author's technique of dividing the twin pairs into same and opposite sex groups could as well have been done from a table of random numbers. In their review of her study, Eaves and Jinks (1972) say, "There is certainly no evidence in Scarr-Salapatek's study that the proportion of genetical variations in either verbal or nonverbal IQ depends on race or social class. In the absence of genotype–environmental interactions for IQ there is little justification for detailed consideration of the particular models suggested by Dr. Scarr-Salapatek (p. 88)."

The dissertation of P. L. Nichols (1970) began as a twin study using data from the N.I.M.H. Collaborative Study (Myrianthopoulos, 1970b), a longitudinal investigation containing a base group of nearly 500 twin pairs. "Unfortunately," Nichols says, "the large confidence intervals found for the twin pair correlations show the difficulty of trying to estimate heritabilities from such a small sample of twins (p. 106)." Results of 61 pairs of Caucasoid and 89 Negroid twins at age 4 years were reported. Later, Nichols (1970) says, "Since the sample of twins who have completed the 7-year exams is too small (much too small) to be useful for estimating subtest heritabilities, an estimate was made from the intraclass correlations of 583 full sib pairs (p. 119)."

The twin aspect of the study was apparently abandoned in favor of a study of the effects of heredity and environment on intelligence test performance in 4- and 7-year-old Caucasoid and Negroid sibling pairs. In the combined analysis, the author used the small sample of twins and test results from a base group of approximately 2000 sibling pairs.

Heritability estimates were made in two ways: *(a)* by computing intraclass correlations of 583 full sibling pairs and *(b)* by determining the *g* loadings of the tests. Correlation between the two estimates of heritability is reported to be .85. Although Nichols' methods are not exactly comparable to those of the three twin studies reviewed, some of his conclusions, which are undeniably pertinent to a study of race differences in heritability of mental test performance, are summarized below:

1. The large within-Negroid family differences suggest that the pattern may result largely from environmental rather than genetic differences between the races.
2. Although the estimates of heritability are indirect, it appears that those subtests with the highest heritability do tend to have the largest Negroid–Caucasoid differences in performance.
3. The data not only suggest that parents' social class differences are responsible for the Negroid–Caucasoid differences in IQ test performance but also offer a fourth line of evidence that IQ differences in children associated with parents' social class differences are largely environmental.

PURPOSE AND SAMPLING PROCEDURES

The design of the Georgia Twin Study is basically a replication of the Cooperative Twin Studies of Vandenberg (1967) and Osborne and Gregor (1967), using a larger sample of twins and only tests of cognitive abilities. Several personality tests of doubtful reliability, such as the Whiteman Test of Social Perception, were used in the Cooperative Studies but were not included in the Georgia Twin Study.

The number of Negroid twins, especially DZs, in the Cooperative Studies was too small to make reliable subgroup comparisons (Jensen, 1973; Loehlin, Lindzey, & Spuhler, 1975). By increasing the sample of all twin pairs from 284 to 427 and by raising the number of Negroid twin sets from 43 to 123, it was hoped that meaningful subgroup comparisons could be made.

In a later section of this chapter, the statistical techniques used in the analysis of the Georgia Twin Study will be described in detail. Suffice it to say here, heritability coefficients will be computed to enable comparisons to be made by age, race, and sex subgroups. The analysis will be given in two parts: *(a)* in which three heritability coefficients will be computed and discussed; *(b)* in which a factor analytic method of comparing subgroups proposed by Arthur Jensen will be utilized. The two parts of the analysis will be reported separately, although in both parts the same pool of subjects and same test battery were used.

Subjects for the Cooperative Twin Study were drawn from the public and private schools in Louisville, Kentucky; Jefferson County, Kentucky; Atlanta, Georgia, city schools; Clarke County, Georgia, public schools; a

TABLE 6.1

Georgia Twin Study Distribution of Like-Sex Twins by Race, Sex, and Zygosity

MZ				DZ			
Race	Males	Females	Total	Race	Males	Females	Total
Negroid	26	50	76	Negroid	14	33	47
Caucasoid	84	87	171	Caucasoid	51	82	133
Total	110	137	247	Total	65	115	180

small number of schools in Indiana. Two hundred eighty-four sets of like-sexed twins were examined in the Cooperative Study. The results have been reported variously by Vandenberg (1967), by Osborne and Gregor (1967), and by Osborne and Suddick (1971).

Twins of the extended sample were drawn from the public schools of Cobb, Fulton, Chatham, Clarke, Walton, and Madison counties in Georgia. There were 190 pairs of twins. Eighty like-sexed pairs were Negroid; 63 like-sexed Caucasoid; 25 boy–girl sets were Negroid, and 22 sets were Caucasoid.

Table 6.1 shows distribution of the 427 pairs of the same-sex twins comprising the Georgia Twin Study. In the combined sample, there are 123 pairs of Negroid twins and 304 pairs of Caucasoid twins. The 47 pairs of unlike-sexed twins were not shown in the table since they will not be used in the analysis. Table 6.2 shows the distribution of twins by age, race, and sex.

Table 6.2

Georgia Twin Study Distribution of Twins by Age, Race, and Sex

	Caucasoid			Negroid			Total			Twin pairs	
Age	Boy	Girl	Total	Boy	Girl	Total	Boy	Girl	Total	Number	Percentage
12		2	2		2	2		4	4	2	.4
13	30	46	76	14	30	44	44	76	120	60	14.1
14	46	52	98	24	36	60	70	88	158	79	18.5
15	54	76	130	18	34	52	72	110	182	91	21.3
16	54	74	128	16	32	48	70	106	176	88	20.6
17	56	52	108	6	24	30	62	76	138	69	16.2
18	22	34	56	2	8	10	24	42	66	33	7.7
19	6	2	8				6	2	8	4	.9
20	2		2				2		2	1	.2
Means	15.59	15.41	15.49	14.78	15.01	14.93	15.41	15.28	15.33		
SDs	1.61	1.56	1.58	1.29	1.50	1.44	1.58	1.55	1.56		
Number twin pairs	135	169	304	40	83	123	175	252	427	427	

Physical Observations, Biometric Measurements, and Questionnaires Used in Zygosity Diagnosis

In order to diagnose twins in the extended sample, classical methods of Verschuer (1925) and statistical techniques developed by Nichols (1965) and Nichols and Bilbro (1966) for use in the National Merit Twin Study were combined with results of personal observations and photographs.

Using the procedure described by Montagu (1945), five standard anthropometric measures were made:

1. Nose length (distance in centimeters between the nasion and the subnasal) was measured with a sliding compass.
2. Face length (distance in centimeters from trichion to gnathion) was measured with a sliding compass.
3. Maximum head length (the distance between the glabella and the farthest point on the midline on the back of the head) was measured with a sliding caliper.
4. Maximum head breadth (the greatest transverse distance of the head usually found over each parietal bone) was measured with the spreading caliper.
5. Head circumference (the distance from the area between the eyebrows around the maximum projection of the occiput) was measured with steel tape.

In addition to the above, the following data were obtained:

1. Standing height in stocking feet was measured in inches.
2. Weight in pounds in street clothes without shoes was determined.
3. Handedness was determined by asking the subject his preferred hand for writing and throwing.
4. Dvorine Pseudo-Isochromatic Plates (Peters, 1954) were administered individually to all twins to determine degree of color blindness.
5. Individual profile and front-view photographs were made of each subject.
6. Project Talent Twin Questionnaire (Schoenfeldt, 1966) was administered to all twins.

Biometric measures were made three times by two trained examiners. The average was used for zygosity diagnosis. The measurement of height was verified by a scale on the color photographs.

Table 6.3 shows the twin physical similarity variables and the derived indexes used in zygosity determination.

Zygosity Determination

Since the publication of a paper by Smith and Penrose (1955), investigators using human twins have relied almost exclusively on serological

TABLE 6.3

Twin Physical Similarity Variables Used for Zygosity Diagnosis

Variable	Coding
Face length in millimeters	Percentage difference
Head length in millimeters	Percentage difference
Head breadth in millimeters	Percentage difference
Head circumference in millimeters	Percentage difference
Height in inches	Percentage difference
Weight in pounds	Percentage difference
Cephalic Index	Percentage difference
Kaup's Index[a]	Absolute difference
Rohrer's Index[b]	Absolute difference
Color of eyes	1 = no difference; 2 = difference in shade only; 3 = different color
Color of hair	1 = no difference; 2 = difference in shade only; 3 = different color
Other characteristics of hair	1 = no difference in rate of hair growth, hairline or pattern of growth, thickness of texture of hair, curliness of hair, or any other difference, including distribution of body hair; 2 = difference in at least one of the "other characteristics."
Nose length in millimeters	Percentage difference
Color blindness I—both normal?	1 = yes; 2 = no (3 or more errors out of a possible 15 diagnosed as color blind)
Color blindness II—both color blind?	1 = yes; 2 = no (3 or more errors out of a possible 15 diagnosed as color blind)
Handedness	1 = both right-handed or both left-handed; 2 = one or both ambidextrous; 3 = one right-handed and one left-handed.
Mistaken by parents when young?	1 = both twins responded frequently; 2 = one frequently, one occasionally; 3 = both occasionally or one frequently and one rarely or never; 4 = one occasionally and one rarely or never; 5 = both rarely or never.

[a] Body weight in grams/height in centimeters^2
[b] Body weight in grams × 100/height in centimeters^3

tests to determine zygosity. Claims of 97% accuracy for MZ diagnosis and 100% accuracy for discordant DZs seemed to have eliminated the need for the standard biometric measurements used in earlier twin research. However, a paper published by the writer (Osborne & Gregor, 1967) that used only blood types for determining zygosity was criticized by readers because reported MZ–DZ proportions did not satisfy Weinberg's rule (1901). (For the theoretical justification of Weinberg's rule, see Cannings, 1969.) It was suggested that there was a significant bias in the like-sexed twin sample against DZ twins or that some DZ pairs were misdiagnosed and classified as MZs. Sample bias was ruled out by the design, which included all like-sexed twins in the participating schools except mentally retarded children and children in special education classes. Weinberg's rule states simply that in an unselected population the number of like-sexed DZ twins is equal to the number of unlike-sexed DZ twins. Accordingly, the proportion of MZ twins is 100 less twice the percentage of opposite-sexed pairs. The rule holds even though the number of male births is slightly higher than that of female births. Using Weinberg's method in their study of over 31 million multiple and single births, Standskov and Edelen (1946) found the MZ percentage of Caucasoid twins to be 34.17 and the MZ percentage of Negroid twins to be 28.89.

In our study of like-sexed twins referred to above, we reported 172 (60.6%)MZ and 112 (39.4%)DZ twins. This is out of line with classical twin studies and with the recent study of Myrianthopoulos (1970a) that used blood type, gross and microscopic placental examination, and fingerprints and palm prints for zygosity determination.

Several other investigators have also relied only on serological tests for the same purpose. The Vandenberg twin studies, published between 1961 and 1965, involved 1140 sets of twins: 277 were diagnosed like-sexed DZs, 478 diagnosed MZs, and 385 boy–girl pairs. Zygosity was determined "exclusively on the results of an extensive battery of serological tests (Vandenberg, 1968, p. 154)." Since Vandenberg used only blood type to determine zygosity and also included boy–girl twins in his study, his data can be used to compare the number of MZ–DZ twins obtained by blood type with the number expected by Weinberg. In Vandenberg's studies, there were 478 (41.9%) pairs of twins concordant for all blood types and thus were automatically called MZs as against an expectation of only 32.5% by the Weinberg rule. Although the rule ignores the problem of differential prenatal and postnatal survival of twins of various types, this alone could not explain the deviation of the percentage from theoretical expectation.

At least two other investigations have relied only on serological tests for zygosity determination. The National Merit Twin Study (Nichols, 1965) involved 1169 sets of twins; Project Talent Twin Study (Schoenfeldt, 1966) involved 493 sets of like-sexed twins. The studies of Nichols and of Schoenfeldt used a computer program to classify twins into mutually exclusive subgroups that were similar with regard to blood diagnosed zygosity. While all twins were not actually blood typed, the effect was the same since the zygosity of the criterion groups was determined by blood typing. For Project Talent, 30% of the blood-diagnosed criterion group was identified as fraternal, 70% identical. For the National Merit Twin Study, the corresponding percentages were 33 and 67. If the percentage of DZ twins was underestimated, that is, if true DZ twins were called MZs, the resulting intraclass *r*s and heritability estimates would be attenuated.

Perhaps the most comprehensive analysis of twin data published since 1955 is that of Myrianthopoulos (1970b). With the cooperation of 14 institutions throughout the United States and the National Institute of Health, Myrianthopoulos studied 615 pairs of twins from among 56,249 pregnancies of known outcome. He did not rely solely on blood types for zygosity determination but also used sex and gross and microscopic examination of the placenta. Fingerprints and palm prints were collected on 113 pairs of like-sexed twins, but they were only used with other supporting evidence for zygosity diagnosis. Due to death and other causes, zygosity was determined for only 498 pairs of twins, of which 316, or 63%, were diagnosed as DZ and 182, or 37%, were called MZ. The expected MZ–DZ percentage in the Myrianthopoulos study, using the Weinberg difference method, were 32.6% MZ and 67.4% DZ.

It is clear from this brief review of four large twin studies that there is no single perfectly reliable method for establishing zygosity for all twin pairs. Even gross and microscopic information about the placenta is not infallible. All monochorionic twins are thought to be MZs. However, not all MZ twins are monochorionic (Corney, Robson, & Strong, 1968). A similar situation holds for serological tests. Twins discordant for any blood type are DZs, but not all pairs concordant for all blood types are necessarily MZs.

Since the level of confidence of MZ–DZ twin diagnosis determines the credibility that can be given to heritability estimates and other statistics comparing twins, in this study, efforts were made to use every practical method for zygosity determination to avoid misclassification of a twin pair. Classical methods of twin diagnosis of Verschuer (1925, 1932) were combined with those of the Automatic Interaction Detector developed at the Survey Research Center at the University of Michigan (Sonquist & Morgan, 1964), two-group discriminant analysis described by Schoenfeldt (1966), and ratings of judges made on the basis of photographs and questionnaire responses to determine zygosity. The techniques were used independently in steps to arrive at our final diagnosis of zygosity.

The present study is a follow-up of the Cooperative Twin Study of Vandenberg (1967) and Osborne and Gregor (1967). The purpose of the follow-up was to enlarge the sample size of the Cooperative Twin Study to enable comparisons to be made by sex and by subpopulation. Zygosity for all of the twins in the Cooperative Study was determined by serological tests made by the Minneapolis War Memorial Blood Bank. The following factors were tested: A, B, O, M, N, S, s, P_1, P_2, Rho, rh′, rh″, Miltenberger, Vermeyst, Lewis, Lutheran, Duffy, Kidd, Sutter, Martin, Kell, Cellano, and occasionally some others. The decision to classify a twin pair of the Cooperative Study as MZ or DZ was based only on the results of serological tests (Osborne & Gregor, 1967). Throughout this paper, the "Cooperative Twin Study" is used when referring to the studies of Vandenberg (1967) and Osborne and Gregor (1967). The Extended Sample refers to the 190 sets of twins tested in 1972 to enlarge subsamples of the Cooperative Study. The Georgia Twin Study refers to the pooled data of the Cooperative Twin Study (N = 284) and the Extended Sample (N = 143 like-sexed pairs and 47 unlike-sexed pairs).

Blood samples were taken for all subjects in the Cooperative Twin Study. Anthropometric measurements were also made for the twins of the Cooperative Study tested in Georgia. While anthropometric data were not used in the original study for zygosity determination, the measurements were coded and stored on master cards. The use of the anthropometric measurements to assist in the zygosity determination for the Extended Twin Sample will be discussed later in this section.

For the present study, 190 additional sets of twins were located and examined with the same battery of cognitive tests that was used in the Cooperative Twin Study, bringing the total number of twin sets for the Georgia Twin Study to 474 pairs. Forty-seven twin sets of the Extended Sample were boy–girl pairs. For the remaining 143 sets of like-sexed twins, it was not practical to take blood samples. Zygosity for this group was determined by combining the classical methods of Verschuer (1932) with the modern multivariate techniques used by Nichols (1965) and Schoenfeldt (1966).

The first step in determining zygosity for the new sample of like-sexed twins was to reexamine the Georgia twins of the Cooperative Twin Study in terms of concordance for blood type, similarity in anthropometric measurements, and derived indexes. The six biometric measures taken at the original testing were: (1) face length, (2) head length, (3) head breadth, (4) head circumference, (5) height, and (6) weight. Three other measures were computed: Cephalic Index, Kaup's Index[1] and Rohrer's Index.[2] These six biometric measures and the three derived indexes made a total of nine separate, but not necessarily independent, criteria for determining twin similarity. On this basis, the 44 sets of like-sexed twins (one twin of the Georgia sample did not take biometric tests) originally diagnosed only by blood type were now reclassified as Similar (MZ), Dissimilar (DZ), or Questionable. (From the tables of intrapair differences of Verschuer (1932) and Dahlberg (1926), the confidence levels of correct classification were obtained.) This set-by-set examination of biometric measures of twins revealed discrepancies when compared with zygosity determined by blood tests alone. For example, a pair of 14-year-old Negroid boys (No. 021) was concordant for all blood tests but differed in

[1] Kaup's Index: body weight in grams/height in centimeters^2

[2] Rohrer's Index of Body Structure: body weight in grams × 100/height in centimeters^3

height by 4 inches and in weight by 59 pounds. On all anthropometric measures, they resembled DZ twins more than MZs.

A pair of 14-year-old Caucasoid twins (No. 043) differed by 5% in height and 12% in weight. Both figures are beyond the .01 level for MZ twins from the Verschuer tables. Other biometric measures, including face length, nose length, and head length, support a DZ diagnosis for set No. 043. All other pairs of Georgia twins were examined carefully on the basis of the six biometric measures, three derived indexes, handedness, color blindness, and color photographs. It was the consensus of three judges that seven twin sets, originally classified as MZs, were actually DZs. Thus, 44 twin sets of the Cooperative Twin Study were reclassified as 25 MZs and 19 DZs. (A forty-fifth twin pair (No. 33) did not take biometric tests and had to be classified on the basis of blood tests alone.) The remaining 239 sets of twins were classified MZ or DZ solely on the basis of blood group data.

For the 143 sets of twins in the Extended Sample, we had the same biometric measures and three derived indexes as for the 44 twin sets of known zygosity. Because group membership or zygosity was known for the 44 pairs, it was appropriate to use the two-group discriminant analysis program described by Schoenfeldt in the Project Talent Twin Study (1966) to maximize the separation of MZ and DZ twins. Discriminant analysis program BMDO 7M (Dixon, 1970) transforms the nine similarity scores of each twin pair of the a priori group into a single discriminant score. This score is the position of a twin pair along the line that best separates the two groups. Based on the mix of the groups at that point on the line, it is possible to classify the twin set as MZ or DZ. The twin pair is allocated to whichever group has the greatest proportion of members at the point in question. By then comparing classification results with known zygosity, the accuracy of predictions based on the discriminant scores can be determined. The classifications made by the discriminant analysis program were set aside to be compared with the classifications made independently by the Automatic Interaction Detector (AID) program (Sonquist & Morgan, 1964).

The next step in zygosity determination was to apply to the 143 pairs of twins of the Extended Sample the criteria of twin physical similarity developed and modified for Project Talent and for the National Merit Scholarship Twin Questionnaire. It will be recalled that the Project Talent Questionnaire was administered as part of the Extended Sample test battery.

The procedure used in developing the AID classification system is explained in detail by Schoenfeldt (1966). The physical feature of height will provide an illustration of how the program operates. On the questionnaire, each twin of each pair was asked his own height as well as the difference in inches between his height and that of his twin. The height in inches for both, plus each twin's estimate of the difference, are the four pieces of data read in by the program. Assuming an individual will have as good or better knowledge of his own height as of the height of another person, even his twin, the most reliable estimate of the difference would be found by simply subtracting the two values, that is, those provided by each twin. This is precisely the first step the computer is programmed to perform. Subsequent steps will check to make sure the computed difference is not substantially deviant from that estimated by each twin. If the differences taken from the questionnaire are within 1 inch of that calculated by the program, the latter becomes the output value. If one or both estimates differ from the computed value by 2 inches, the output will be the average of the three differences in height, that is, those provided by the twins and the computed value.

From the classification chart shown in Figure 6.1, one can easily follow each classification step. Using the AID Program, twins of the Extended Sample were classified as either MZ or DZ.

The next step was to compare the two independent classifications, the one made by discriminant analysis and the one made by the AID Program. There is some overlap in the variables used in the two systems. However, it should be pointed out that the biometric measurements used in discriminant analysis were actual physical measurements made at the same time the psychological tests were administered, while the input for the AID Program was a self-report questionnaire completed by the subjects subsequent to the psychological testing.

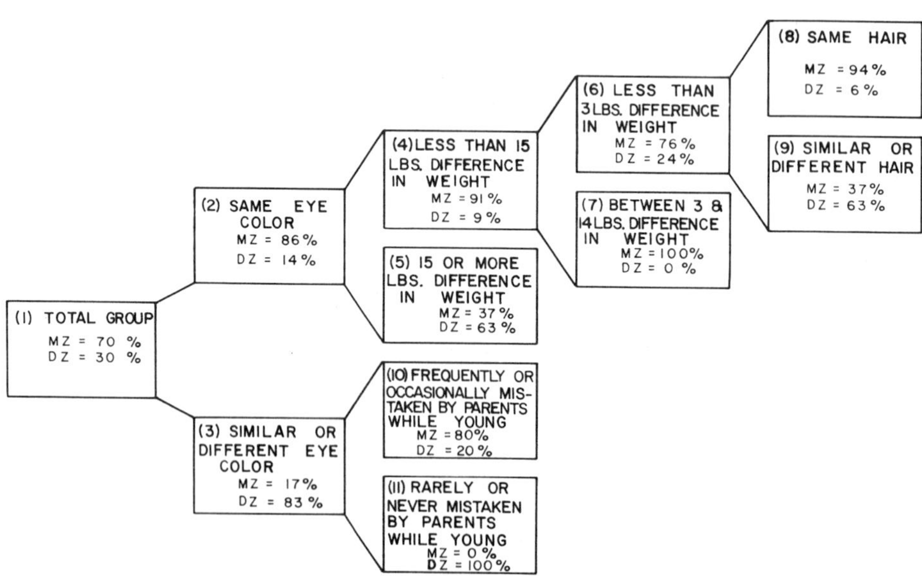

Figure 6.1 Classification structure produced by using AID with twins of known zygosity.

Of the 143 twin pairs, 96 pairs (61 MZs and 35 DZs) were classified the same by both the AID Program and 9 variable discriminant analysis. Those twins identified the same way by both programs were classified accordingly for this study.

To classify the remaining 47 pairs of like-sexed twins, a second discriminant analysis was run with the 96 twin sets on which agreement could be reached in the a priori group and the 47 unclassified sets in the test group. For the second discriminant analysis, 17 variables were used. To the 9 variables from the first discriminant analysis program were added: (1) nose length, (2) eye color, (3) hair color, (4) other hair differences, (5) color blindness (two variables), (6) mistaken identity variables from the Project Talent Questionnaire, and (7) handedness (Table 6.3). The classifications made by the second discriminant analysis were again set aside to be compared with those made by independent judges.

Three judges classified the 47 pairs of twins as MZ or DZ, using front and profile photographs, statements of likenesses and differences made by the twins, and the twins' self-report of zygosity. Classifications made by the 17 variable discriminant analysis program and by judges were concordant for 16 DZ pairs and 20 MZ pairs. The two systems were discordant on 11 sets. The 36 pairs of twins, diagnosed the same by the second discriminant analysis program and by the judges, were identified accordingly for this study.

Of the 143 pairs of like-sexed twins in the Extended Sample, 11 pairs remained in the doubtful classification. To reach a decision on zygosity of these 11 doubtful pairs, the complete files, except psychological test results, were examined for classification by the principal investigator with the following results:

Twin Pair No. 233: These 16-year-old Caucasoid girls were called MZ by the discriminant analysis program and DZ by AID. The girls were exactly the same height but different by 14.5% in weight. Differences in head length and breadth were also significant. One sister was right-handed, the other ambidextrous. The twins reported they were rarely misidentified. They believe they are DZ. Twin A says, "There is no resemblance. Everything is unlike." Final classification, DZ.

Twin Pair No. 277: In terms of biometric measurements, these 14-year-old Negroid girls appear to be identical. They are the same height. Head length and head breadth are also the same. There are only slight differences in the other physical measurements. However, Twin A is right-handed; Twin B is not. The test for color blindness probably convinced the investigator. Final classification, DZ.

Twin Pair No. 282: These 14-year-old Negroid boys were classified MZ by AID and DZ by the discriminant analysis program. Weight difference was 6%; face length difference, 8%. Twin A was color blind; Twin B was not. A is left-handed; B is right-handed. Final classification, DZ.

Twin Pair No. 284: These 15-year-old Negroid girls were not classified the same way by the computer programs. Examination of their files convinced the investigator they were DZ. The twins say they are fraternal. A is left-handed; B is right. Both say they do not look alike. Both twins say that Twin A is darker skinned and heavier; but also both say that their noses, mouths, and eyes look alike. Teachers, parents, and friends sometimes mistake one for the other. Differences in head length and breadth both are significant at the .01 level from Verschuer tables. Final classification, DZ.

Twin Pair No. 309: This pair of 17-year-old Caucasoid girls was classified DZ by AID and MZ by discriminant analysis. The girls differ by 8% in height and 27% in weight. Twin A is right-handed; B is left-handed. The attending physician said they were DZ, and the girls believe they are fraternal. Final classification, DZ.

Twin Pair No. 317: These are 14-year-old Caucasoid girls. A says she knows she is an MZ twin; B is just as confident she is DZ because the attending physician said they were DZ. In the questionnaire, B said their noses were not alike. This, in fact, is the case since their noses differ in lenght by 9%. Height difference is 5%; weight, 13%. A is right-handed; B is left-handed. Final classification, DZ.

Twin Pair No. 347: This pair of 14-year-old Caucasoid girls says their attending physician said they are identical. However, Twin B says, "We look nothing alike." A's hair is brown; B's is auburn. They never, or only rarely, are mistaken by teachers and parents. Differences in nose length, face length, head length, and height all support final diagnosis of DZ.

Twin Pair No. 362: These 13-year-old Negroid boys "know we are identical," but Twin A says that B's hair grows faster than his. They are only occasionally mistaken by teachers, friends, and parents. Differences in five biometric measurements, height, weight, head breadth, nose length, and face length, convinced the investigator of the final DZ classification.

Twin Pair No. 373: These 13-year-old Negroid girls know they are fraternal. A is right-handed; B is left-handed. They are rarely mistaken by friends, teachers, or parents. A's hair is lighter and thinner than B's. Both twins report their faces, legs, and heads to be different. The AID Program called them DZ; the discriminant analysis, MZ. Final classification, DZ.

Twin Pair No. 375: These 17-year-old Caucasoid girls say they are identical but rarely misidentified. They indicate their noses, fingers, hands, stomachs, and busts to be similar. The discriminant analysis program classifies the girls as DZ; the AID Program as MZ. Rohrer's Index of Body Structure and Kaup's Index both support the diagnosis of DZ. Differences in nose length and face length confirm the final DZ classification.

Twin Pair No. 379: These 16-year-old Negroid boys say they look alike and know they are identical because their attending physician said they were. They are seldom misidentified. Face-length difference is the only biometric measure that supports a DZ diagnosis. Other measurements are within MZ limits. Final classification, MZ.

The addition of the twins in the Extended Sample to those of the Cooperative Twin Study brings the total number of like-sexed twin pairs in the Georgia Twin Study to 427. The age range was from 12 to 20. There were 175 twin boys, 272 twin girls. Negroids made up 29% of the sample, Caucasoids accounted for 71%. (Table 6.2 gives the distribution by age, race, sex, and zygosity of the 427 pairs of twins of the Georgia Twin Study.)

Psychological Tests

The psychological tests used in the Georgia Twin Study were selected to represent, insofar as possible, the unique primary mental abilities identified independently by Thurstone (1938), Cattell (1957), and Guilford (1967). Global IQ tests, which may include specific learned achievements and tests of short-term memory, which Jensen calls "Level I" and which are found in the Stanford–Binet and Wechsler scales, were not considered in this study. The primary abilities tested are called "Level II" abilities by Jensen (1973) and "gc" by Cattell (1971).

Jensen says Level II abilities include mental manipulation and transformation of information in order to arrive at a satisfactory output. Level II is much the same as what Spearman termed *g* (Jensen, 1973).

Cattell's crystallized general mental ability "gc" shows itself heavily in such primary mental abilities as verbal factor, numerical ability, reasoning, mechanical information, and experimental judgment (Cattell, 1971).

In this section, the 12 tests used in the new study are described in detail and identified by author or publisher. Sample items are given and the method of scoring is explained.

On the Calendar Test, developed by Remondino (1962), the examinee is asked to mark the exactness of 50 sentences pertaining to the relationship of the days of the week true or false. In a factor analysis, Remondino found that this test loaded on the Number factor. The Calendar Test, scored number right minus number wrong, yields a single test score. Following are examples of the types of questions asked:

If today is Sunday, then tomorrow will be Monday.	T	F
If yesterday was Wednesday, then today is Saturday.	T	F

The Cube Comparisons Test was developed from Thurstone's Cubes. Each item presents two drawings of a cube. Assuming no cube can have two faces alike, the subject has to decide whether the two drawings *can* represent the same cube or *must* represent different cubes. The instructions indicate that the task can be performed *(1)* by mentally turning one of the cubes so that its face is oriented in the same way as the like face of the second cube and then comparing the sides one by one or *(2)* by noting whether two faces that are side by side have the same letters or numbers in the same position relative to one another. The process of obtaining the answers by the second method consists largely of verbal reasoning, although it does require a "static" awareness of three-dimensional relations as opposed to a more "dynamic" moving around of the blocks in space. Cube Comparisons, scored number right minus number wrong, yields two-part scores and a total score (French, Ekstrom, & Price, 1963).

The Simple Arithmetic Test, taken from an unpublished study by Mukherjee (1963), contains seven parts, each consisting of a number of simple arithmetical problems. Part 1 contains 15 problems; Part 2, 20 problems; and parts 3–7 each contains 25 problems. Speed is an important factor since the examinee is allowed only 2 minutes per test. This is a multiple-choice test with five alternatives for each problem. Complexity of problems decreases from Part 1 to Part 7. There are five choices for each item on the Arithmetic Test, scored number right minus one-fourth number wrong. The seven subtests are scored to obtain the total score. Examples contained in each part are given below:

Part 1: $4(77 + 39 - 4)/7 =$ 60 68 74 64 84
Part 2: $5(69 + 18 - 3) =$ 420 400 410 415 425
Part 3: $69 + 25 - 9 =$ 85 95 90 89 80
Part 4: $640 \div 5 =$ 120 128 88 136 126
Part 5: $8 \times 91 =$ 738 728 732 739 737
Part 6: $19 - 7 =$ 12 13 14 15 16
Part 7: $83 + 17 =$ 90 110 100 109 101

The Wide Range Vocabulary Test, which was adapted from a Cooperative Vocabulary Test (French *et al.*, 1963) is a five-choice synonym test with items ranging from very easy to very difficult. Scoring is by the formula: number right minus one-fourth number wrong. There are no part scores. Samples of the items follow:

JOVIAL: 1. refreshing 2. scare 3. thickset 4. wise 5. jolly
DULLARD: 1. peon 2. duck 3. braggart 4. thief 5. dunce

The Surface Development Test is adapted from Thurstone. In this test, the subject has to imagine or visualize how a piece of paper can be folded to form some kind of object. Each item consists of a drawing of a piece of paper that can be folded on the dotted lines to form the object drawn at the right. The subject is to imagine the folding, to figure out which of the lettered edges on the object are the same as the numbered edges on the piece of paper at the left, and to identify the letters of the answers in the number spaces at the far right. He is told that the side of the flat piece marked with the X will always be the same as the side of the object so marked. This task apparently requires mental movement of the parts of the pattern and probably cannot be performed by verbal reasoning only. The test, scored number right, yields two-part scores and a total score (French *et al.*, 1963).

Each item of the Form Board Test (French *et al.*, 1963) presents five shaded drawings of pieces, some or all of which can be put together to form a figure presented in outline form. The task is to indicate which of the pieces, when fitted together, will form the outline. The test is scored as number right. The two parts are added to yield the total score.

The Self-Judging Vocabulary Test, developed by Heim (1965), contains two parts. The first part contains 128 words, each of which is to be marked with an A, B, or C. (A = I know this word and could explain it to someone unfamiliar with it, B = I am doubtful as to what this word means, C = I have never seen this word before and have no idea what it means.) The second part of the test consists of the first 80 words of the 128-word list presented as a multiple-choice test with six choices. The second part of the test combines the advantages of the multiple-choice and creative-answer techniques by allowing the examinee, who thinks he knows the word but dislikes the six choices offered, to write his answer in his own words below the six choices. The test is scored number right minus one-fifth number wrong. In this study, only the second part of the test is used. Examples of test items are given below:

AUTHENTIC: 1. writer 2. to allow 3. respectful 4. a bargain 5. antique 6. genuine

VERSATILE: 1. of varied activities 2. pouring out 3. form of poetry 4. having masculine vigor 5. intense 6. kind of turnstile

The Paper Folding Test was suggested by Thurstone's Punched Holes. For each item, successive drawings illustrate two or three folds made in a square sheet of paper. A drawing of the folded paper shows where a hole is punched in it. The subject selects one of five drawings to show how the sheet would appear completely unfolded. While it is probable that the problems can be solved more quickly by imagining the folding and unfolding, they can also be solved by verbal reasoning. The latter, however, is more likely to lead to incorrect answers. The items are scored number right minus one-fourth number wrong. The two subtests are summed to obtain the total score (French *et al.*, 1963).

In the Object Aperture Test, a test of spatial visualization developed by Philip H. DuBois and Goldine C. Gleser (1948), a three-dimensional object is shown at the left, followed by outlines of five apertures or openings. The subject is to imagine how the object looks from all directions, then to select from the five apertures outlined the opening through which the solid object would pass directly if the proper side were inserted first. This usually requires the subject to turn mentally the object into other positions. The test is scored number right minus one-fourth number wrong. It yields two-part scores that are added for the total score.

The Identical Pictures Test was adapted from Thurstone. For each item, the subject is asked to check which of five geometrical figures or pictures in a row is identical to the given figure at the left end of the row. The test is scored number right minus one-fourth number wrong. Two subtests are summed for the total score (French *et al.*, 1963).

The Newcastle Spatial Test, developed by I. McFarlane Smith and J. S. Lawes (1959) for the National Foundation for Educational Research in England and Wales, consists of six different subtests ranging in difficulty from simple recognition of selections of regular solids to the more complex problems of surface development.

Subtest 1 consists of 10 sets of drawings in which the end views and middle section of a solid object (in the order end, middle, end) are shown. The subject is to determine which one of 12 solid objects on the opposite page fits each set of drawings. It appears that this test requires some idea of perspective drawing but not strongly developed spatial ability.

Subtest 2 requires the subject to indicate which of four choices is a view from above of the solid model shown at the left of the row. This test also seems to call for only a modest amount of spatial visualization.

In each item for Subtest 3, the subject is given three sides of a cube in a flat pattern and a drawing of a solid cube, part of which is shaded. The subject is to draw lines on the pattern to indicate where he would cut to remove the parts shown shaded on the solid model. One would probably use spatial visualization to solve this problem, although it seems possible to perform the task by verbal reasoning.

In Subtest 4, each item shows a block of wood. The subject is to imagine a cut made where shown by the dotted lines and to indicate which of the three drawings on the right shows the shape of the cut face. It appears that for this task no highly developed ability to visualize three-dimensional objects is needed.

In each item on Subtest 5, there is a drawing of a solid object, called Shape, and a place to copy it, called Framework. The subject is to put circles around the crosses in the Framework that could be joined to make the Shape. It is not necessary to visualize the shape in three dimensions to copy it. In fact, the task may be easier if one regards the Shape as a flat pattern and merely counts units of distance.

Each item in Subtest 6 shows a model built from the shapes shown next to it. The subject is required to indicate the number of times each Shape was used to make the model. Although one could rely largely on verbal reasoning to solve these problems, visualization would probably contribute to speed of solution. For each subtest, the score is the number of correct answers. The total score is the sum of the six subtest scores.

The Spelling Achievement Test was taken from the Metropolitan Achievement Test (Allen, Bixler, Connor, and Graham, 1946). In this test, each word was pronounced by the examiner, used in a sentence, and then pronounced again. The student was then instructed to write the word. The test, consisting of 60 words, was administered to small groups of subjects by trained examiners in accordance with standard procedures. There are no parts scores. The total score is the number of words spelled correctly. Examples are given below:

garage	I keep my car in a garage.	garage
instructor	One who teaches is an instructor.	instructor
tuberculosis	Tuberculosis is a serious lung disease.	tuberculosis

The 12-test battery produced 29 individual scores that will be factor analyzed after a method suggested by Jensen (personal communication, 1974) before heritability ratios are computed. In a separate analysis, formulas used in the earlier studies of Vandenberg (1970) and Osborne, Gregor, and Miele (1968) will be used for computing heritability ratios for the 12 individual tests and for various combinations of the tests that yield verbal, spatial, perceptual speed, and full-scale IQs. The formulas used in the analyses will be described in the next section.

INHERITANCE OF GENERAL AND SPECIFIC MENTAL ABILITIES

The primary aim of this chapter was to replicate the Cooperative Twin Studies of Vandenberg (1970) and Osborne *et al.* (1968) with a larger sample of twins, especially black DZs. The same classical methods of genetic analysis of human twin data were used. It was expected that the larger sample of twins would permit more reliable heritability comparisons to be made by age, race, and sex than in the previous studies.

Since the Cooperative Studies were published, behavioral geneticists have found the heritability formulas of Holzinger, Nichols, and Vandenberg less powerful indexes of heritability than some newer multivariate techniques. One serious criticism of Holzinger's H and Nichols' HR is that one is not a monotonic function of the other, and neither is a monotonic function of h^2 (Jensen, 1972). Jensen found Vandenberg's F ratio faulty as an index of heritability because F is a linear function of H.

Since the variance ratio F is an essential step in computing h^2, h^2 cannot be presumed to differ significantly from zero if F is not significant (Jensen, 1972).

Other behavioral geneticists have developed more comprehensive and sophisticated approaches to the genetic analysis of human twin data. Cattell's Multiple Abstract Variance Analysis (MAVA) assesses the importance of correlation between genetic and environmental influences within the family as well as within the culture (Cattell, 1971). The biometric genetic approach of Jinks and Faulker (1970) attempts to go beyond the other methods to an assessment of the kinds of gene action and mating systems operating in the population.

Only the three classical heritability formulas used in the Cooperative Twin Studies will be applied in the first part of this section. The second part involves a factor analytic approach before applying the heritability formulas (Jensen, personal communication, 1974). It is hoped that the biometric genetic methods of Jinks and Faulker (1970) can be applied to the present data.

Holzinger's H index (Holzinger, 1929) is the ratio of half the heritability variance to the variance within sets of fraternal twins.

$$H = \frac{r\text{MZ} - r\text{DZ}}{1 - r\text{DZ}}$$

Nichols (1965) developed the HR coefficient for analysis of the National Merit Twin Study data. It is the ratio of hereditary variance to variance due to heredity and environment common to both twins of a set.

$$HR = \frac{2\,(r\text{MZ} - r\text{DZ})}{r\text{MZ}}$$

Vandenberg's (1965) F ratio compares within pair variance of DZ twins with that of MZs, and significance is tested by Fisher's F test.

$$F = \frac{\sigma^2 w\text{DZ}}{\sigma^2 w\text{MZ}}$$

Heritability Coefficients for Negroid and Caucasoid Twins: Classical Methods of Analysis

After standardizing test raw scores for age and race, the Holzinger, Nichols, and Vandenberg heritability ratios were computed for the 12 separate tests described in the previous section. The F tests shown in the tables refer to the DZ within pair variance/MZ within pair variance. To test the significance of the difference between two Fs, the Fs are transformed to a unit normal variate after the method described by Paulson (1942). Paulson's U statistic is entered in a Z table to determine its probability level.

Table 6.4 gives the three heritability ratios, intraclass *r*s, within pair variances for MZ and DZ twins and the *U* statistic for determining the significance of difference between *F*s. The striking thing about the results shown in Table 6.4 is the wide range of heritability ratios for both groups, suggesting that mental abilities represented by the 12 tests are not uniform in their genetic and environmental characteristics. For example, the tasks required in the Simple Arithmetic Test turned out to be highly heritable for both races. Not only are the *F* tests significant for Negroids and Caucasoids, but the intraclass *r*s are all high; the correlations for Negroids are slightly greater than for Caucasoids. For Caucasoids, 8 of the 12 *F* ratios are significant at the .05 level or better; for Negroids, four are significant. On the other hand, there is one test, Form Board, which yields a negative *H* value; that is, the intraclass correlation for DZ black twins is greater than for MZs. In all other cases, rMZ − rDZ is in the expected direction, that is, rMZ > rDZ.

Three other tests deserve mention. The Spelling Test, the Newcastle Spatial Test, and Identical Pictures yield significant *F* ratios for the Caucasoids at the .01 level and at the .05 level for Negroids even though the number of Negroid DZ twins is only 47. As a rule of thumb, Loehlin *et al.* (1975) say with typical values of IQ correlations and 50 pairs in each group, a standard error of the index of broad heritability is approximately .23. For 500 pairs in each group, it is .07. I would add this is about the likelihood of an investigator locating 500 twin pairs for each of eight subgroups.

The next question to be asked of Table 6.4 is, Are any of the between race variance ratios significant? The answer is yes. Four of the 12 individual tests yield *U* values significant at the .01 level. This means that the heritability ratios for two verbal tests, Spelling and Heim Vocabulary, and two spatial tests, Cube Comparison and Surface Development, are different for Negroid and Caucasoid twins. All other individual tests show insignificant variance ratio differences.

The 12 tests, standardized for age and race (mean 100; standard deviation 15), were averaged to get a composite score that would give equal weight to the individual tests. Heritability ratios for the composite or total scores are shown in the last two rows of Table 6.4. When all 12 tests are combined into a general mental test score, *F* ratios for both Negroids and Caucasoids are significant at the .01 level, but there is no difference in the *U* statistic; that is, when the 12 tests are equally weighted and combined into a general mental ability score, there is no difference in variance ratios between the two races.

Since, by design, tests in the battery represented the broad spectrum of specific primary mental abilities (number, space, verbal, and perceptual speed), it would have been remarkable if all tests had reflected the same degree of within pair variance for both races. On the other hand, group-

TABLE 6.4

Heritability Coefficients on 12 Mental Tests for Negroid and Caucasoid Twins: Classical Methods of Analysis

	MZ		DZ					Within-pair variance			
Variable	*r*	*N*	*r*	*N*	*T*(Cor)	*H*	*HR*	*MZ*	*DZ*	*F*	*U*[c]
Calendar											
Caucasoid	.48	170	.40	130	.91	.14	.36	119.09	125.78	1.06	
Negroid	.54	76	.42	46	.78	.20	.43	112.74	103.40	.92	.65
Cube Comparison											
Caucasoid	.43	168	.29	129	1.34	.19	.65	117.54	176.04	1.50[a]	
Negroid	.28	76	.10	47	.95	.19	1.26	190.22	137.36	.72	3.68[b]
Surface Development											
Caucasoid	.72	171	.36	133	4.65	.57	1.02	62.44	138.94	2.23[b]	
Negroid	.48	76	.25	47	1.40	.31	.95	110.72	167.98	1.52	3.29[b]
Wide Range Vocabulary											
Caucasoid	.52	171	.23	133	2.95	.38	1.12	106.98	165.88	1.55[b]	
Negroid	.43	76	.22	47	1.25	.27	.98	128.17	161.81	1.26	1.80
Form Board											
Caucasoid	.59	168	.44	133	1.77	.27	.52	98.86	112.01	1.13	
Negroid	.21	75	.33	47	−.68	−.18	−1.18	183.05	135.02	.74	1.86
Arithmetic											
Caucasoid	.80	168	.53	133	4.29	.57	.68	44.26	109.97	2.49[b]	
Negroid	.84	76	.65	47	2.37	.55	.46	31.91	88.91	2.79[b]	1.57
Heim Vocabulary											
Caucasoid	.85	169	.57	132	5.24	.66	.66	33.81	89.13	2.64[b]	
Negroid	.76	75	.57	47	1.86	.45	.51	57.41	82.78	1.44	4.49[b]
Paper Folding											
Caucasoid	.55	169	.45	133	1.07	.17	.34	101.27	118.08	1.17	
Negroid	.45	76	−.02	47	2.59	.45	2.07	135.04	173.27	1.28	.00
Object Aperture											
Caucasoid	.49	168	.39	132	1.05	.16	.41	114.04	134.95	1.18	
Negroid	.39	76	.17	47	1.24	.26	1.13	141.67	154.88	1.09	.67
Identical Pictures											
Caucasoid	.76	164	.55	128	3.20	.47	.56	56.73	90.57	1.60[b]	
Negroid	.51	72	.32	47	1.20	.28	.75	99.57	162.63	1.63[a]	.96
Spelling											
Caucasoid	.85	169	.54	132	5.57	.68	.73	32.67	99.66	3.05[b]	
Negroid	.79	76	.58	47	2.14	.50	.53	49.22	82.91	1.69[a]	4.70[b]
Newcastle Spatial											
Caucasoid	.78	158	.60	125	2.91	.45	.47	46.02	95.45	2.07[b]	
Negroid	.85	75	.44	47	4.13	.74	.96	39.97	74.06	1.85[a]	1.92
Subtest mean											
Caucasoid	.85	171	.62	133	4.58	.61	.54	14.09	35.71	2.53[b]	
Negroid	.88	76	.51	47	4.18	.75	.85	10.64	31.34	2.95[b]	1.48

[a] $p < .05$.
[b] $p < .01$.
[c] difference in *U*s

ing tests of similar factor structure or combining several short tests into one composite score or general factor score should produce a more reliable measure of mental ability than a specific test alone if for no other reason than that the composite test represents a larger sample of mental test performance than the specific test.

Factor analysis of the 12 individual tests produced three distinctly separate factors: *(1)* verbal factor, made up of Calendar, Wide Range Vocabulary, Heim Vocabulary, Spelling, and Arithmetic tests; *(2)* spatial factor, made up of Cube Comparison, Surface Development, Form Board, Paper Folding, Object Aperture, and Newcastle Spatial tests; and *(3)* perceptual speed factor, represented by only one test, Identical Pictures. Since the 12 individual tests were standardized for age and race with a mean of 100 and a standard deviation of 15, the derived factor scores were called IQs. *F* ratios for each of the factor IQs and the full-scale IQ, which represents the average of the three factors, are shown in Table 6.5. With the exception of the spatial IQ, all variance ratios are significant for both races. For Caucasoids, all *F* ratios are significant at the .01 level. Full-scale IQ variance ratios are significant at the .01 level for Negroids, but verbal IQ and perceptual speed reach only the .05 level.

The variance ratios for verbal and spatial IQs are significantly larger for Caucasoids than for Negroids. It should be remembered that the four

TABLE 6.5

Heritability Coefficients for Factor IQ Tests for Negroid and Caucasoid Twins: Classical Methods of Analysis

	MZ		DZ					Within-pair variance			
Variable	*r*	*N*	*r*	*N*	*T*(Cor)	*H*	*HR*	*MZ*	*DZ*	*F*	U^c
Verbal IQ											
Caucasoid	.82	171	.59	133	4.21	.57	.57	24.29	52.54	2.16^b	
Negroid	.83	76	.64	47	2.19	.52	.45	22.44	43.28	1.93^a	2.17^a
Spatial IQ											
Caucasoid	.81	171	.57	133	3.98	.55	.59	24.25	53.98	2.23^b	
Negroid	.77	76	.45	47	2.80	.58	.84	28.01	39.82	1.42	3.55^b
Perceptual Speed IQ											
Caucasoid	.76	164	.55	128	3.20	.47	.56	56.73	90.57	1.60^b	
Negroid	.51	72	.32	47	1.20	.28	.75	99.57	162.63	1.63^a	.96
Full Scale IQ											
Caucasoid	.85	171	.60	133	4.68	.62	.58	15.49	36.57	2.36^b	
Negroid	.80	76	.34	47	3.90	.70	1.15	14.84	47.79	3.22^b	.73

[a] $p < .05$.
[b] $p < .01$.
[c] difference in *U*s

individual tests with significant Negroid–Caucasoid F differences are included in these factor IQs. The perceptual speed factor shows no significant difference in F ratios between races; neither does the full-scale IQ in which all three factor IQs are weighted equally and combined.

In Table 6.6, the races are combined, and heritability comparisons are made by sex. For the most part, there are no big surprises since the same subjects are represented as in Table 6.4. Within-pair variances are significant at the .01 level for both boys and girls on the Arithmetic, Spelling, Surface Development, and Heim Vocabulary tests. There are four tests on which variance ratios were different for boys and girls. On the Cube Comparison, Identical Pictures, and Object Aperture tests, F ratios were significant for boys only. The Wide Range Vocabulary F ratio was significant only for girls. Heritability ratios for the means of the 12 subtests are significant at the .01 level for both sexes.

Only three of the 12 boy–girl F ratios are significantly different. The boys' within-pair variance was significantly greater on two spatial tests. The girls' was greater on one. For all other tests, including Arithmetic, Spelling, and Vocabulary, in which heritability differences might be expected to be different, the U statistic shows no significant difference.

In Table 6.7, the races are combined, and heritability comparisons are made by sex for the three factor IQs and the full-scale IQ. All variance ratios for factor IQs are significant at the .01 level except the single-test factor, perceptual speed. The full-scale factor IQ variance ratio is significant at the .01 level for both boys and girls.

The differences in F ratios for the verbal and spatial IQs are insignificant for male and female comparisons. On the one-test factor, perceptual speed, sex difference in the F ratio is significant. Full-scale factor IQs show highly significant Fs for both boys and girls, but the within-pair variance difference between the sexes is insignificant.

A Factor Analytic Method of Comparing Negroid and Caucasoid Twins

In this section, twin study data are factor analyzed before heritability coefficients are computed. The first step was to eliminate from the base group those individuals who did not have scores on all 29 subtests. Seventy-seven subjects were dropped, leaving 540 Caucasoids and 237 Negroids.

For the initial factor analysis, the Negroid and Caucasoid groups were combined, and the first principal component was obtained for the total group. The groups were then separated by race, and the first principal component was computed for each race.

TABLE 6.6

Heritability Coefficients on 12 Mental Tests by Sex: Classical Methods of Analysis

	MZ		DZ					Within-pair variance			
Variable	*r*	*N*	*r*	*N*	*T*(Cor)	*H*	*HR*	*MZ*	*DZ*	*F*	*U*[c]
Calendar											
Male	.52	110	.36	63	1.25	.25	.62	125.38	117.01	.93	
Female	.48	136	.42	113	.54	.10	.23	110.45	121.56	1.10	−.84
Cube Comparison											
Male	.45	108	.28	64	1.18	.23	.73	123.51	185.44	1.50[a]	
Female	.29	136	.20	112	.70	.11	.60	153.41	154.44	1.01	1.80
Surface Development											
Male	.73	110	.44	65	2.81	.51	.79	76.30	157.23	2.06[b]	
Female	.56	137	.23	115	3.13	.43	1.19	78.09	140.47	1.80[b]	.06
Wide Range Vocabulary											
Male	.52	110	.31	65	1.61	.31	.81	96.68	132.40	1.37	
Female	.48	137	.19	115	2.55	.35	1.21	127.01	183.14	1.44[a]	−.59
Form Board											
Male	.60	108	.54	65	.51	.12	.19	114.22	97.38	.85	
Female	.30	135	.32	115	−.14	−.02	−.11	133.35	129.68	.97	−.54
Arithmetic											
Male	.80	109	.48	65	3.63	.62	.81	40.88	124.80	3.05[b]	
Female	.82	135	.60	115	3.45	.53	.52	40.04	92.98	2.32[b]	.48
Heim Vocabulary											
Male	.85	108	.50	64	4.31	.69	.81	41.44	116.55	2.81[b]	
Female	.80	136	.62	115	2.86	.47	.45	40.77	71.27	1.75[b]	1.61
Paper Folding											
Male	.49	109	.34	65	1.12	.22	.61	130.65	162.19	1.24	
Female	.54	136	.36	115	1.80	.28	.68	96.59	115.70	1.20	−.03
Object Aperture											
Male	.48	109	.35	64	.92	.19	.51	128.83	198.66	1.54[a]	
Female	.37	135	.23	115	1.19	.18	.76	117.64	107.64	.92	2.44[a]
Identical Pictures											
Male	.70	105	.34	63	3.12	.54	1.02	63.78	149.97	2.35[b]	
Female	.68	131	.57	112	1.48	.27	.34	74.63	87.40	1.17	3.01[b]
Spelling											
Male	.86	108	.41	64	5.40	.77	1.05	35.00	119.13	3.40[b]	
Female	.79	137	.59	115	3.05	.48	.51	40.01	81.98	2.05[b]	1.59
Newcastle Spatial											
Male	.79	99	.58	61	2.45	.50	.54	55.17	92.43	1.68[a]	
Female	.81	134	.53	111	4.15	.60	.69	35.87	88.05	2.46[b]	−2.64[b]
Subtest mean											
Male	.87	110	.63	65	3.73	.65	.55	13.41	38.56	2.88[b]	
Female	.84	137	.57	115	4.56	.64	.64	12.73	32.32	2.54[b]	−.29

[a] $p < .05$.
[b] $p < .01$.
[c] difference in *U*s

TABLE 6.7

Heritability Coefficients for Factor IQ Tests by Sex: Classical Methods of Analysis

	MZ		DZ					Within-pair variance			
Variable	*r*	*N*	*r*	*N*	*T*(Cor)	*H*	*HR*	*MZ*	*DZ*	*F*	*U*[b]
Verbal IQ											
Male	.85	110	.53	65	4.06	.67	.74	22.76	58.21	2.56[a]	
Female	.80	137	.63	115	2.81	.46	.42	24.49	45.55	1.86[a]	.88
Spatial IQ											
Male	.83	110	.63	65	2.87	.55	.50	24.99	53.80	2.15[a]	
Female	.74	137	.45	115	3.59	.52	.79	25.74	48.30	1.88[a]	.01
Perceptual Speed IQ											
Male	.70	105	.34	63	3.12	.54	1.02	63.78	149.97	2.35[a]	
Female	.68	131	.57	112	1.48	.27	.34	74.63	87.40	1.17	3.01[a]
Full Scale IQ											
Male	.85	110	.55	65	3.92	.66	.70	15.18	49.01	3.23[a]	
Female	.83	137	.54	115	4.41	.62	.69	15.38	34.12	2.22[a]	.97

[a] $p < .01$.
[b] difference in *U*s

Using weights determined by the factor analyses, three factor scores were assigned to each subject: *(a)* one based on weights from the total group, *(b)* one based on weights from the Caucasoid group, and *(c)* one based on weights from the Negroid group. Own-race determined factor scores were then intercorrelated with opposite-race determined scores and total group scores. The idea here was to get a good general factor score for the entire test battery.

To determine if the first principal component factor scores are measuring the same mental factor in the two races, own-race determined factor scores were intercorrelated with cross-race and total group factor scores. All *r*s were .99+, suggesting that whatever mental factor is measured in the Caucasoid group is the same as that measured in the Negroid group and in the total sample.

To get an estimate of the "reliability" of this method of cross-racial comparisons, the two racial groups were split in half at random. Twins of a pair were always assigned to the same group to avoid spuriously high correlations. The first principal component was then obtained for each of the four new subgroups.

Three factor scores based on the factor analysis were assigned to each subject in the four subgroups. Factor scores obtained for own-within race subgroup, opposite-within race subgroup, and total racial group were intercorrelated. The idea here was to test the method of comparing the same group, opposite group, and total group factor scores without introducing the variable of race. Similar factor weights across groups and high intercorrelations would suggest that the first principal component of this complex battery of mental tests is measuring the same general factor in both subgroups of each race. This is exactly what we found. Correlations for own subgroup, opposite subgroup, and total racial group factor scores were .99+ for Negroids and for Caucasoids.

Table 6.8 gives the factor loadings for all 29 subtests for each of the seven groups: total group, Negroid and Caucasoid groups separately, and

TABLE 6.8

Loadings of the General Factor on 29 Subtests by Various Subgroups

		Caucasoid group			Negroid group		
Subtest[a]	Combined total group	Random Subgroup A	Random Subgroup B	Total	Random Subgroup A	Random Subgroup B	Total
1	.63	.64	.65	.64	.57	.60	.58
2	.48	.45	.53	.49	.58	.34	.46
3	.50	.54	.57	.56	.37	.29	.33
4	.63	.63	.74	.68	.48	.47	.48
5	.67	.69	.71	.71	.60	.54	.57
6	.47	.46	.55	.51	.23	.50	.37
7	.60	.63	.68	.66	.49	.35	.42
8	.54	.56	.59	.58	.57	.29	.43
9	.53	.54	.55	.54	.37	.58	.48
10	.55	.58	.49	.54	.47	.68	.58
11	.72	.73	.70	.72	.70	.79	.74
12	.71	.71	.69	.70	.71	.77	.74
13	.62	.65	.58	.61	.58	.70	.64
14	.66	.66	.62	.64	.77	.74	.75
15	.65	.65	.61	.63	.69	.76	.72
16	.70	.67	.71	.69	.73	.72	.72
17	.62	.66	.67	.66	.65	.39	.52
18	.57	.55	.61	.58	.58	.49	.53
19	.36	.40	.42	.41	.23	.24	.24
20	.42	.46	.55	.50	.32	.11	.21
21	.39	.34	.51	.43	.26	.27	.27
22	.46	.41	.59	.51	.40	.26	.32
23	.59	.60	.51	.55	.72	.69	.70
24	.60	.57	.67	.62	.62	.47	.54
25	.53	.52	.55	.53	.50	.52	.51
26	.65	.58	.70	.65	.68	.62	.65
27	.62	.59	.73	.66	.55	.46	.50
28	.68	.69	.71	.70	.68	.57	.62
29	.60	.57	.57	.56	.70	.68	.69

[a] See text for identification of test.

the four randomly selected subgroups. Similarity of the seven groups with respect to the factor loadings is remarkable. Arithmetic tests (9–15) yield especially high loadings across all groups. Spelling (23) and the two vocabulary tests (6 and 16) also load heavily on the first principal component.

As an independent check of the validity of the first principal component factor scores, the three scores obtained from the analyses were correlated with results from a standard IQ test, Primary Mental Abilities (PMA; Science Research Associates, 1962). It will be recalled that the PMA was included in the test battery for the Extended Twin Sample. For Caucasoids, the PMA correlates .85 with both own race and opposite race factor scores; for Negroids, .82 with own race and .81 with opposite race factor scores. These *r*s are significant and approach the test–retest reliability coefficients for the PMA.

From the above cross-race correlations, it is clear that the same general factor is being measured in each group separately and in the composite group when the two races are combined. When the races are split at random and factor analyzed, the high intercorrelations of the resulting factor scores indicate the significant reliability of the first principal com-

TABLE 6.9

Heritability Ratios for Factor Scores Based on First Principal Component Analysis of Own Race, Opposite Race, and Total Group

	MZ		DZ					Within-pair variance			
Score	cor	*N*	cor	*N*	*T*(Cor)	*H*	*HR*	*MZ*	*DZ*	*F*	U^b
				Factor weights from own racial group							
Caucasoid	.85	141	.63	115	3.89	.58	.50	3517.4	9335.4	2.65^a	
Negroid	.91	70	.58	46	4.45	.79	.73	1619.3	6398.3	3.95^a	.31
Total	.91	211	.71	161	5.78	.67	.43	2887.7	8496.3	2.94^a	
				Factor weights from opposite racial group							
Caucasoid	.85	141	.64	115	3.83	.58	.50	2839.1	7696.9	2.71^a	
Negroid	.91	70	.56	46	4.46	.79	.77	1955.7	7340.8	3.75^a	.62
Total	.90	211	.71	161	5.48	.65	.42	2546.0	7595.1	2.98^a	
				Factor weights from total group							
Caucasoid	.85	141	.63	115	3.88	.58	.50	3302.0	8822.3	2.67^a	
Negroid	.91	70	.56	46	4.48	.79	.76	1838.9	7049.7	3.83^a	.47
Total	.86	211	.62	161	5.61	.65	.57	2816.6	8315.9	2.95^a	

[a] $p < .01$.
[b] difference in *U*s

ponent as a basis for the "cross-racial" correlations. The first principal component yields a general mental ability factor that is indistinguishable between races. Total group principal component factor scores correlate highly with an independent measure of IQ—.85 for Caucasoids and .82 for Negroids.

Satisfied that the mental test factors generated by the first principal component analysis were stable and represented the same factor in each race, the groups were reassembled as twins for the final step in the analysis. Classical heritability ratios were applied to the nine factor scores derived from own race, opposite race, and total group factor analyses.

Heritability ratios for total, Caucasoid, and Negroid groups are shown in Table 6.9. In the top third of the table, factor scores of the subjects' own racial group were used to compute the heritability ratios. The results are clear, and the *F* ratios for all comparisons are significant at the .01 level. In the center of the table, opposite race factor scores were used. There is no apparent change in heritability ratios from the same race factor scores. When the total group factor weights were used (lower third of Table 6.9), the results were indistinguishable from those obtained from own and opposite race analyses.

When a general mental factor, not unlike Spearman's *g*, is used to compute heritability ratios, not only are the *F* ratios highly significant for own race, other race, and total group factor scores, but there is no significant difference between the heritability ratios of the two races.

SUMMARY AND CONCLUSIONS

While granting heritability of 40–80% for IQ in Caucasoid samples, environmentalists Mercer and Brown (1973) and Lewontin (1975) say we do not know if similar heritability coefficients would occur in Negroid or Mexican–American groups. There are no data.

Even with no data available, Feldman and Lewontin (1975) tell us that no valid statistical method exists for separating variation that is the result of environmental fluctuation from variation that is the result of genetic segregation. Estimates of heritability in the broad sense are nearly equivalent to no information at all for any serious problem of human genetics.

The present study was begun well before the futility of the use of the variance analysis in human genetics was discovered by Feldman and Lewontin (1975). In fact, the Georgia Twin Study is a replication of earlier comparative studies, using a larger sample of Negroid and Caucasoid twins. The primary purpose of this investigation was to compare heritability estimates and variance ratios of Negroid and Caucasoid twins on a comprehensive battery of tests with several tests or subtests representing each of the primary mental abilities.

Classical methods of Holzinger, Nichols, and Vandenberg were applied to 427 sets of like-sexed twins ranging in age from 12 to 20; 304 pairs were Caucasoid, 123 Negroid. Obviously, the ideal of 500 subjects in each subgroup was not attained, nor is it likely to be in the foreseeable future with the ever-increasing number of federal and professional restrictions on the use of human subjects in research involving psychological tests. Acknowledging the small sample limitations and some other less than ideal experimental conditions for the study, I shall report the most consistent trends that emerge from the comparative study of Negroid and Caucasoid twins using classical heritability methods.

Primary mental abilities represented by the 12 separate tests comprising the battery show a wide range of heritability, suggesting that mental abilities represented by the tests are not uniform in their genetic and environmental characteristics. Some mental tasks yield no significant heritability ratios for either Caucasoids or Negroids. For arithmetic, spelling, and some spatial tests, the variance ratios are significant for both races. On two spatial and two vocabulary tests, F ratios are significant for Caucasoids only. On no test is the variance ratio significant for Negroids only.

On the basis of Lyon's (1961) work and later research, Maccoby and Jacklin (1974) have predicted sex differences in MZ twins since female identical twins can be somewhat more unlike than male identical twins with respect to any characteristic carried on the X chromosome but not so unlike as fraternals. To the extent that the X chromosome is implicated in a wide range of psychological attributes, then female MZs should show less congruence than male MZs. Maccoby and Jacklin's (1974) prediction is supported by our data. On 9 of the 12 individual tests, on all factor IQs, and on the total subtest score, male MZ intraclass rs were greater than female MZ rs.

The classical heritability ratios show a consistent trend for mental characteristics to be more heritable for boys than girls. This is especially true for the spatial abilities. The trend holds for individual tests and verbal, spatial, and perceptual speed IQs based on combinations of the individual tests.

When examined individually, the 12 tests, or 29 subtests, produce a wide range of both intraclass correlations and heritability estimates resulting from what Spearman might have called the seemingly unscientific course of throwing very miscellaneous tests into a common hotchpot (Jenkins & Patterson, 1961, p. 263).

Pooling the tests, as we have, in various combinations does not necessarily yield an average of the person's ability. "What the pooling does is to make the influences of the many specific factors more or less neutralize each other so that the eventual result would become the approximate measure of *g* alone (Jenkins & Patterson, 1961, p. 263)." This is exactly what we found. In whatever way all individual tests or subtests were pooled, by simply averaging the 12 standard scores, combining the factor IQs to get a full-scale IQ, or by using weighted scores determined from the first principal component factor analysis of either race singly or both races combined, the results were the same. Heritability variance ratios for both Negroids and Caucasoids were significant at the .01 level. In no case was the difference between variance ratios of the races significant.

Mental characteristics measured by the pooled scores of our comprehensive test battery correlate significantly with a standard IQ score and show congruent heritability patterns for Negroids and Caucasoids.

REFERENCES

Allen, R. D., Bixler, H. H., Connor, W. L., and Graham, F. B. *Spelling achievement test; metropolitan achievement test.* Yonkers on Hudson, New York: World Book, 1946.

Cannings, C. A discussion of Weinberg's rule on the zygosity of twins. *Annals of Human Genetics,* 1969, *32* (4), 403–406.

Cattell, R. B. A universal index for psychological factors. *Psychologia,* 1957, *1,* 74–85.

Cattell, R. B. *Abilities: Their structure, growth, and action.* Boston: Houghton Mifflin, 1971.

Corney, G., Robson, E., & Strong, S. Twin zygosity and placentation. *Annals of Human Genetics,* 1968, *32,* 89–96.

Dahlberg, G. *Twin births and twins from a hereditary point of view.* Stockholm: Tidens tryckeri, 1926.

Dixon, W. J. *Biomedical computer programs.* Berkeley: Univ. of California Press, 1970.

DuBois, P. H., & Gleser, G. Object-aperture test. *American Psychologist,* 1948, *3,* 363.

Eaves, L. J., & Jinks, J. L. Insignificance of evidence for differences in heritability of IQ between races and social classes. *Nature,* 1972, *240,* 84–88.

Feldman, M. W., & Lewontin, R. C. The heritability hang-up. *Science,* 1975, *190,* 4220.

French, J. W., Ekstrom, R. B., & Price, L. A. *Manual for a kit of reference tests for cognitive factors.* Princeton, New Jersey: Educational Testing Service, 1963.

Guilford, J. P. *The nature of human intelligence.* New York: McGraw-Hill, 1967.

Heim, A. W. Self-judging vocabulary test. *Journal of Genetic Psychology,* 1965, *72,* 285–294.

Holzinger, K. J. The relative effect of nature and nurture influences on twin differences. *Journal of Educational Psychology,* 1929, *20,* 241–248.

Jenkins, J. J., & Patterson, D. G. *Studies in individual differences.* New York: Appleton, 1961.

Jensen, A. R. *Genetics and education.* New York: Harper, 1972.

Jensen, A. R. *Educability and group differences.* New York: Harper, 1973.

Jinks, J. J., & Fulker, D. W. Comparison of the biometrical genetical, CPS and classical approaches to the analysis of human behavior. *Psychological Bulletin,* 1970, *73,* 311–349.

Lewontin, Richard C. Race and intelligence. In Ashley Montagu (Ed.), *Race and IQ.* New York: Oxford Univ. Press, 1975. Pp. 174–191.

Loehlin, J. C., Lindzey, G., & Spuhler, J. N. *Race differences in intelligence.* San Francisco: Freeman, 1975.

Lyon, M. F. Gene action in the X-chromosome of the mouse *(Mus musculus L.)*. *Nature*, 1961, *190*, 372.

Maccoby, E. E., & Jacklin, C. N. *The psychology of sex differences*. Stanford, California: Stanford Univ. Press, 1974.

Mercer, J. R., & Brown, W. C. Racial differences in I.Q.: Fact or artifact? In Carl Senna (Ed.), *The fallacy of I.Q.* New York: Joseph Okpaku Publishing, 1973. Pp. 56–113.

Montagu, M. F. A. *An introduction to physical anthropology*. Springfield, Illinois: Charles C. Thomas, 1945.

Mukherjee, B. N. Simple arithmetic test. Unpublished Ph.D. thesis, Univ. of North Carolina, 1963.

Myrianthopoulos, N. C. A survey of twins in the population of a prospective collaborative study. *Acta Geneticae Medicae et Gemellogia*, 1970, *19*, 644–648. (a)

Myrianthopoulos, N. C. An epidemiologic survey of twins in a large prospectively studied population. *American Journal of Human Genetics*, 1970, *22*, 611–629. (b)

Nichols, P. L. The effects of heredity and environment on intelligence test performance in 4 and 7 year old white and negro sibling pairs. Unpublished Ph.D. thesis, Univ. of Minnesota, 1970.

Nichols, R. C. The national merit twin study. In S. G. Vandenberg (Ed.), *Methods and goals in human behavior genetics*. New York: Academic Press, 1965. Pp. 231–243.

Nichols, R. C., & Bilbro, W. C., Jr. The diagnosis of twin zygosity. *Acta Genetica et Statistica Medica*, 1966, *16*, 265–275.

Osborne, R. T., & Gregor, A. J. Racial differences in inheritance ratios for tests of spatial ability. Paper presented to the Instituto Internacional de Sociologia, XXII Congreso, Madrid, October 1967.

Osborne, R. T., & Gregor, A. J. Racial differences in heritability estimates for tests of spatial ability. *Perceptual and Motor Skills*, 1968, *27*, 735–739.

Osborne, R. T., Gregor, A. J., & Miele, F. Heritability of factor V: Verbal comprehension. *Perceptual and Motor Skills*, 1968, *26*, 191–202.

Osborne, R. T., & Miele, F. Racial differences in environmental influences on numerical ability as determined by heritability estimates. *Perceptual and Motor Skills*, 1969, *28*, 535–538.

Osborne, R. T., & Suddick, D. E. Blood type gene frequency and mental ability. *Psychological Reports*, 1971, *29*, 1243–1249.

Paulson, E. An approximate normalization of the analysis of variance distribution. *Annals of Mathematical Statistics*, 1942, *13*, 233–235.

Peters, G. A., Jr. The new dvorine color perception test. *The Optometric Weekly*, November 11, 1954, 1801–1803.

Remondino, C. Calendar test. *Revue de Psychologie Applique*, 1962, *12*, 62–81.

Scarr-Salapatek, S. Race, social class, and IQ. *Science*, 1971, *174*, 4016, 1285–1295.

Schoenfeldt, L. F. The project talent twin study—A preliminary report. Univ. of Pittsburgh and American Institutes for Research. A paper presented at the Second Invitational Conference on Human Behavior Genetics, Louisville, May 1966.

Science Research Associates, Inc. *Primary mental abilities examiner's manual*. (Rev. ed.) Chicago, Illinois: Thelma Gwinn Thurstone, 1962.

Smith, I. McFarlane, & Lawes, J. S. *Newcastle spatial test*. Bedford, England: Newnes Educational Publishing, 1959.

Smith, S. M., & Penrose, L. S. Monozygotic and dizygotic twin diagnosis. *Annals of Human Genetics*, 1955, *19*, 273–289.

Sonquist, J. A., & Morgan, J. N. *The detection of interaction effects*. Monograph No. 35, Survey Research Center, Institute for Social Research, Univ. of Michigan, 1964.

Standskov, H. H., & Edelen, E. W. Monozygotic and dizygotic twin birth frequencies in the total, the "white" and the "colored" U.S. populations. *Human Genetics*, 1946, *31*, 438–446.

Thurstone, L. L. *Primary mental abilities.* Chicago: Chicago Univ. Press, 1938.

Vandenberg, S. G. *Methods and goals in human behavior genetics.* New York: Academic Press, 1965.

Vandenberg, S. G. *A twin study of spatial ability.* Research Report from the Louisville Twin Study, Child Development Unit, Department of Pediatrics, Univ. of Louisville School of Medicine, Report No. 26, April, 1967.

Vandenberg, S. G. *Progress in human behavior genetics.* Baltimore: Johns Hopkins Press, 1968.

Vandenberg, S. G. A comparison of heritability estimates of U.S. negro and white high school students. *Acta Geneticae Medicae et Gemellogia,* 1970, *19,* 280–284.

Verschuer, O. V. Anthropologische Studien in ein-und zweieiigen Zwillingen. *Zeitschrift für induktive abstammungs und vererbungslehre,* 1925, *41,* 115–119.

Verschuer, O. V. Die biologischen Grundlagen der menschlichen Mehrlingsforschung. *Zeitschrift für induktive abstammungs und vererbungslehre,* 1932, *61,* 147–205.

Weinberg, W. Beitrage zur Physiologie und Pathologie der Mehrlingsgeburten beim Menschen. *Pflüger's Archiv für die gesamte Physiologie des Menschen und der Tiere,* 1901, 88, 346–430.

7

Sex Linkage: A Biological Basis for Greater Male Variability in Intelligence

ROBERT G. LEHRKE

Brainerd State Hospital, Brainerd, Minnesota

INTRODUCTION

An examination of modern textbooks on individual differences discloses no major attempt during the past half century to demonstrate a sex difference in average intelligence. In fact, should such an attempt have been made using standard intelligence tests, its efforts probably would have been doomed to failure since the tests most frequently used have been designed to eliminate (or at least minimize) such differences. Terman (1925), in his revision of the Binet test, simply deleted test items that were "relatively less fair to one sex than the other," and this practice has apparently continued in the design of intelligence tests up to the present. For a discussion of this topic, see Masland, Sarason, and Gladwin (1958, pp. 260–264). More recent data are summarized in Chapter 3 of Maccoby and Jacklin's (1974) book, *The Psychology of Sex Differences*.

The practice is understandable since it has been a goal of such tests to tap basic, essentially genetic, individual differences. The genes relating to intellectual functioning being the same for the two sexes, it would probably be illogical to design a test of general intelligence in which there were consistent sex differences.

However, the hypothesis of greater male *variability* in biological traits (including mental functioning) is still widely accepted. Barton Childs (1965, p. 803) said, " . . . the female is less frequently represented at the extremes of a distribution of measurements of quantitative expressions than is the male." Lionel Penrose (1963, p. 186) referred this greater male variability specifically to intelligence. "The larger range of variation in males than in females for general intelligence is an outstanding phenomenon."

This certainly is not a new concept. Havelock Ellis (1904, p. 425) discussed it early in this century. "We have, therefore, to recognize that in men, as in males generally, there is an organic variational tendency to diverge from the average, in women, as in females generally, an organic tendency, notwithstanding all their facility for minor oscillations, to stability and conservatism, involving a diminished individualism and variability." He also refers this greater variability to mental processes (Ellis, 1904, p. 426). "It is undeniably true that the greater variational tendency of the male is a psychic as well as a physical fact . . . "

On the other hand, the authors of two outstanding texts on individual differences play down the hypothesis of greater male variability. Anne Anastasi (1958, p. 629) says, "There is, however, no evidence in these surveys for a greater male variability [in intelligence], nor for a greater frequency of boys at the upper IQ levels."

Leona Tyler (1965, p. 251) essentially agrees. "The hypothesis of the greater variability of the male will probably persist as long as no really decisive evidence shows up to disprove it. However, it does not rate as an important theoretical concept in current research on sex differences."

The persons taking pro and con positions on the question seem to use different interpretations of essentially similar data in arriving at their conclusions. As pointed out in an earlier paper (Lehrke, 1972b, p. 626), "It would take only a little change in emphasis and a closer examination of certain studies to turn Dr. Anastasi's discussion in her outstanding textbook to a strong argument *for* the idea of greater male variability." The same thing is true for the comparable materials in Dr. Tyler's book.

Determinants of which viewpoint a person accepts are undoubtedly highly complex, but a single, very simple one is obvious. In the small sample cited, all those accepting the hypothesis of greater male variability have been males, all those rejecting it, females.

Is it that the empirical evidence is so incomplete or so inconclusive that the sex of the interpreter swings the balance? Probably not. What is lacking is not so much a preponderance of empirical evidence as a logical (or biological) reason that males should be *expected* to be more variable—or less stable, to use Ellis's term, if that is more pleasing to the ladies. Lacking such a logical basis for the phenomenon, it is not hard for those so inclined to decide that the demonstrated differences are due to cultural rather than intrinsic factors.

However, a biological basis for such greater male variability in several major aspects of intelligence seems to have been stumbled upon in the course of a study of several families that displayed obviously sex-linked mental retardation with no known physiological correlates. Essentially what happened was that a theory of sex linkage (or X-linkage) of major intellectual characteristics arose as the virtually obligatory outcome of analysis of the data presented in these earlier studies (Lehrke, 1968, 1974). Greater male variability is, in essence, what would be expected if there are major genes for intelligence on the X-chromosomes of humans.

The families described were mainly from the United States. More recently, researchers in Australia have described many more such families (Turner, Engisch, Lindsay, & Turner, 1972; Turner & Turner, 1974; Turner, Turner, & Collins, 1970, 1971). They gave the condition the eponym, *Renpenning's Syndrome*, after the Canadian physician who participated in the reporting of two such families (Renpenning, Gerrard, Zaleski, & Tabata, 1962; Dunn, Renpenning, Gerrard, Miller, Tabata, & Federoff, 1963). Opitz (Kaveggia, Opitz, & Pallister, 1972, for example) prefers the eponym *Martin–Bell Syndrome*, which probably should have priority since it refers to an English kinship described in 1943. Obviously, the phenomenon of nonspecific sex-linked mental retardation is geographically widespread.

An understanding of why X-linkage should result in greater male variability requires only very elementary genetics. Sexual differentiation in higher animals depends on the sex chromosome complement—two X-chromosomes for females, an X and a Y for males. The X-chromosome in man is of medium size, containing about 5 or 6% of the genetic material in a haploid set of human chromosomes. On the basis of recognized traits, it seems to carry about that same proportion of genetic information, including known genes affecting every major body system. The Y-chromosome, on the other hand, is one of the smallest chromosomes and, as far as is known, carries only the genetic instructions for maleness.

The autosomes (non-sex chromosomes) operate in pairs. That is, homologous portions of the two sets generally summate in some fashion to determine the genotype. However, since males have only one X-chromosome, and since the Y-chromosome is homologous for little, if any, of the genetic material on the X-chromosome, some method of dosage compensation is necessary if the non-sex genes on the X-chromosome are to be expressed at comparable levels in the two sexes.

The currently accepted explanation of how this occurs is that proposed by Mary Lyon (1962). According to the Lyon Hypothesis, shortly after fertilization, one of the two X-chromosomes in each cell of female conceptions is inactivated, and that same X-chromosome (maternal or paternal) remains relatively inert in descendants of that particular cell. Obviously, then, a chromosomally normal female is a mosaic of cells in which either the maternal or paternal X-chromosome is active. Such mosaicism has been repeatedly demonstrated for a variety of X-linked traits. See Davidson (1964) and Thompson (1965) for examples.

One result of this mosaicism is that extreme manifestation of any gene on the X-chromosome is less likely to appear in females since the effect of a particularly deviant allele would probably be moderated by that of the homologous gene remaining active in another line of cells. A female would show an abnormal sex-linked recessive trait at full strength only in the case that she received the same mutant allele from both parents.

On the other hand, males, lacking a second X-chromosome, would have a particular gene active in all cells and would show an X-linked trait at full force, whether deleterious, beneficial, or neutral. In a case in which there are just two alleles at a locus, one normal with a frequency in the population of .6 and one a recessive variant with a frequency of .4, 40% of the males would show the variant trait. However, a female would receive the variant from both parents only .4 × .4, or 16% of the time.

For rarer traits, such as deleterious ones are likely to be, the discrepancy between males and females would be even more marked. For example, if a recessive gene on the X-chromosome had a population frequency of .05, the corresponding trait would show up in 5% of males but only in about a quarter of 1% of females. This is what is noted (among Caucasoids) for the sex-linked trait of red–green color blindness.

Such a simple model is appropriate for such nonlethal conditions as color blindness. The two-allele model does show that there are more males who are deviant from the normal pattern when a recessive X-linked gene is involved. The same thing can be shown to be true for a continuously variable trait like intelligence as long as X-linked genes are involved, but it will be necessary to go to a more complex (though still greatly simplified) model to demonstrate it.

Let us assume a gene on the X-chromosome that has six alleles, that is, six variations, each resulting in a different extent of neuronal development relating to some aspect of intelligence. Such multiplicity of alleles is fairly common among genetic traits. We will assume that allele No. 1 leads to defective functioning for this trait, and No. 6 to the highest level, with other alleles intermediate in proportion to their numerical position.

Further, let us assume that these alleles occur in the population with a frequency distribution similar to the expansion of the binomial, in proportions 1:5:10:10:5:1, with the intermediate values being the most frequent. This is not unrealistic, although it is undoubtedly oversimplified. Severely retarded individuals would not often reproduce, thus keeping the frequency of allele No. 1 low. It is also unfortunately true in our society that individuals with high intelligence tend to marry later in life and to have small families, both of which would tend to reduce the frequency of allele No. 6.

For this gene on the X-chromosome, the number of males at each level of functioning would be in the same proportion as the gene frequency. The distribution for 1024 males is shown in Figure 7.1.

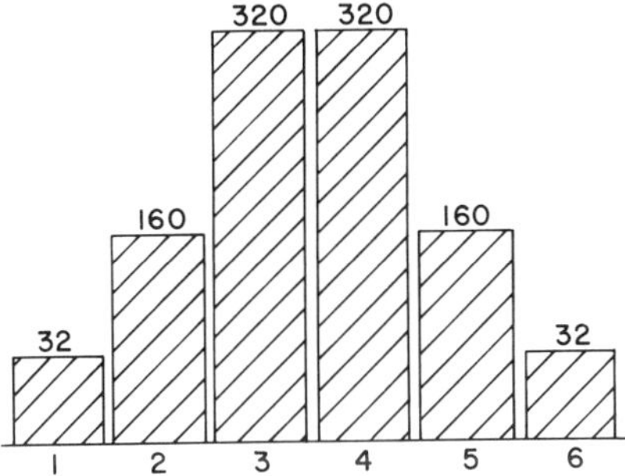

Figure 7.1 Hypothetical distribution of phenotypes among 1024 males for an X-linked recessive gene with six alleles.

For females, the level of functioning would depend on the average of two alleles, according to the Lyon Hypothesis. Thus, for a female to function at the lowest level (i.e., 1), it would be necessary that she had inherited allele No. 1 from both parents. Since that allele has a frequency of $\frac{1}{32}$, females would have two such alleles only $(\frac{1}{32})^2$ of the time, or one in 1024, if there is no selective mating.

At the same time, average values would be more common for females since a low value on one X-chromosome would probably be offset by a higher value on the other. The same effect, basically regression to the mean, would be expected for unusually high values. The expected distribution for 1024 females, on the same baseline as that for males, is shown in Figure 7.2.

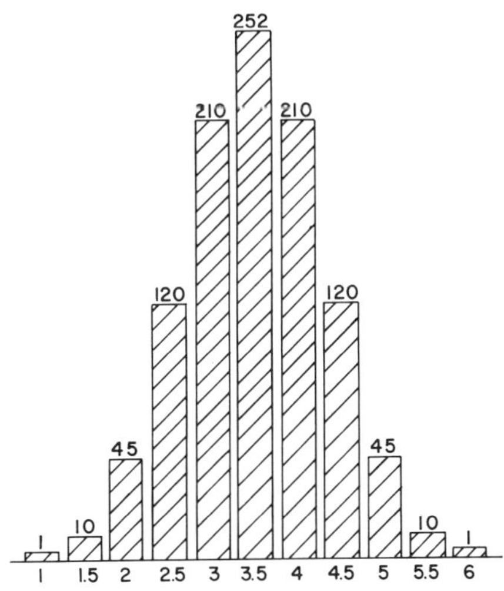

Figure 7.2 Hypothetical distribution of phenotypes among 1024 females for an X-linked recessive gene with six alleles.

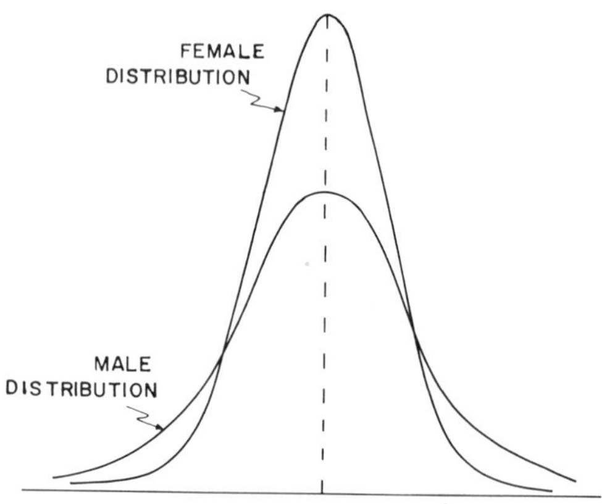

Figure 7.3 Distributions from Figures 7.1 and 7.2, superimposed and expressed as curves.

When the two distributions are compared, it can be seen that there are more males at the extremes, more females in the middle. This result, shown as superimposed curves, is apparent in Figure 7.3, with a greater variance (though not necessarily a greater range) for males than for females.

It should be noted that Figure 7.3 is probably a reasonably valid representation at the high end of the scale. However, the single events (chromosomal aberrations, genetic and other diseases, trauma, etc.) that cause the more severe levels of retardation are seldom sex-linked. Thus, the greatest sex differences would be apparent in the moderate, mild, and borderline levels (Heber, 1961) of retardation; and, in fact, Turner, Collins, and Turner (1971) have shown a very distinct group of males who are moderately retarded without apparent physiological basis.

Consequently, a comparative distribution like that of Figure 7.4 is probably a more accurate representation. (Differences between the

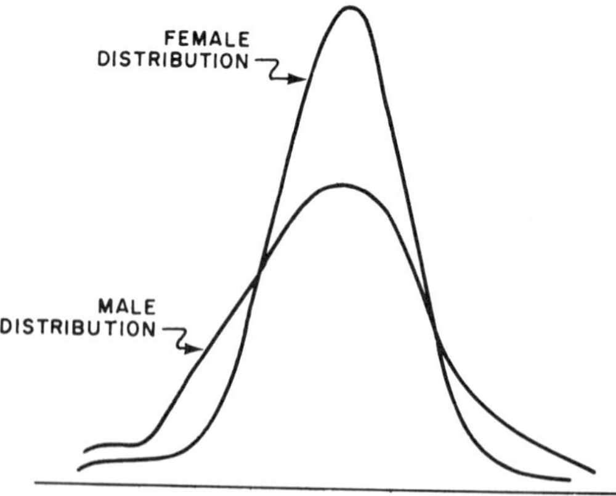

Figure 7.4 Probable distributions of IQ in the population, by sex, with differences exaggerated for clarity.

curves have been exaggerated for graphic purposes.) When the male and female distributions are lumped together, the result is very similar to the combined distribution of intelligence test scores as shown, for example, in Jensen (Fig. 2; 1969a and 1969b, p. 25), suggesting that the hypothetical model fits closely with the actual distribution of intelligence in the community.

COMPARATIVE RATES OF RETARDATION

Institutional data (Table 7.1) over recent years have consistently shown the excess of male retardates that would be expected under the hypothesis of greater male variability in intelligence. There have been some objections to the use of such figures as evidence of an actual greater prevalence of male retardation since, as Anastasi (1972, p. 620), Nance and Engel (1972, p. 623), and many others have pointed out, there could be a sex difference in the proportions of male and female retardates admitted to institutions.

However, as Penrose once said in relation to the Colchester study (Penrose, 1938), such an admission or selection bias can operate in either direction. It is the author's own opinion (Lehrke, 1972a, p. 612; 1972b, p. 627), based on years of experience in state institutions, that the bias has

TABLE 7.1
Residents and Charges at End of Year of U.S. Public Institutions for the Retarded, by Sex

Year	Institutions reporting	Number of males	Number of females	Male–female	Percentage excess males[a]
1950	95 of 96	64,116	58,041	6,075	10
1951	94 of 95	65,458	59,050	6,408	11
1952	95 of 96	61,886	56,025	5,861	10
1953	97 of 98	64,534	58,079	6,455	11
1954	97 of 97	68,257	60,930	7,327	12
1955	98 of 99	76,308	67,073	9,235	14
1956	99 of 100	80,012	69,113	10,899	16
1957	96 of 99	82,523	70,829	11,694	17
1958	98 of 102	84,220	73,587	10,633	14
1959	102 of 106	87,337	75,665	11,672	15
1960	105 of 108	96,672	82,852	13,820	17
1961	110 of 113	98,255	84,406	13,849	16
1962	118 of 124	101,496	85,483	16,013	19
1963	126 of 128	107,795	90,617	17,178	19
1964	126 of 135	100,995	84,462	16,533	20
1965	135 of 143	107,709	88,937	18,772	21

Source: From National Institute of Mental Health publications listed in references.
[a] Calculated (M − F)/F.

TABLE 7.2

Residents and Charges of U.S. Public Institutions for the Retarded, by Sex and Degree of Retardation, 1950–1962

Year	Degree of retardation	Number of males	Number of females	Male–female[a]	Percentage of male excess[b]
1950—63 of 96 institutions	Idiot	5,407	5,104	303	6
	Imbecile	11,097	10,385	712	7
	Moron	6,921	7,030	−109	−2
	Unclassified	1,662	1,411	251	18
1951—57 of 95 institutions	Idiot	6,550	6,052	498	8
	Imbecile	12,610	11,849	761	6
	Moron	7,240	7,880	−640	−8
	Unclassified	1,081	807	274	35
1952—64 of 96 institutions	Idiot	8,055	7,228	827	11
	Imbecile	14,534	13,360	1,174	9
	Moron	7,993	8,464	−471	−6
	Unclassified	1,781	1,291	490	38
1953—69 of 98 institutions	Idiot	9,201	8,272	929	11
	Imbecile	15,753	14,457	1,296	9
	Moron	8,563	8,764	−201	−2
	Unclassified	1,735	1,331	404	30
1954—73 of 97 institutions	Idiot	10,780	9,587	1,193	12
	Imbecile	17,705	16,298	1,407	9
	Moron	9,343	9,259	84	1
	Unclassified	3,052	2,412	640	27
1955—77 of 99 institutions	Idiot	12,378	11,004	1,374	12
	Imbecile	20,590	18,672	1,918	10
	Moron	11,236	10,449	787	8
	Unclassified	3,601	2,729	872	32
1956—86 of 100 institutions	Idiot	14,381	12,228	2,153	18
	Imbecile	26,757	23,377	3,380	14
	Moron	14,920	13,219	1,701	13
	Unclassified	4,775	3,977	798	20
1957—79 of 99 institutions	Idiot	14,942	12,831	2,111	16
	Imbecile	27,651	23,963	3,688	15
	Moron	15,172	13,334	1,838	14
	Unclassified	2,592	2,213	379	17
1958—87 of 102 institutions	Idiot	17,123	14,813	2,310	16
	Imbecile	29,792	25,521	4,271	17
	Moron	16,139	13,876	3,263	24
	Unclassified	2,846	2,398	448	19
1959—86 of 106 institutions	Idiot	17,275	14,687	2,588	18
	Imbecile	29,721	25,014	4,707	19
	Moron	15,967	13,384	2,583	19
	Unclassified	2,493	2,115	378	18
1960—78 of 108 institutions	Idiot	17,123	14,769	2,354	16
	Imbecile	31,186	25,987	5,199	20
	Moron	15,793	13,430	2,363	18
	Unclassified	2,257	1,991	266	13

TABLE 7.2 *(Continued)*

Year	Degree of retardation	Number of males	Number of females	Male–female[a]	Percentage of male excess[b]
1961—81 of 113 institutions	Idiot	17,555	15,136	2,419	16
	Imbecile	31,478	26,297	5,181	19
	Moron	15,650	12,908	2,742	21
	Unclassified	2,458	2,176	282	13
1962—91 of 124 institutions	Idiot	18,426	15,948	2,478	16
	Imbecile	32,682	26,831	5,851	22
	Moron	16,131	13,112	3,019	23
	Unclassified	2,902	2,514	388	15

Source: Data from National Institute of Mental Health publications listed in references.
[a] Minus (−) sign means excess of females.
[b] Calculated (M − F)/F Minus (−) sign indicates excess of females.

generally been in the opposite direction. That is, retarded females have been institutionalized more frequently in proportion to their numbers in the community as a means of controlling their fertility. Heber (1970, p. 24), for instance, shows a greater number of female admissions to state institutions in the early part of the century, and this tendency seems to have continued, though only for the higher levels of retardation, until the early 1950s. In Table 7.2, for instance, the only noted excesses of females in institutions were at the moron level, during the years 1950–1953.

Community studies, however, have consistently shown a greater incidence of male retardation, going back to the 1890 U.S. Census (Johnson, 1897, p. 26). A few such studies are summarized in Table 7.3. Others are given in Abramowicz and Richardson (1975).

As shown in the table, these community surveys of the prevalence of retardation in many parts of the world have tended to report a greater excess of males than even the most recent institutional figures. In addition to the studies in Table 7.3, there are scores of others that have reached the author's attention. Among these, there are only two groups (actually, subgroups) that have shown an excess of females. One is the age 0–10 group in Akesson's Western Sweden study, reported by Abramowicz and Richardson (1975), in which there was reported a 9% excess of girls. This was more than offset by the male excess at the higher age levels. For instance, in the age 10–20 group, there were 53% more males. The other such situation is in the 0–49 IQ group from Imre's (1968) report of the "Rose County" study. Here there was a 55% greater frequency of females. However, when the entire retarded population was included, that is, IQs below 69, there was about a 7% excess of males. Furthermore, among the adults, in which ascertainment was excellent,

TABLE 7.3

Male–Female Ratios from Some Community Studies of the Prevalence of Mental Retardation

Reference	Population included	Number of males	Number of females	Percentage of male excess	Comments
Berg (1966)	Retarded persons in Denmark.	11,493	8,966	28	
British Columbia Department of Health Service (1971)	Retarded persons in British Columbia, 1970.	4,699	3,471	35	Virtually total ascertainment
	All levels less borderline.	3,459	2,738	26	
British Columbia Department of Health Service (1972)	Retarded persons in British Columbia, 1971.	4,841	3,585	35	Virtually total ascertainment
	All levels less borderline.	3,551	2,820	26	
Hasan (1972, p. 62)	Retarded persons seen at a diagnostic center in Karachi, Pakistan.	634	376	69	
Imre (1967, p. 154)	Retarded persons age 1–59, IQ to 69, in "Rose" County, Md.	7.68%	7.21%	7	See text for further data and comments.
Jensen (1971, p. 154)	Data from Lemkau and Imre (1966) study of "Rose" County, Md. Persons ages 20–59 with subnormal functioning or education and WAIS Verbal IQ below 70.				A large study, including 7475 adults, and very high ascertainment.
	White population	1.69%	1.02%	68	
	Negro population	21.57%	16.49%	31	
Johnson (1897)	U.S. Census for 1890.	52,940	42,631	24	
	Colored population.	5,788	4,786	21	
New York State Department of Mental Hygiene (1955)	Retarded persons in Onondaga Co., N.Y.	2,075	1,118	86	

Reed and Reed (1965)	Minnesota families over several generations.	867	583	49	
Richardson and Higgins (1964)	Retarded children in Alamance Co., N.C.	10.0%	5.9%	69	
Scottish Council for Research in Education (1949)	Scottish children, age 11 at time of test.				
	1932 study	608	465	31	
	1947 study	3,191	1,810	76	
Socialstyrelsen (1972)	Residents in Swedish facilities for the retarded.	9,098	6,799	34	
Sterner (1967)	Retarded children in Vasternorrland Co., Sweden.	517	298	73	
Stomma and Wald (1972)	Residents of homes for low grade retarded children in Poland in 1970.	4,888	3,825	28	
Turner and Turner (1974)	Retarded children, IQ's 30–55 in New South Wales, Australia.	1,335 (.31%)	1,010 (.25%)	32	Virtually complete ascertainment
Verbraak (1975, p. 664)	Retarded persons of all ages in the Netherlands.				Highly sophisticated sampling technique covering 4 provinces, population 2.2 million, and extrapolated to total population
	Severely handicapped (to IQ 50).	25,618	18,623	38	
	Per 1000 of same sex.	4.07	2.93	39	
	Mildly handicapped (to IQ 80).	56,505	33,972	66	
	Per 1000 of same sex.	10.6	6.4	66	
Wunsch (1951)	Retarded persons reported to state registry in R.I.	3,706	2,970	25	

TABLE 7.4

Children and Adults with Mental Retardation, by Sex and Degree of Retardation, in British Columbia at Year End 1971[a]

Degree of retardation[b]	Under 21 (children)				21 and over (adults)				Total			
	Male	Female	Total	Male excess (%)	Male	Female	Total	Male excess (%)	Male	Female	Total	Male excess (%)
Borderline	730	414	1,114	76	560	351	911	60	1,290	765	2,055	69
Mild	447	302	749	48	380	346	726	10	827	648	1,475	28
Moderate	563	421	984	34	233	220	453	6	796	641	1,437	24
Severe	275	191	466	44	247	184	431	34	522	375	897	39
Profound	222	165	387	35	143	130	273	10	365	295	660	24
Unspecified	678	549	1,227	23	363	312	675	16	1,041	861	1,902	21
Total	2,915	2,042	4,957	43	1,926	1,543	3,469	25	4,841	3,585	8,426	35

Source: From British Columbia Department of Health Services and Hospital Insurance (1972).

[a] Total population: Males 1,100,400—prevalence of retardation, .0044; females 1,084,200—prevalence of retardation, .0033; total 2,184,600—prevalence of retardation, .0039.

there were 68% more males in the Caucasoid retardate group and 31% more in the Negroid group. (See the "Jensen" heading in Table 7.3.)

In general, the few substantial surveys showing an excess of males of less than 10% have involved severely subnormal persons, among which group (as previously was mentioned) the male excess would be expected to be smaller; older populations in which the greater mortality rate for males would start to show its effect; or groups in which community educational deficits have resulted in extremely high levels of psychometrically retarded persons of both sexes, thus decreasing proportional differences.

The surveys in Table 7.3 spread out in time over 80 years, in space from Australia to Scandinavia, and in level of retardation from profound to borderline. It is hardly likely that they all show the same type of sex bias in determining which individuals are counted as retarded; and even if some such bias were present, it is hardly conceivable that it would reach such extreme levels as to account for the differences seen in most studies.

Perhaps the best evidence that ascertainment bias does not account for these differences is found in the annual surveys of the incidence of handicapping conditions in the Province of British Columbia. (See Table 7.4.) Regarding these surveys, Dr. James R. Miller (personal communication, 1972) said, "I believe these data could be used to refute the comments of some of your critics because as I indicated this is close to 100% ascertainment in this population and certainly there would be no bias that I am aware of which would result in this peculiar sex ratio." Even if one

TABLE 7.5

Prevalence of Mentally Retarded in the IQ Range 30–55, Born between January 1955 and December 1964 in New South Wales, Australia

Retarded boys ascertained	Boys that age in New South Wales	Prevalence
1,335[a]	431,520	$\frac{1,335}{431,520} = 3.1/1,000$
Retarded girls ascertained	**Girls that age in New South Wales**	**Prevalence**
1,010[b]	412,910	$\frac{1,010}{412,910} = 2.5/1,000$
Male excess = 32.2%		

Source: From Turner and Turner, Table 1, 1974.
[a] 196 Down's syndrome.
[b] 219 Down's syndrome.

eliminates the borderline classification, in which the chances of selection bias would seem to be the greatest, there still remain 26% more retarded males than females. Institutional data for the Province show about the same excess of males, 34%, as in the community, 35%.

Another survey with virtually complete ascertainment is reported by Turner and Turner (1974). Among New South Wales children born during the 10 years 1955–1964, there were 32% more boys than girls with IQs in the range of 30–55. These data are summarized in Table 7.5.

It is rather surprising that none of the critics of the author's earlier papers have mentioned one possible source of bias. That is, boys, during the earlier school years, do tend to mature more slowly than girls. Thus, they may be more frequently placed in special classes and consequently

TABLE 7.6

Enrollment in Public Secondary Day Schools of Educable Mentally Retarded Children, by Sex and Region: 1969–1970

	Male	Female	Total[a]	Percentage male excess
Region I: (Mideast) Del., Md., N.J., N.Y., Pa., D.C.	28,414	16,383	44,797	73
Region II: (Southeast) Ala., Ark., Fla., Ga., Ky., La., Miss., N.C., S.C., Tenn., Va., W.Va.	31,082	17,613	48,695	76
Region III: (Southwest) Ariz., N.Mex., Okla., Tex.	8,375	5,083	13,458	65
Region IV: (Far West) Alaska, Calif., Hawaii, Nev., Oreg., Wash.	9,698	10,158	19,856	−5
Region IV: less Cal. (See text.)	3,333	2,379	5,712	40
Region V: (Rocky Mountains) Colo., Idaho, Mont., Utah, Wyo.	2,015	1,276	3,291	57
Region VI: (Plains) Iowa, Kans., Minn., Mo., Nebr., N.Dak., S.Dak.	5,393	3,298	8,691	61
Region VII: (Great Lakes) Ill., Ind., Mich., Ohio, Wis.	14,751	7,393	22,144	99
Region VIII: (New England) Conn., Maine, Mass., N.H., R.I., Vt.	3,708	2,488	6,196	49
United States	103,436	63,692	167,128	62
United States, less Calif. (See text.)	97,071	55,913	152,984	74

Source: From Becker (1972).

[a] Data compiled from questionnaires sent to chief school officers in each state and from census populations in state directories of special education.

considered to be retarded. However, by the high school level, both sexes have ordinarily attained their growth spurt and should be developmentally at par.

Therefore Table 7.6, showing enrollment in secondary level special education classes, is of interest if one makes the obvious assumption that special class placement correlates with low intelligence. The original data, gathered by Becker (1972) in connection with the planning for a vocational interest test for special class students, shows one obvious anomaly. Area Division IV, unlike the others, shows an excess of females.

The discrepancy arose because California, the most populous state in the division, had no data on special class enrollment by sex and therefore reported the same sex division as in all classes, that is, 55% females, 45% males. California, like the rest of the country, probably has more boys than girls in special classes, although the actual proportions are uncertain. Eliminating the California figures from the calculations, the table shows that the rest of Region IV has a 40% excess of males in secondary special classes; and the country, exclusive of California, has a 74% excess of males. These figures, of course, are only approximate.

As Nance and Engel (1972) infer, those figures showing the greater numbers of retarded males might mean merely that there is a greater amount of pathology, including sex-linked disorders, affecting males' intellectual functioning, unless it can also be shown that there are more males at the high end of the distribution of IQs. For many reasons, data indicative of such an excess are less plentiful.

STUDIES OF BRIGHT CHILDREN

However, there is some evidence from psychometric studies that a real difference does exist. Perhaps the best, to date, arose from the California studies of exceptionally bright children. Regarding this study, Terman (1925, p. 54) pointed out, " . . . the facts we have presented are in harmony with the hypothesis that exceptionally superior intelligence occurs with greater frequency among boys than among girls." The actual ratio from Terman's high-IQ group (computed from Terman 1925, Table 2, p. 39) was 813 boys to 592 girls, or 37% more males; $\chi^2 = 34.5$, 1 *df;* $p < .001$.

Anastasi (1958, p. 628) suggests that the excess of males in this study was due to selection bias on the part of the teachers.

> It should be noted that the children in the California study were located in large part through teachers' recommendations . . . It is thus likely that the excess of boys in the California group resulted from the effect of sex stereotypes on teachers' judgments. Perhaps a girl with a high IQ was more often regarded by her teachers simply as a "good pupil," while a boy with the same IQ was judged to be "brilliant."

However, Dr. Anastasi missed out on a comparison that suggests that the bias was actually in the opposite direction. The original group, selected, as she says, largely through teachers' recommendations, consisted of 857 boys and 671 girls. However, after actual testing, there remained 813 boys with IQs over 140, or 95% of the original group, but only 592 girls, or 88% of the original group. Thus, when it came to objective scores, the teachers more frequently overestimated the intelligence of the girls.

Obviously, more studies should be done along these lines. One resource would be data from graduate school selection tests such as the Miller Analogies Test and the Graduate Record Examination. According to the hypothesis of greater male variability, the male–female ratio in the top quartile for such tests should be greater than for the bottom quartile. Up until 1973, at least, the publishers of the Miller Analogies Test had no record of studies of this type (Hall, personal communication, 1973), nor does the manual (Psychological Corp., 1970) have separate norms for males and females. One small study reported in the manual (p. 16) did give separate scores for males and females, which did favor the males, but it would seem impossible to rule out extraneous selection factors.

Actual data on sex differences in variance are extremely limited. The manual for the Stanford–Binet, Form LM, gives nothing at all, and the texts on the Wechsler tests are not much better. There is a suggestion of greater male variability on the Wechsler Adult Intelligence Scale since on 10 of 11 subtests the variance is greater for males (Wechsler, 1958). Clark (1958) compared the two sexes' performance on a large battery of academic tests. He was able to demonstrate that average scores favored neither sex, but on 44 of the 56 tests the variance for boys was greater than that for girls ($\chi^2 = 18.3$, 1 *df*; $p < .001$).

In a summary of eight rather large surveys, which used a variety of cognitive ability tests, Jensen (1971, p. 145) reported that Caucasoid males had, on an average, a 13% greater variance, and Negroid males a 23% greater variance, than females of the same race. The 13% difference in the white group's variance turns out to be exactly the same as that reported (as a minimum) by Roberts, Norman, and Griffiths (1945) for Scottish schoolchildren.

In spite of the fact that the group tests used were truncated in range, particularly at the low end, both of the Scottish studies (see Table 7.3) showed significantly higher variances for boys than for girls (data from Anastasi, 1958, p. 459; Gruenberg, 1964, pp. 285–288). For the 1932 survey, the standard deviations were 15.92 and 15.02; and for 1947, 16.68 and 15.44, for boys and girls, respectively. Anastasi (1972, p. 620) points out that these differences are significant because of the large numbers of subjects involved, presumably implying that while they are *statistically* significant, they are not *practically* significant.

The practical importance of a small difference in *average* IQ might, indeed, be slight. However, the effect of small differences in standard deviation increases with distance from the mean.

Suppose that the standard deviation of an intelligence test, nominally 16 points (as for the Binet), should actually be 16.5 points for males and 15.5 points for females. This is 13% greater variance, which is actually a little less difference than was noted in the 1947 Scottish study. On this basis, there would be expected to be 37% more males than females with IQs below 68, and the same would be true for IQs above 132. The actual ratio, from a table of normal probabilities, would be 0.026:0.019. In effect, a very slight difference in variance can result in marked differences at the tails of the normal curve. Such differences would be predicted under the hypothesis of sex linkage of intellectual traits.

Of course, evidence of greater male variability is not, per se, proof of X-linkage of major intellectual traits. Other hypotheses could undoubtedly be devised to explain the phenomenon or even to explain it away. It is necessary, therefore, to find other lines of evidence.

FAMILY SIMILARITIES AND X-LINKAGE

One such line of evidence would be whether family similarities in IQ are of the degree one might expect of an X-linked trait. Bayley (1966, p. 102), in evaluating a somewhat different hypothesis, has provided the necessary data for such a test.

From a collection of intelligence test scores for family groups compiled by Outhit (1933), Bayley selected families in which there were IQs for both parents, plus a son and a daughter. What one would expect, if there are major genes relating to intelligence on the X-chromosome, is that the correlations of test scores for mother–daughter, father–daughter, and mother–son would be somewhat similar. In each case, the parent and child have one X-chromosome in common. The correlations between fathers and sons should be lower since they have no X-chromosome in common; and the brother–sister correlation should be intermediate since they have an X-chromosome in common half the time.

To quote Bayley (1966):

> The resulting correlations . . . are, again, higher for the daughters than the sons. The mother–daughter *r* is .68, the father–daughter *r* is .66; the mother–son *r* is .61, and the father–son *r* is .44. The fact that the parents in this sample are the same for both sons and daughters makes this a potentially crucial test. Tests of significance (the brother–sister *r* is .55) indicate that these differences in father–child correlations approach significance at the .05 level [p. 102].

In other words, the order of size of correlations is exactly what might be expected of an X-linked trait. If one takes into account that the rank order of the first three correlations (for mother–daughter, father–daughter, and mother–son) is not critical, there are 6 out of 120 possible permutations that fit the hypothesis. This makes the Bayley data significant at the .05 level as a test of the hypothesis of X-linkage of intelligence.

Additional information from the original source increases the importance of the differences. According to Outhit (1933, p. 43) the correlation between mothers and fathers in this study was .741 ± .042. In other words, the entire correlation between father and son, who have no X-chromosome in common, could be accounted for by the product of the correlation between the parents and that between mother and son (.74 × .61 = .45). Obviously this is an overstatement since there must be non-sex-linked genes and environmental factors involved as well. However, these additional data do increase both the statistical and practical significance of Bayley's study in regard to the present hypothesis.

As would be predicted from the hypothesis, there is evidence that retardation of women is more frequently associated with retardation of their children than is the case for men. Some such evidence, based on the Reed and Reed (1965) study, is presented in Lehrke (1972a, pp. 615–616) and Lehrke (1974, pp. 23–26). Ahern and Johnson (1973) also did an analysis of the Reed and Reed data with results that are compatible with the hypothesis of X-linkage of intelligence, but because the hypothesis they were testing is slightly different, the results are not given in a way that can be applied to the theory of sex linkage. In general, a child of a retarded mother is more than twice as likely to be retarded as is the child of a retarded father. This is as might be expected since the mother passes on an X-chromosome to all of her children, the father only to his daughters. However, in this regard, it is important to take into account Bessman's (1972) theory of prenatal maternal influence associated with metabolic deficiencies.

In a study with probands from an institutional retarded population, Priest, Thuline, LaVeck, and Jarvis (1961) found that in sibships in which more than one person was retarded, those with only boys affected outnumbered those with only girls affected by 29–15. To quote the authors, "One is tempted to speculate concerning the influence of sex linkage to account for the predominance of families with only males affected (p. 44)."

Other studies by Wortis, Pollack, and Wortis (1966); by Wright, Tarjan, and Eyer (1959); and by Turner *et al.* (1972) also show, in addition to greater numbers of male retardates, a tendency for these male retardates to cluster in certain families. In effect, not only are males more likely to be retarded in general, but there is a familial tendency to male retardation that would be hard to explain on any other basis than X-linkage.

For example, out of the 58 families with more than one retarded child in the Wright, Tarjan, and Eyer (1959) study, there were 22 families with a total of 52 retarded males, compared with 5 families with 10 retarded females. Families with retardates of both sexes numbered 31, with 66 males and 51 females. The retarded totaled 118 males to 61 females, while the normal sibs numbered 77 males and 81 females.

If there were no tendency to familial incidence of male retardation, we would expect that the combined male–female ratio in families in which all retardates were of one sex would be the same as that in families with retardates of both sexes. By each of several tests of the distribution, this was not the case. Perhaps the critical test is whether the numbers of retarded persons in families with male only, male and female, and female only retardates conform to the expected $p^2 + 2pq + q^2$, where p is the proportion of males (118 of 179) and q the proportion of females (61 of 179). For this analysis, $\chi^2 = 31.7$, $2\,df$; $p < .001$. In other words, the study shows a strong tendency for male retardation to run in families.

Under the Lyon Hypothesis, it would be expected that those women who have one of their X-chromosomes bearing genes for low intelligence would be duller than their sisters who do not. While they might only occasionally be retarded, they would show some effect of the defective gene, that is active in half of their cells.

For the present, there seems to be no data that would make possible a statistical test of this, but Turner *et al.* (1972, p. 326) report that in some families the obligate carriers do seem duller than their presumptive non-carrier sisters. This partial expression in heterozygous females could very well account for the presence of an occasional female retardate in families that otherwise fit the pattern of X-linked recessive transmission, for instance, pedigree KMD-2 in Reed and Reed's (1965) collection.

Elucidation of an hypothesis as to the nature of the X-linked gene or genes involved will have to be left for a work now in preparation.[1] However, X-linkage of certain other intellectual traits is now well established. Stafford (1961) demonstrated sex linkage of a gene related to spatial perception, and further evaluation by Bock and Kolakowski (1973) seems to show that the X-linked gene may be one that essentially enhances this ability in those males who carry the appropriate gene or those females who are homozygous for it. Jensen's paper in this volume discusses the matter in greater detail.

There is some evidence that the gene (or genes) related to general intelligence is, like that for spatial perception, one that enhances some substrate of basic skills. It could do this by controlling not the structure or metabolic processes of the brain but the rate and duration of neuronal growth, particularly in the temporal area. Mutation of this X-linked gene

[1] In a book now being prepared by this author, it is hypothesized that the major basis of evolution of *homo erectus* to *homo sapiens* was a neotenous mutation that extended the period of neuronal development in certain critical areas of the brain until adult years. Probably the period of such growth for modern man's most recent ancestors was about 5 years, compared with 3 years for the higher apes.

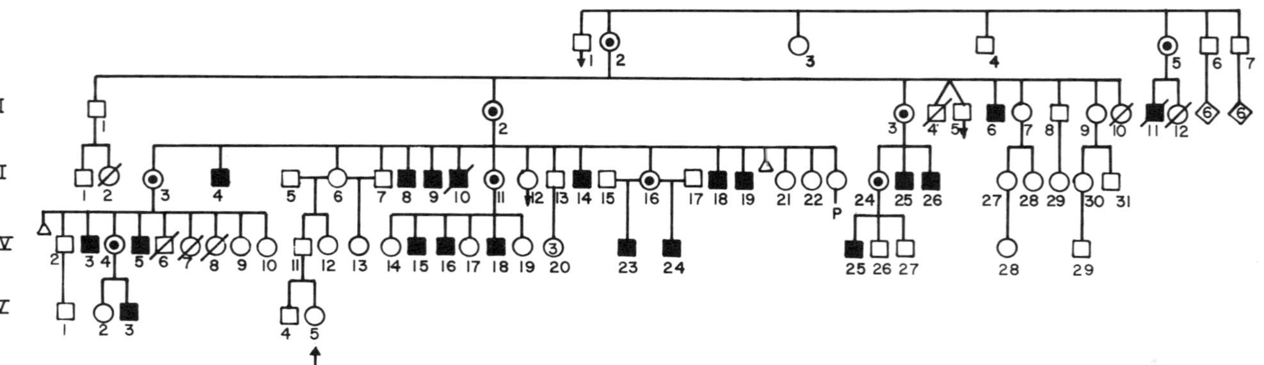

Figure 7.5 Pedigree of a family with nonspecific X-linked mental retardation. (From Lehrke, R. G.: *X-Linked Mental Retardation and Verbal Disability*. In Bergsma, D. [Ed.]: Birth Defects *Original Article Series*, Vol. X, No. 1. Florida: Symposia Specialists for the National Foundation–March of Dimes, White Plains, N.Y., 1974.)

(possibly a throwback to a primitive state) could limit verbal ability without necessarily affecting other areas of functioning. For instance, in the family with X-linked mental retardation shown in Figure 7.5, 11 of the 20 retardates were able to take Wechsler intelligence tests; and in all cases the verbal IQ was lower than the performance IQ. Regarding the other 9 retarded males, there was frequent mention of verbal, language, or speech deficits.

There is evidence from studies such as those of Blewett (1954) and Vandenberg (1962) that it is specific areas of ability rather than global intelligence that are inherited. To quote Blewett, "The fact that h^2 (heritability) values for all the total scores are considerably lower than for Verbal, Reasoning, and Fluency test scores [of the Primary Mental Abilities (PMA) Battery] may be interpreted as support for the view that these abilities are determined to a greater extent by heredity than is the general factor to which they give rise (p. 930)."

THE ABSENCE OF PHYSICAL CORRELATES

McKusick (1971) lists about two dozen sex-linked disorders in which mental defect in some or all cases is secondary to structural or metabolic disorder involving the brain. No such defect has shown up in the reports of persons identified only by their presence in pedigrees of nonspecific sex-linked mental retardation. Dr. Gillian Turner, who has studied scores of such cases, is very emphatic about such lack of an organic basis. In a letter to me, she said, "In your lecture you say that there is no way of picking individuals with the X-linked form of retardation clinically. I think you can. I suspect that all those [males with an IQ below 50] who are physically normal are X-linked."

The absence of physical correlates is further borne out by institutional data, which show that the excess of males is greatest in those categories for which there is no demonstrable basis for retardation. For instance, for the year 1964 there were in state institutions for the retarded 2176 males to 2127 females in American Association on Mental Deficiency (AAMD) Category IV (Heber, 1961), *Mental retardation associated with diseases and conditions due to disorder of metabolism, growth, or nutrition* (National Institute of Mental Health, 1966). In other words, in this category, in which there is a demonstrable physiological basis for retardation, males did not appreciably outnumber females.

However, in Category VIII, *Mental retardation due to uncertain (or presumed psychological cause* . . . , there were 22,357 males to 17,896 females, an excess of 25%. Category VIII included 24% of reported populations of institutions for that year, but it accounted for 36% of the excess of males.

What we apparently are seeing, then, is essentially that there are genes, or groups of genes, related to the development of certain very important aspects of intelligence, which are on the X-chromosome. These genes seem to have a strong effect on IQ at all levels from severe retardation to genius.

Because of the absence of physical signs, it is not possible to identify many of the persons whose retardation is due to these defective X-linked genes either in the community or in institutions. Equally important, it is obvious that in most cases such sex-linked genes are not, by themselves, solely responsible for a person's retardation. Rather, these genes, in their less severe forms, interact with other marginal or defective genes, plus unfavorable environment, to produce cultural–familial mental retardation. The same combination could affect many females, either homozygous or heterozygous for the X-linked genes, but the effect would be more likely to be clearly apparent in males. However, the presence of these females would increase the relative importance of X-linked factors in the entire population.

In effect, most of the male excess of retardation would be found in the category of cultural–familial or "garden variety" retardation, in which group most persons are borderline or mildly retarded. There would, as Turner and Turner (1974) have pointed out, also be a substantial number among those moderately and severely retarded. Many of these latter would be from families in which there is a pedigree pattern of sex-linked mental defect.

Because of the X-linked component in cultural–familial retardation, it is probable that the greater the prevalence of mental subnormality in a population, the greater will be the excess of retarded males. This will undoubtedly need more study, but there is some evidence for the idea among the studies listed in Table 7.3.

For instance, the Richardson and Higgins study (1964, p. 1824) covers a population with a 7.9% incidence of retardation. The excess of males is 69%. In the Reed and Reed (1965) data, the incidence goes down to 2.7% and the male excess to 49%. The British Columbia population, excluding the borderline category, has less than a 0.5% incidence and a 26% excess of males. Other studies in the table for which incidence is known fit into the general pattern, although the correlation is not perfect.

At the high end of the IQ scale, the hereditary pattern of X-linkage might be less apparent. For one thing, there is the matter of selective mating. The adage, "Like father, like son," might be true, as far as IQ is concerned, mainly because an intelligent man would be smart enough to select (and win) an intelligent woman to be the mother of his children, or vice versa. For instance, in the Outhit (1933, p. 43) study already cited, the correlation between parents was .742; in the Reed and Reed (1965, p. 57) study, it was .464, using 10-point intervals. Also, the brighter fathers would tend to provide a better social and environmental background for their children, thus introducing an important, but nongenetic, variable.

SUMMARY AND SOCIAL IMPLICATIONS

Overall, the implications of greater male-variability in intelligence due to X-linkage are of considerable social importance. Facilities and programs for the retarded must be planned with the expectation of a greater number of males and with the understanding that to a great extent the reasons for this excess are genetic rather than environmental. At the other end of the scale, it is highly probable that basic genetic factors rather than male chauvinism account for at least some of the difference in the numbers of males and females occupying positions requiring the highest levels of intellectual ability.

Because of the importance of these and other implications of the theory, it is essential that further research on the theory of X-linkage of major intellectual traits and its implications be carried out. It must be recalled that the theory, as is usually the case, was designed to fit the data at hand. Later reports, including this one, have added little in the way of new approaches or new data; but what there has been has tended to add to the theory's plausibility. Nonetheless, further studies along many lines will be necessary for the theory's thorough evaluation and to bring forth all of its implications.

As several critics, especially Anastasi (1972), Nance and Engel (1972), and Wittig (1976), have made clear, each bit of supporting data and each line of evidence can be interpreted in such a fashion as to lead to different conclusions. However, to date, no one has come up with an alternative theory that fits *all* of the data. By the principle of parsimony, then, it seems logical to accept the theory as a basis for further research.

Finally, for the benefit of the ladies, it must be stressed that in the theory as set forth there is no inference that males have a greater *range* of intelligence than females. All that is implied is that as the score or rating deviates from the mean in either direction, the male–female ratio in numbers increases. The very brightest women could very well be on par with the very brightest men—there would just be fewer of them.

Nowhere is this more evident than among the authors of studies most essential to the development of the X-linkage theory. Had it not been for the contributions of such outstanding women as Anne Anastasi, Nancy Bayley, Mary Lyons, Marion Outhit, Margaret Thompson, Gillian Turner, and Leona Tyler, among others, this chapter would not have been possible.

Dr. Samuel Johnson said it most succinctly. When asked which were more intelligent, men or women, he replied, "Which man? Which woman?"

REFERENCES

Abramowicz, H. K., & Richardson, S. A. Epidemiology of severe mental retardation in children: Community studies. *American Journal of Mental Deficiency*, 1975, *80*, 18–39.

Ahern, F. M., & Johnson, R. C. Inherited uterine inadequacy: An alternate explanation for a portion of cases of defect. *Behavior Genetics*, 1973, *3*, 1–12.

Anastasi, A. *Differential psychology*. (3rd ed.) New York: Macmillan, 1958.

Anastasi, A. Four hypotheses with a dearth of data: Response to Lehrke's "A theory of X-linkage of major intellectual traits." *American Journal of Mental Deficiency*, 1972, *76*, 620–622. (Also available in the National Foundation–March of Dimes reprint series, RS-186.)

Bayley, N. Developmental problems of the mentally retarded child. In I. Philips (Ed.), *Prevention and treatment of mental retardation*. New York: Basic Books, 1966. Pp. 85–110.

Becker, R. B. Personal communication of data compiled by the Project in Vocational Interest, Columbus (Ohio) State School and Institute, 1972.

Berg, E. Andssvage problemets omfang i Danmark. *Socialt Tidsskrift (Kobenhaun)*, 1966, *5–6*, 150–162.

Bessman, S. F. Genetic failure of amino acid "justification": A common basis for many forms of metabolic, nutritional, and "nonspecific" mental retardation. *Journal of Pediatrics*, 1972, *81*, 834–842.

Blewett, D. B. An experimental study of the inheritance of intelligence. *Journal of Mental Science*, 1954, *100*, 922–923.

Bock, R. D., & Kolakowski, D. Further evidence of sex-linked major-gene influence on human spatial visualizing ability. *American Journal of Human Genetics*, 1973, *25*, 1–14.

British Columbia Department of Health Services and Hospital Insurance. *Registry for handicapped children and adults: Annual report, 1970*. Vancouver: British Columbia Dept. of Health Services, 1971.

British Columbia Department of Health Services and Hospital Insurance. *Registry for handicapped children and adults: Annual report, 1971*. Vancouver: British Columbia Dept. of Health Services, 1972.

Childs, B. Genetic origin of some sex differences among human beings. *Pediatrics*, 1965, *35*, 798–812. (Also available in the National Foundation—March of Dimes reprint series, RS-39.)

Clark, W. W. Research findings on mental sex differences and their implications for education. Paper read at American Association of School Administrators, Cleveland, Ohio, 1958.

Davidson, R. G. The Lyon Hypothesis. *The Journal of Pediatrics*, 1964, *65*, 765–775. (Also available in the National Foundation—March of Dimes reprint series, RS-39.)

Dunn, H. G., Renpenning, H., Gerrard, J. W., Miller, J. R., Tabata, T., & Federoff, S. Mental retardation as a sex-linked defect. *American Journal of Mental Deficiency*, 1963, *67*, 827–848.

Dutch National Association for the Care of the Mentally Retarded, and Bishop Bekkers Institute. *The structure of Dutch services for the mentally retarded*. Author, P. O. 415, Utrecht, the Netherlands, 1973.

Copies of the National Foundation–March of Dimes reprints are available (in limited supply) from The National Foundation–March of Dimes Professional Educational Department; Box 2000, White Plains, N.Y. 10602.

Ellis, H. *Man and woman: A study of human secondary sexual characters.* (4th ed.) London: Walter Scott Publ., 1904.
Gruenberg, E. M. Epidemiology. In H. A. Stevens & R. Heber (Eds.), *Mental retardation.* Chicago: Univ. of Chicago Press, 1964. Pp. 259–306.
Hall, D. H. (Senior Research Associate, Test Division, The Psychological Corp.) Personal communication, 1973.
Hasan, K. Z. Mental retardation in Pakistan. In D. A. A. Primrose (Ed.), *Proceedings of the Second Congress of the International Association for the Scientific Study of Mental Retardation.* Warsaw, Poland: Polish Medical Publishers, 1972. Pp. 61–73.
Heber, R. A manual on terminology and classification in mental retardation. *American Journal of Mental Deficiency Monograph Supplement,* 1961.
Heber, R. *Epidemiology of mental retardation.* Springfield, Illinois: Thomas, 1970.
Imre, P. D. The epidemiology of mental retardation in a S. E. rural USA community. In B. W. Richards (Ed.), *Proceedings of the First Congress of the International Assoc. for the Scientific Study of Mental Deficiency.* Reigate (Surrey), Great Britain: Michael Jackson Publ. 1968. Pp. 550–655.
Jensen, A. R. *Environment, heredity, and intelligence.* Cambridge, Mass.: Harvard Educational Review, 1969. (a)
Jensen, A. R. How much can we boost IQ and scholastic achievement? *Harvard Educational Review,* 1969, *39,* 1–123. (b)
Jensen, A. R. The race × sex × ability interaction. In R. Cancro (Ed.), *Intelligence: Genetic and environmental influences.* New York: Grune and Stratton, 1971. Pp. 107–161.
Johnson, G. E. Contribution to the psychology and pedagogy of feeble-minded children. *Journal of Psycho-Asthenics,* 1897, *2,* 26–32.
Kaveggia, E. G., Opitz, J. M., & Pallister, P. D. Diagnostic/genetic studies in severe mental retardation. In D. A. A. Primrose (Ed.), *Proceedings of the Second Congress of the International Association for the Scientific Study of Mental Deficiency.* Amsterdam: Swets and Zeitlinger, 1972. Pp. 305–312.
Lehrke, R. G. *Sex-linked mental retardation and verbal disability.* Ph.D. dissertation, University of Wisconsin-Madison, 1968. Ann Arbor, Michigan: University Microfilms Order No. 68–15,999.
Lehrke, R. A theory of X-linkage of major intellectual traits. *American Journal of Mental Deficiency,* 1972, *76,* 611–619. (a) (Also available in the National Foundation–March of Dimes reprint series, RS-186.)
Lehrke, R. Response to Dr. Anastasi and to the Drs. Nance and Engel. *American Journal of Mental Deficiency,* 1972, *76,* 626–631. (b) (Also available in the National Foundation–March of Dimes reprint series, RS-186.)
Lehrke, R. G. *X-linked mental retardation and verbal disability.* In The National Foundation–March of Dimes, Birth Defects Original Article Series, Vol. 10, No. 1. Miami, Florida: Symposia Specialists, 1974.
Lyon, M. P. Sex chromatin and gene action in the mammalian X chromosome. *American Journal of Human Genetics,* 1962, *14,* 135–148.
Maccoby, E. E., & Jacklin, C. N. *The psychology of sex differences.* Stanford, California: Stanford Univ. Press, 1974.
Masland, R. L., Sarason, S. B., & Gladwin, T. *Mental subnormality.* New York: Basic Books, 1958.
McKusick, V. A. *Mendelian inheritance in man.* (3rd ed.) Baltimore: Johns Hopkins Press, 1971.
Nance, W. E., & Engel, E. One X and four hypotheses: Response to Lehrke's "A theory of X-linkage of major intellectual traits." *American Journal of Mental Deficiency,* 1972, *76,* 623–625. (Also available in the National Foundation–March of Dimes reprint series, RS-186.)
National Institute of Mental Health. *Patients in mental institutions, 1950 and 1951.*

U.S.P.H.S. publication No. 356. Washington, D.C.: U.S. Government Printing Office, 1954.

National Institute of Mental Health. *Patients in mental institutions, 1952. Part I: Public institutions for mental defectives and epileptics.* U.S.P.H.S. Publication No. 483. Washington, D.C.: U.S. Government Printing Office, 1956. (a)

National Institute of Mental Health. *Patients in mental institutions, 1953. Part I: Public institutions for mental defectives and epileptics.* U.S.P.H.S. Publication No. 495. Washington, D.C.: U.S. Government Printing Office, 1956. (b)

National Institute of Mental Health. *Patients in mental institutions, 1954. Part I: Public institutions for mental defectives and epileptics.* U.S.P.H.S. Publication No. 523. Washington, D.C.: U.S. Government Printing Office, 1957.

National Institute of Mental Health. *Patients in mental institutions, 1955. Part I: Public institutions for mental defectives and epileptics.* U.S.P.H.S. Publication No. 574. Washington, D.C.: U.S. Government Printing Office, 1958.

National Institute of Mental Health. *Patients in mental institutions, 1956. Part I: Public institutions for mental defectives and epileptics.* U.S.P.H.S. Publication No. 632. Washington, D.C.: U.S. Government Printing Office, 1969.

National Institute of Mental Health. *Patients in mental institutions, 1957. Part I: Public institutions for mental defectives and epileptics.* U.S.P.H.S. Publication No. 715. Washington, D.C.: U.S. Government Printing Office, 1960. (a)

National Institute of Mental Health. *Patients in mental institutions, 1958. Part I: Public institutions for mental defectives and epileptics.* U.S.P.H.S. Publication No. 781. Washington, D.C.: U.S. Government Printing Office, 1960. (b)

National Institute of Mental Health. *Patients in mental institutions, 1959. Part I: Public institutions for mental defectives and epileptics.* U.S.P.H.S. Publication No. 820. Washington, D.C.: U.S. Government Printing Office, 1961.

National Institute of Mental Health. *Patients in mental institutions, 1960. Part I: Public institutions for the mentally retarded.* U.S.P.H.S. Publication No. 963. Washington, D.C.: U.S. Government Printing Office, 1963.

National Institute of Mental Health. *Patients in mental institutions, 1961. Part I: Public institutions for the mentally retarded.* U.S.P.H.S. Publication No. 1091. Washington, D.C.: U.S. Government Printing Office, 1964. (a)

National Institute of Mental Health. *Patients in mental institutions, 1962. Part I: Public institutions for the mentally retarded.* U.S.P.H.S. Publication No. 1143. Washington, D.C.: U.S. Government Printing Office, 1964. (b)

National Institute of Mental Health. *Patients in mental institutions, 1963. Part I: Public institutions for the mentally retarded.* U.S.P.H.S. Publication No. 1222. Washington, D.C.: U.S. Government Printing Office, 1965.

National Institute of Mental Health. *Patients in mental institutions, 1964. Part I: Public institutions for the mentally retarded.* U.S.P.H.S. Publication No. 1452. Washington, D.C.: U.S. Government Printing Office, 1966.

National Institute of Mental Health. *Patients in mental institutions, 1965. Part I: Public institutions for the mentally retarded.* U.S.P.H.S. Publication No. 1597. Washington, D.C.: U.S. Government Printing Office, 1967.

New York State Department of Mental Hygiene, Mental Health Research Unit. *Technical report: A special census of suspected referred mental retardation, Onondaga County, N.Y.* Syracuse, New York: Syracuse Univ. Press, 1955.

Outhit, M. C. A study of the resemblance of parents and children in general intelligence. *Archives of Psychology*, 1933, *149:*1–60.

Penrose, L. S. *A clinical and genetic study of 1280 cases of mental defect.* London: His Majesty's Stationery Office, 1938.

Penrose, L. S. *The biology of mental defect.* (Rev. ed.) New York: Gruen and Stratton, 1963.

Priest, J. H., Thuline, H. C., LaVeck, C. D., & Jarvis, D. B. An approach to genetic factors in mental retardation. *American Journal of Mental Deficiency,* 1961, *66,* 42–50.

Psychological Corp. *Manual: Miller Analogies Test.* New York: Psychological Corp., 1970.

Reed, E. W., & Reed, S. C. *Mental Retardation: A family study.* Philadelphia: Saunders, 1965.

Renpenning, H., Gerrard, J. W., Zaleski, W. A., & Tabata, T. Familial sex-linked mental retardation. *Canadian Medical Association Journal,* 1962, *87,* 954–956.

Richardson, W. P., & Higgins, A. C. A survey of handicapping conditions and handicapped children in Alamance County, North Carolina. *American Journal of Public Health,* 1964, *54,* 1817–1830.

Roberts, J. A. F., Norman, R. M., & Griffiths, R. On the difference between the sexes in the dispersion of intelligence. *British Medical Journal,* 1945, *1,* 727–730.

Scottish Council for Research in Education. *The trend of Scottish Intelligence.* London: Univ. of London Press, 1949.

Socialstyrelsen. *Living standard in Swedish facilities for the mentally retarded. Special report, May 15, 1972.* Stockholm: Socialstyrelsen, 1972.

Stafford, R. E. Sex differences in spatial visualization as evidence of sex-linked inheritance. *Perceptual Motor Skills,* 1961, *13,* 428.

Sterner, R. Note on the number of retarded children in Vasternorrland County, Sweden. Paper presented at International League of Societies for the Mentally Handicapped, Stockholm, Sweden, 1967.

Stomma, D., & Wald, I. *The influence of medical rehabilitation on the general performance of institutionalized mentally retarded children.* Warsaw, Poland: Psychoneurological Institute, 1972.

Terman, L. M. *Genetic studies of genius,* Vol. 1. Stanford, California: Stanford Univ. Press, 1925.

Thompson, M. W. Genetic implications of heteropyknosis of the X-chromosome. *Canadian Journal of Genetics and Cytology,* 1965, *7,* 202–213. (Also available in the National Foundation—March of Dimes reprint series, RS-76)

Turner, G., Collins, E., & Turner, B. Recurrence risk of mental retardation in sibs. *The Medical Journal of Australia,* 1971, *1,* 1165–1166.

Turner, G., Engisch, B., Lindsay, D. S., & Turner, B. X-linked mental retardation without physical abnormality (Renpenning's Syndrome) in sibs in an institution. *Journal of Medical Genetics,* 1972, *9,* 324–330.

Turner, G., & Turner, B. X-linked mental retardation. *Journal of Medical Genetics,* 1974, *11,* 109–113.

Turner, G., Turner, B., & Collins, E. Renpenning's Syndrome–X-linked mental retardation. *Lancet,* August 15, 1970, 365–366.

Turner, G., Turner, B., & Collins, E. X-linked mental retardation without physical abnormality: Renpenning's Syndrome. *Developmental Medicine and Child Neurology,* 1971, *13,* 71–78.

Tyler, L. E. *The psychology of human differences.* (3rd ed.) New York: Appleton, 1965.

Vandenberg, S. G. The hereditary abilities study: Hereditary components in a psychological test battery. *American Journal of Human Genetics,* 1962, *14,* 220–237.

Verbraak, P. A new prevalence figure for mental retardation in the Netherlands. In D. A. A. Primrose (Ed.), *Proceedings of the Third Congress of the International Association for the Scientific Study of Mental Retardation.* Warsaw: Polish Medical Publishers, 1975. Pp. 664–673.

Wechsler, D. *The measurement and appraisal of adult intelligence.* (4th ed.) Baltimore: Williams and Wilkins, 1958.

Wittig, M. A. Sex differences in intellectual functioning: How much of a difference do genes make? *Sex Roles: A Journal of Research*, 1976, *2*, 63–74.

Wortis, H., Pollack, M., & Wortis, J. Families with two or more mentally retarded or mentally disturbed siblings: The preponderance of males. *American Journal of Mental Deficiency*, 1966, *70*, 745–752.

Wright, W. S., Tarjan, G., & Eyer, L. Investigation of families with two or more mentally defective siblings. *American Journal of Diseases of Children*, 1959, *97*, 445–463.

Wunsch, W. L. The first complete tabulation of the Rhode Island Mental Deficiency Record. *American Journal of Mental Deficiency*, 1951, *55*, 293–312.

8

Own-Race Preference and Self-Esteem in Young Negroid and Caucasoid Children

AUDREY M. SHUEY
Randolph–Macon Woman's College

INTRODUCTION

For more than 25 years, the view that low self-esteem characterizes Negroids in the United States has been generally accepted by competent writers. This low self-esteem supposedly distinguishes Negroid from Caucasoid Americans and is responsible for slower Negroid school progress, presumed lack of ambition, and lower cognitive abilities as measured by various aptitude and achievement tests. (Improvement of self-concept and a sense of personal worth, a major goal of Head Start, was listed by 46% of the 104 participating officials interviewed as the most important objective of their centers [Granger *et al.*, 1969].) We propose to review critically the available pertinent research on self-esteem of Negroid and Caucasoid children between 3 and 8 years of age and to present the studies in comparable tabular form.

In describing these researches, we have differentiated *race-preference studies* from *self-esteem studies* per se. While many investigators have not connected race-preference findings with self-esteem, some have done so, and other scholars have assumed "prefer my race, prefer me," and "reject my race, reject me." (Carried a little further, preference for *own-sex figures, own-religion* symbols, and the like, could be assumed to measure self-esteem in older schoolchildren.) Thus, Clark (1955) writes,

"Younger children, as we have seen, tend to express their self-hatred by concrete and direct rejection of brown skin color (p. 50)." Similarly, Vontress (1970, p. 193) indicates that the Negroid child, even at four years, knows that he is racially different and tends to despise his group and hate himself for being a member of this group. (See Kardiner and Ovesey (1951), Karon (1958), and Bronfenbrenner (1967) for other expressions of this familiar view.)

In reviewing these researches, we have looked for factors other than race that may be related to *own-race preference* or self-esteem, such as sex and age of child, socioeconomic level, IQ, section of country, race of interviewer, and so on. We have grouped the studies into four tables according to the type of test or instrument employed: (1) Dolls and Puppets, (2) Photographs, Sketches and Drawings, and a Coloring Test, (3) Constructs and Symbols, and (4) Questionnaire and Rating Scales. The first two purport to measure race preference, the latter two self-esteem. In each table, we have included wherever possible: (1) *Author* and *Date* of publication, (2) *Location* of study, (3) Name and brief description of *Instrument* used, (4) *Subjects*, including number of each race, age, sex, grade in school, and method of selection, (5) *Results* as reported by the author or combined by reviewer from author's data, and (6) *Comments of Author*. The text has been primarily centered around the material as presented in the tables.

DOLLS AND PUPPETS

Mary Ellen Goodman (1946) was one of the first to use dolls as a means of studying racial self-identification and preference among young children. Her report emphasizes interpretation and only briefly describes her materials, subjects, procedure, and results. Some 27 3- and 4-year-old children of both races, attending the Ruggles Street Nursery School in Boston, served as subjects. It appears that 21 of them responded to the first set of baby dolls and the question *Which looks most like you when you were a baby?* Sixteen of them responded to the same question when shown a second set of baby dolls. Shown a third set of dolls dressed to represent nursery school children of the subject's own sex, 25 responded to *Which looks most like you?* Negroid and Caucasoid subjects were reported to comprise approximately half of each group. Correct identifications were made by 40% of the Negroid subjects and by 80% of the Caucasoids in each situation. We are informed without supporting evidence that the Negroid–Caucasoid difference in self-identification bears no apparent relation to age, sex, or IQ.

The children were given four different opportunities to indicate their esthetic preferences with respect to the three sets of dolls shown in the identification test. Fifty to 70% of the 15–20 subjects who responded were Negroids. "Nevertheless 70% to 90% of each responding group preferred the white doll." Negroid preference for the white doll was reported to be as great as the Caucasoid preference for this doll.

Kenneth and Mamie Clark (1947) employed several techniques in studying the racial identification and preferences of Negroid children. Four dolls, identical except in skin and hair color, were presented to each child. For half of the subjects, the dolls were presented in the order: white, colored, white, colored; for the other half the order was reversed. The children were asked to respond to eight requests by choosing *one* of the dolls and giving it to the experimenter. The first four were designed to reveal racial preferences:

(1) *Give me the doll that you like to play with—like best.*
(2) *Give me the doll that is a nice doll.*
(3) *Give me the doll that looks bad.*
(4) *Give me the doll that is a nice color.*

All children interviewed were young Negroids, ages 3–7; 134 of them attended segregated nursery or public schools in Arkansas, and 119 were enrolled in racially mixed nursery or public schools in Massachusetts. The Clarks indicated that results for a few children who showed generalized negativism were not included. Total doll preferences were tabulated according to age (in years), North or South, and observed skin color (light, medium, or dark).

The Negroid children on the whole showed a preference for the white dolls and tended to reject the brown. Thirty-six percent of the 253 children chose their own-race (brown) doll in response to the three combined requests of: *play with—like best, nice doll,* and *nice color,* as opposed to 60% who chose the white doll. (This difference is significant (χ^2 = 6.58, df = 1, p = <.01).) Conversely, 50% of the subjects selected their own-race doll when asked for the doll that *looks bad,* as against 17% who selected the white doll, likewise a highly significant difference.

The younger Negroid subjects (3–4 years) identified with one of the colored dolls no more than 50% of the time, whereas the older children (6–7 years) selected their own-race doll as looking like themselves 78% of the time (Determined by reviewer from the Clarks' Table 2). Not unpredictably, the younger subjects likewise showed less preference for their own-race doll (28.7%) than the older children (42.0%), as determined by the proportion of their responses to the three combined positive requests of play with–like best, nice doll, and nice color (Difference significant at the .05 level of confidence, computed by reviewer).

The subjects attending segregated schools in the three Arkansas cities showed a slight and insignificant preference for their own-race dolls over the children attending the racially mixed schools of the Massachusetts city, as evidenced by their responses to the three combined requests (41% and 31.7%, respectively; see Table 8.1).

TABLE 8.1

Dolls and Puppets

Author and date	Location	Instrument	Subjects		
			N		Age
Goodman, M. E. (1946)	Boston, Mass.	Two sets of baby dolls; one set of dolls dressed to represent nursery school children of sex of subject.		27 50–70% were Negroids.	3–4
Clark, K. B., and M. P. (1947)	Hot Springs, Pine Bluff, Little Rock, Ark., and Springfield, Mass.	Two brown dolls with black hair, 2 white dolls with yellow hair. All in diapers.	Negroid (Ark.)	Negroid (Mass.)	
			18	13	3
			19	10	4
			12	34	5
			39	33	6
			46	29	7
			134	119	
Radke, M. J., and Trager, H. G. (1950)	Philadelphia, Pa.	Plywood form boards of men, women, clothes, houses.	Negroid	Caucasoid	
			90	152	Kindergarten 2

<table>
<tr><th>Subjects
Method of selection</th><th>Results</th><th>Comments of author</th></tr>
<tr><td>Not given; Ruggles Street Nursery School "where 27 children were studied . . ."; four opportunities for a child to indicate esthetic preference with respect to baby dolls and those representing own age; only 15–20 responded to the 4 situations.</td><td><table><tr><th colspan="2">Preference for white doll</th></tr><tr><td>All subjects</td><td>70–90%</td></tr></table></td><td>About ¾ of both Negroid and Caucasoid children preferred white doll as prettier.</td></tr>
<tr><td>All subjects Negroid. Ark. subjects were in segregated nursery and public schools with no experience in racially mixed schools; Mass. subjects in mixed nursery and public schools. Examined all present except a few who showed "generalized negativism."</td><td><table><tr><th colspan="4">Choices of all subjects</th></tr><tr><th>Request</th><th>Brown doll</th><th>White doll</th><th>?[a]</th></tr><tr><td>Play with- best</td><td>32%</td><td>67%</td><td>1%</td></tr><tr><td>Nice doll</td><td>38%</td><td>59%</td><td>3%</td></tr><tr><td>Looks bad</td><td>59%</td><td>17%</td><td>24%</td></tr><tr><td>Nice color</td><td>38%</td><td>60%</td><td>2%</td></tr></table><table><tr><th colspan="4">Choice of own-race doll[b] to play with, according to age of subjects</th></tr><tr><th>3–5</th><th>6–7</th><th>t</th><th>p</th></tr><tr><td>28.7%</td><td>42.0%</td><td>2.17</td><td><.05</td></tr></table></td><td>Data indicate a basic knowledge of racial differences in Negroid children 3–7 years of age in Northern and Southern communities tested.</td></tr>
<tr><td>From 6 public schools in 1 school district; largely lower-middle to upper-lower socioeconomic levels. Selected every third name on class lists. Subjects interviewed by experimenter of own race.[c]</td><td><table><tr><th colspan="4">Preference for own race[d] indicated by choice of:</th></tr><tr><th>Subjects</th><th>Doll</th><th>Costume</th><th>House</th></tr><tr><td>Negroid</td><td>57%</td><td>78%</td><td>33%</td></tr><tr><td>Caucasoid</td><td>89%</td><td>54%</td><td>77%</td></tr></table></td><td>White doll preferred by 89% of Caucasoid subjects; Negroid subjects chose black doll 57% of time. Majority from both races gave poor houses to black doll.</td></tr>
</table>

TABLE 8.1 *(Continued)*

Author and date	Location	Instrument	Subjects: N			Age
				Negroid	Caucasoid	
Stevenson, H. W., and Stewart, E. C. (1958)	Austin, Tex.	Four small dolls made of soft, flesh-colored plastic.		23	25	3
				13	25	4
				22	25	5
				17	25	6
				20	25	7
				95	125	
Gregor, A. J., and Mc-Pherson, D. A. (1966)	Southern metropolitan area	Two dolls, one white with blue eyes, other chocolate with brown eyes; both in diapers		Negroid	Caucasoid	
			Boys	38	45	
			Girls	54	38	
				92	83	6–7

<table>
<tr><th>Subjects
Method of selection</th><th>Results</th><th>Comments of author</th></tr>
<tr>
<td>Subjects enrolled in 10 private segregated nursery and elementary schools. Chosen at random.[e] Evenly divided by sex; testing by experimenter of own race.</td>
<td>Choice of own-race doll[f] as playmate, by:
<table>
<tr><th colspan="4">Race of subjects</th></tr>
<tr><th>Negroids</th><th>Caucasoids</th><th>t</th><th>p</th></tr>
<tr><td>45%</td><td>68%</td><td>3.43</td><td><.001</td></tr>
</table>
<table>
<tr><th colspan="5">Age and race of subjects</th></tr>
<tr><th></th><th>3–5</th><th>6–7</th><th>t</th><th>p</th></tr>
<tr><td>Negroid</td><td>41.8%</td><td>50.3%</td><td>.81</td><td>—</td></tr>
<tr><td>Caucasoid</td><td>58.7%</td><td>82.0%</td><td>2.75</td><td><.01</td></tr>
</table>
</td>
<td>Greater frequency of own-race rejection by Negroid subjects seen in lower proportion choosing own-race for playmate.</td>
</tr>
<tr>
<td>Caucasoid and Negroid children, ages 6–7, in first grade, in racially homogeneous schools. No report of number of schools or classes surveyed, nor whether all present at 2 ages were interviewed.</td>
<td>Choice of own-race doll to play with,[g] according to:
<table>
<tr><th colspan="4">Race of subjects</th></tr>
<tr><th>Black</th><th>White</th><th>t</th><th>p[h]</th></tr>
<tr><td>56.8%</td><td>85.5%</td><td>4.17</td><td><.001</td></tr>
</table>
<table>
<tr><th colspan="5">Sex and race of subjects</th></tr>
<tr><th></th><th>Male</th><th>Female</th><th>t</th><th>p</th></tr>
<tr><td>Negroid</td><td>69.2%</td><td>48.0%</td><td>2.02</td><td><.05</td></tr>
<tr><td>Caucasoid</td><td>84.5%</td><td>86.7%</td><td>—</td><td>—</td></tr>
</table>
</td>
<td>The 2 groups in this Deep South area gave evidence of secure self-system; both identified with own group.</td>
</tr>
</table>

TABLE 8.1 *(Continued)*

Author and date	Location	Instrument	Subjects	N: Negroid	N: Caucasoid	Age
Greenwald, H. J., and Oppenheim, D. B. (1968)	New York City and New Rochelle, N.Y.	Three dolls (dark brown, mulatto,[i] and white) identical except in skin color. All in diapers.		Negroid	Caucasoid	
			Boys	21	21	
			Girls	18	15	
				39	36	3–5
Asher, S. R., and Allen, V. (1969)	Newark, N.J., and vicinity	Three pairs of puppets, identical except for skin and hair color.		Negroid	Caucasoid	
			Boys	96	77	
			Girls	85	71	
			Unidentified	5	7	
				186	155	3–8

Subjects: Method of selection	Results	Comments of author
Subjects interviewed in both integrated and nonintegrated nursery schools; most were of ages 4 or 5, a few were age 3. No mention of number of schools or whether all children present were tested.	Choice of Negroid subjects; Choice of Caucasoid subjects (see tables below)	Negroid subjects were judged to be from lower and middle classes, while Caucasoid subjects all from middle class.

Choice of Negroid subjects

Question	Brown doll	Mulatto doll	White doll	?
Play with	28%	13%	56%	3%
Good doll	35%	15%	50%	—
Bad doll	21%	59%	10%	10%
Nice color	31%	8%	56%	5%

Choice of Caucasoid subjects

	Brown doll	Mulatto doll	White doll	?
Play with	22%	4%	63%	11%
Good doll	20%	3%	69%	8%
Bad doll	26%	51%	3%	20%
Nice color	18%	8%	71%	3%

Subjects: Method of selection	Results	Comments of author
From nursery schools, neighborhood centers, pre-school programs and play street run by city.	Choice of own-race doll to play with,[j] according to: (see tables below)	Social-class data for Negroid children suggest enhanced status may lead to increased inferiority feeling.

Choice of own-race doll to play with,[j] according to:

Race of subjects

Negroid	Caucasoid	t	p
27.3%	75.0%	8.77	<.001

Sex and race of subjects

	Male	Female	t	p
Negroid	22.3%	32.7%	1.57	—
Caucasoid	78.3%	72.0%	—	—

TABLE 8.1 *(Continued)*

Author and date	Location	Instrument	Subjects: N			Age
				Negroid	Caucasoid	
Crooks, R. C. (1970)	Halifax, Nova Scotia	Dolls identical except skin and hair color. Two were brown with black hair; 2 white with blond hair.	*E*	17	17	
			C	17	17	
				34	34	4–5
				Negroid	Caucasoid	
Hraba, J., and Grant, G. (1970)	Lincoln, Nebr.	Dolls identical except skin and hair. Two medium brown with curly black hair; 2 white with fair hair. All same eye color.[f]		89	71	4–8
				Negroid	Caucasoid	
Durrett, M. E., and Davy, A. J. (1970)	San Jose, Calif.	Two girl and 2 boy dolls of soft plastic. Black and white dolls same except for color of skin and hair.	Boys	11	14	
			Girls	14	16	
				25	30	4

<table>
<tr><th>Subjects
Method of selection</th><th>Results</th><th>Comments of author</th></tr>
<tr>
<td>All subjects from interracial neighborhood with low income, overcrowding and large families. Experimental group had 1 year enriched preschool program; *C* group, none.</td>
<td>Choice of own-race doll to play with,[k] according to:
<table>
<tr><th colspan="4">Race of subjects</th></tr>
<tr><th>Negroid</th><th>Caucasoid</th><th>t</th><th>p</th></tr>
<tr><td>34.0%</td><td>66.7%</td><td>2.70</td><td>.01</td></tr>
</table>
<table>
<tr><th colspan="5">Preschool program and race</th></tr>
<tr><th></th><th>E</th><th>C</th><th>t</th><th>p</th></tr>
<tr><td>Negroid</td><td>49.0%</td><td>18.0%</td><td>1.91</td><td>—</td></tr>
<tr><td>Caucasoid</td><td>55.0%</td><td>78.3%</td><td>1.44</td><td>—</td></tr>
</table>
</td>
<td>Necesssary to attack problem of prejudice in very young children.</td>
</tr>
<tr>
<td>60% of eligible Negroids 4–8 in city's public schools; 71 Caucasoids drawn at random from mixed classrooms: all in kindergarten or grades 1–2.</td>
<td>Choice of own-race doll to play with,[m] according to:
<table>
<tr><th colspan="4">Race of subjects</th></tr>
<tr><th>Negroid</th><th>Caucasoid</th><th>t</th><th>p</th></tr>
<tr><td>64.3%</td><td>67.0%</td><td>—</td><td>—</td></tr>
</table>
</td>
<td>Respondents randomly assigned to Negroid and Caucasoid interviewers.</td>
</tr>
<tr>
<td>85 4- to 5-year-old subjects including Negroid, Anglo- (Caucasoid) and Mexican–American in prekindergarten interracial schools. All from lower socioeconomic families.</td>
<td>Choice of own-race doll as playmate,[n] according to:
<table>
<tr><th colspan="4">Race of subjects</th></tr>
<tr><th>Negroid</th><th>Caucasoid</th><th>t</th><th>p</th></tr>
<tr><td>48%</td><td>83.3%</td><td>2.71</td><td>.01</td></tr>
</table>
<table>
<tr><th colspan="5">Sex and race of subjects</th></tr>
<tr><th></th><th>Male</th><th>Female</th><th>t</th><th>p</th></tr>
<tr><td>Negroid</td><td>27%</td><td>64%</td><td>1.85</td><td>—</td></tr>
<tr><td>Caucasoid</td><td>86%</td><td>81%</td><td>—</td><td>—</td></tr>
</table>
</td>
<td>Negroid subjects showed least own-group preference. Although positive changes have been noted (since 1958) many Negroids, especially boys, need help in developing own-race acceptance.</td>
</tr>
</table>

TABLE 8.1 *(Continued)*

Author and date	Location	Instrument	Subjects			
				N		Age
Harris, S., and Braun, J. R. (1971)	Media and Chester, Penn.	Two pairs of puppets of about subject's age, one pair male and one female. Within each pair, they were same except for skin and hair color. Black puppet medium brown face and black hair; white had light skin and light hair.		Negroid		
				60		7–8
Fox, D. J., and Jordan, V. B. (1973)	New York, N.Y.	Eight dolls. Each presentation consisted of 2 dolls of medium-brown skin, brown eyes, and dark hair; 2 white dolls with blue eyes and blond hair; all 4 of same sex as subject and identically dressed.		Negroid	Caucasoid	
			Boys	180	180	
			Girls	180	180	
				360	360	5–7

[a] Table here condensed from the Clark's Tables 5 and 6. Clarks have not separated their subjects according to sex. ? indicates no choice was recorded.

[b] Present writer combines responses to *play with-like best, nice doll,* and *nice color;* here and elsewhere.

[c] These subjects also tested on Race Barrier and Non-Barrier pictures. See Table 8.2, this chapter.

[d] Differences between preferences for own race as indicated by choice of preferred doll, choice of costume, and choice of house all highly significant; $p < .01$ (determined by reviewer).

[e] From class rolls in elementary schools. However, all in nursery schools included if within selected age range. These 220 children were also tested on drawings of Negroid and Caucasoid children at play. See Table 8.2.

[f] Calculations by reviewer.

[g] Combining responses to *play with, like best, nice doll,* and *nice color.*

[h] Reviewer is responsible for combining requests in various studies as well as the resulting calculations.

[i] A white doll had been painted a light grayish-brown by a hospital employee to simulate mulatto color.

[j] *Play with, nice doll,* and *nice color.* For the 3–5-year-olds used one pair of puppets appearing about 2-years-old; for the 6–8-year-olds, used two puppets of same sex as subject and estimated by authors to represent children of about 11 years of age.

[k] *Play with, nice doll,* and *nice color* combined.

[l] Personal communication from J. Hraba.

[m] Yet 36% of Negroid and 34% of Caucasoid subjects indicated that the dolls selected to represent their races *looked bad.*

Subjects: Method of selection	Results	Comments of author
All 30 Negroid 7–8-year-old pupils in the six second and third grade classes of middle-class suburban schools. In lower-class inter-city school, every fifth Negroid from class lists of two grades.	Choice of own-race doll to play with[o] Subjects H + B | Other Negroids[p] | t | p 68.7% | 54.1% | 2.18 | .05	Absence of significant difference between middle and lower class subjects suggests that former are as ethnocentric as latter.
Each racial group comprised of equal numbers of 5-, 6-, and 7-year-old boys and girls, half in integrated and half in segregated schools;[q] all subjects born in New York City, English-speaking, living with both natural parents.	Choice of own-race doll to play with:[r] according to: *Race of subjects* Negroid | Caucasoid | t | p 58.0% | 70.2% | 3.42 | <.001 *Sex and race of subjects* | Male | Female | t | p Negroid | 54.2% | 62.0% | 1.48 | — Caucasoid | 74.0% | 65.5% | 1.76 | —	No significant difference in choices of segregated and integrated Negroid children.

[n] Followed procedure of Stevenson and Stewart (1958), only *preference for playmate* question used to determine own-race preference.

[o] We have combined percentages on *play with* item with *nice puppet* and *nice color*. Only 21% of Harris and Braun's subjects chose the brown puppet as looking *bad*. All subjects and experimenters were black.

[p] *Other Negroids* comprising 516 subjects, ages 6–8, responding to similar requests as reported by: Clark and Clark (1947), Stevenson and Stewart (1958), Gregor and McPherson (1966), and Fox and Jordan (1973). Reviewer's calculations.

[q] The three Negroid segregated schools in Manhattan, Brooklyn, and Queens; the three Caucasoid segregated schools in Manhattan, Brooklyn, and Staten Island. Of the integrated schools, two in Brooklyn, one in Queens, and one in the Bronx.

[r] These percentages comprising responses to *play with, nice doll, like best,* and *nice color;* all calculations by present writer; material from authors' Tables 1–4. See Fox and Jordan for responses of Chinese and additional Caucasoid subjects to photographs of Chinese and Caucasoid children.

The children of medium or dark skin color showed a slightly greater preference for the brown dolls than the subjects of light skin, as indicated by their choices to the three combined requests.[1] A majority of the children, regardless of skin color, chose the white doll.

The Clarks' doll studies fully support the finding that Negroid children between 3 and 7 years of age tend to prefer white dolls with yellow hair to brown dolls with black hair. Responses to play with–like best, nice doll, and nice color were similar, with few response refusals. To the request to show a doll that looks bad, a significantly large percentage failed to respond. Older, Southern, and children of medium and dark skin may perhaps react more favorably to own-race dolls than younger, Northern, and light-skinned children. Unfortunately, age, regional, and skin-color variables seem to be inextricably entwined. Without access to the raw data of this poineer work, no definite conclusions can be reached at this point.

Radke and Trager (1950) interviewed 90 Negroid and 152 Caucasoid children in kindergarten, first, or second grades of six Philadelphia public schools in a largely lower-middle or upper-lower class district. Test materials consisted of plywood form boards with cutout stylized figures of men and women, plywood clothes to fit the figures, and plywood forms of houses. In the test situation, a pair of form boards (two men dolls for the boy subjects and two women dolls for the girls, identical except for skin color) was presented to the subject. The costume insets included duplicate sets of dress-up, work, and shabby clothes. The houses represented run-down tenements and neat one-family houses of red brick with a front lawn and trees visible in the background.

The children were asked to put one dress (or suit) on each doll. Questions followed, such as, *What would this man be doing wearing these clothes?* The children were also asked, *Which man (woman) do you like the best? Why?* The experimenter then presented the houses, determining in like manner which the child liked best and the one the Negroid man lives in and why. These were followed by the same questions while the child was attending to the Caucasoid figure.

Fifty-seven % of the Negroid children liked the brown doll best, whereas 89% of the Caucasoid subjects liked the white doll best. Seventy-eight % of the Negroid children assigned the dress-up costume to the doll of their own race, in contrast to 54% of the Caucasoids who assigned a dress-up costume to the white doll. Only 30% of both racial groups assigned a dress-up costume to a doll of the other race.

While the selection of the *doll* and clothes for the *preferred doll* may suggest some positive Negroid self-esteem, the giving of a *good house* to

[1] $t = 1.57$, calculated by reviewer. The mean percentage of own-race doll choices of the 79 *dark subjects* (dark brown to black) was 37.7; that of 127 *medium subjects* (light brown to dark brown), 38.3; that of the 46 *light subjects,* nearly white), 28.3.

the white doll by 60% of the Negroid children and to the *brown doll* by only 33% may be irrelevant to attitudes toward self. A sample of six photographs of Negroid and Caucasoid dwellings included in the authors' publication suggests a striking similarity between the houses of the two groups, with the poorest of the three Negroid houses being more dilapidated than the poorest of the Caucasoids. Unfortunately, the Negroid children who assigned the brown doll to the poor house were seldom articulate.

As a part of their research project involving 220 Negroid and Caucasoid children enrolled in private segregated nursery and elementary schools in Austin, Texas, Stevenson and Stewart (1958) used four plastic dolls. Two of them were Caucasoid, and two were "modified to create Negro dolls." The children, all interviewed individually, were shown two dolls of their own sex, identically dressed, and asked, Which one looks most like you?[2] and Which would you rather play with?

Own-race choices to *rather play with* varied with race and age. Sixty-eight % of the Caucasoids versus 45% of the Negroids chose own-race dolls, the difference being significant at the .001 level of confidence. However, 41.8% of the 3–5-year-old Negroids versus 50.3% of the 6–7-year-old Negroids chose own-race dolls for playmates, the difference being insignificant; whereas comparable age groups of the Caucasoid subjects preferred own-race dolls as playmates 58.7 and 82.0% of the time, the difference between them being significant at the .01 level of confidence.

Gregor and McPherson (1966) gave a variant of the Clarks' dolls test to 175 Negroid and Caucasoid first-grade children. In a city referred to as "Southern," subjects were asked to choose between a *white, fair-haired, blue-eyed,* and a *chocolate-colored, black-haired, and brown-eyed doll.* Each child, tested alone by a member of his race, was given nine requests, the first five designed to produce affective responses to his own racial group. The Caucasoid children gave evidence of marked own-race preference, an average of 85.5% of them choosing the white doll. The Negroid children showed some in-group preference but significantly less than that of the Caucasoids, with 56.8% selecting the brown doll. Compatible with these own-race preferences are the responses to the fifth request. Whereas 92% of the white subjects indicated that the brown doll *looks bad,* only 9% of the Negroid children selected the brown doll as looking bad, the other 91% making no meaningful response. The refusal of such a large percentage of Negroid children to select either doll as *bad* suggested to the authors an interpretation offered by Kardiner and Ovesey (1951) and Karon (1958) that Southern Negroids "accommodate,"

[2] At all age levels, the proportion of own-race correctly identified was greater for Caucasoids than for Negroids. From 3 to 7 years, the respective percentages of Caucasoids selecting own-race doll as more like themselves were: 52, 72, 88, 96, and 96; for Negroids the percentages were: 43, 33, 33, 53, and 85.

that is, refrain from expressing openly hostility toward the Caucasoid community. This explanation could be applied equally to the boys and girls of the Gregor-McPherson study since 89% of the former and 93% of the latter made no interpretable response to looks bad.

In responding to the combined requests of play with, like best, nice doll, and nice color, the Negroid girls were significantly less likely to select the brown doll than were the Negroid boys; in contrast, the percentages of Caucasoid boys and girls who chose a white doll were very close together, 84.5 and 86.7.

Greenwald and Oppenheim (1968) gave a variant of the dolls test to 39 Negroid and 36 Caucasoid children in integrated and nonintegrated nursery schools in New York City and New Rochelle. In addition to the traditional brown and white dolls, they included a mulatto doll. This doll was least frequently chosen by both Caucasoids and Negroids as one to play with and as a good doll or nice color but was designated as a *bad doll* by more than half of all the children.

Asher and Allen (1969) replicated the Clarks' research with 341 Negroid and Caucasoid children between the ages of 3 and 8 in New Jersey. Subjects were separated into middle and lower classes according to parental occupation, 167 and 175, respectively, but the number in each class was not reported by race. Brown and white puppets, rather than commercial dolls, were used.[3] Subjects were asked to choose a puppet to *play with, nice, bad,* and *nice color.* Approximately 27% of the Negroid subjects and 75% of the Caucasoids chose own-race puppets to the combined items of *play with, nice puppet,* and *nice color,* the difference, of course, being highly significant. Asher and Allen compared their findings on the Negroid subjects with those of the Clarks, the two groups of subjects being all Negroid, Northern, urban, of the same age range, and examined by a Negroid examiner. Using chi-square tests, on none of the four questions did the difference reach statistical significance.

Roland Crooks (1970) closely followed the procedure initiated by the Clarks in order to evaluate a preschool program on racial attitudes and self-respect in a poor and crowded interracial neighborhood of Halifax, Nova Scotia. Subjects in the experimental group had just completed one school year of an enriched preschool program (with equal numbers of Negroid and Caucasoid children) that emphasized racial differences and discussed these differences "in favorable terms." Stress was placed on the development of self-respect, especially in Negroid children. The Negroid and Caucasoid control group, of the same age range and similar family backgrounds, had no preschool training.

Crooks essentially duplicated the Clarks' eight requests. He found that

[3] The authors say, curiously, as do Harris and Braun (1971) that the puppets were placed in a *prone position* before the child. They obviously mean in a *supine position.*

nearly twice as many Caucasoids as Negroids chose their own-race dolls in response to the combined requests—play with, nice doll, and nice color ($t = 2.70$, $p = .01$). A trend appeared, although not reaching the .05 level of significance, for the Negroid experimental group to have *more* own-race choices and for the Caucasoid experimental group to have *fewer* than their respective controls. This tendency for both experimental groups to choose the brown doll more frequently than did their controls seems to have been due mainly to reactions to the *nice color* item. Other children, that is, Negroid and Caucasoid non-preschool controls, preferred a *white doll* as a *nice color*.

Hraba and Grant (1970) replicated the Clarks' work with Negroid and Caucasoid public school children from 4 to 8 years of age in Lincoln, Nebraska. They reported Caucasoid own-race preferences of 67% versus Negroid own-race preferences of 64.3%, the difference between them being insignificant. Relative to these findings, the authors note that a black-pride campaign had been conducted by local organizations within the Negroid Lincoln communities that may have caused an increase in own-race preferences among their Negroid subjects. Later, Hraba (1972), in discussing this work, observed that some subjects may have been especially diplomatic when interviewed by a member of the other race. Analyzing their results, he concluded that the higher the ethnocentrism of the Negroid subjects, the *less likely* were they to have been interviewed by a Caucasoid examiner.

We have calculated the mean percentages of own-race choices from the 10 other studies reviewed in this section. Of the 975 Caucasoid and 1234 Negroid 3–8-year-old children examined, 75% of the former and 47% of the latter preferred own-race dolls.

Durrett and Davy (1970) used white and brown dolls with 55 Negroid and Caucasoid children, as well as 30 Mexican-Americans, in public interracial prekindergarten schools in San Jose, California. Correct answers to *looks like me* were given by 76% of the Negroid and by 97% of the Caucasoid children. Forty-eight % of the Negroids and 83% of the Caucasoids preferred own-race dolls as playmates, the difference being significant at the .01 level of confidence. In both racial groups, sex differences proved to be insignificant.

Harris and Braun (1971) tested 60 Negroid children in the second and third grades of two Pennsylvania integrated schools with puppets placed in "prone position" before the subjects. Thirty were drawn from a suburban school and 30 from the inner city. These 7- and 8-year-olds were judged to be of at least normal intelligence. Nearly 69% of the Negroid subjects chose their own-race doll to the combined requests of play with, nice puppet, and nice color. This number compares favorably to the 54.1% of own-race choices of other Negroid subjects of similar age range reported in five dolls studies. Harris and Braun indicated that no significant differences were found between socioeconomic classes or between the sexes.

The Piers–Harris Children's Self-Concept Scale (Piers, 1969), an 80-item questionnaire, was read aloud to these same Negroid children by the senior author. The median of the Piers–Harris Scale for the subjects who had preferred the brown puppet in at least three of the four request items was higher than the median attained by those who had made two or fewer choices for the brown puppet (63.06 versus 55.5, $p < .01$). Since the self-concept medians for both classes were reported to be well above the norms established by Piers and Harris, and since the puppet test showed a preference for brown over white dolls, Harris and Braun concluded that their Negroid subjects had a viable and secure self-esteem and apparently valued the traits of their own group.

Fox and Jordan (1973) describe in a monograph a carefully designed and executed research program for studying self-esteem of young Negroid and Caucasoid children. Three hundred sixty Negroid and 360 Caucasoid children, aged 5–7 years in 10 public schools of New York City, served as subjects. Each racial group comprised equal numbers of 5-, 6-, and 7-year-old boys and girls; half of each race, age group, and sex attended integrated and half-segregated schools. A segregated school was defined as one having at least 85% of its enrollment of a particular race. All integrated schools contained, by definition, at least 40% of each of the two races. Estimate of the socioeconomic levels of the schools was attempted. The Federal Elementary and Secondary Education Act (E.S.E.A.) guidelines provided by the New York City Board of Education revealed that two of the integrated schools were in designated poverty areas and two were not. Two of the segregated Negroid and one of the segregated Caucasoid schools were in E.S.E.A. designated poverty areas.

Each subject was randomly selected from the school record cards of children of a given race and sex who were born in New York City, English speaking, living with both natural parents, and at the time of interviewing within 3 months of a particular age level. Each child was tested individually by an examiner of his own race. Two of the five Caucasoid examiners were male, while the Negroid examiners were all female.

Eight flexible brown and white dolls, 15 inches high with identical features, half of them dressed as boys and half as girls, were utilized. The dolls were placed in alternating racial group sequence, subjects being instructed to point to the doll that answered the question best. The first five question-requests followed the Clarks' preference system: play with, nice, bad, like best, and nice color. The small number of children who did not respond to any question was not included in the sample.

As will be observed in the table, 58% of the Negroid and 70.2% of the Caucasoid subjects selected an own-race doll to the combined requests of play with, nice doll, like best, and nice color. The difference proved to be significant at the .001 level.[4] Neither race showed significant sex differences in choice of favored doll. However, there is a suggestion of greater own-race preference on the part of Negroid girls and Caucasoid boys. In order to determine the effect of *age* on own-race preference, we have combined Fox and Jordan's data on their 6- and 7-year-olds (older subjects) and compared them with those obtained on the 5-year-olds (younger subjects). Within each race, a higher percentage of the older children preferred their own-race doll; the age-group differences, however, were small and insignificant.

Of interest is the relation between racial composition of the schools and the percentage of own-race doll choices. No observable difference was found between the integrated and segregated Negroids; however, the segregated Caucasoids own-race choices (play with, nice doll, like best, and nice color combined) proved to be significantly higher than those of the integrated Caucasoids ($t = 2.18$, $p < .05$). Negroid interviewers were instructed to classify the skin color of each Negroid subject during the testing period as *light, medium,* or *dark*. Of the 360 subjects, 81 were deemed light, 137 medium, and 142 dark. Some 43.4% of the light, 58.5% of the medium, and 62.8% of the dark children preferred their own-race doll to the four combined requests, the only significant difference being between own-race preferences of the light and dark groups.[5]

PHOTOGRAPHS, SKETCHES AND DRAWINGS, AND A COLORING TEST

We have examined eight research reports that describe the use of photographs, drawings, or sketches of Negroid and Caucasoid children in the study of racial preferences. Summaries of these have been included in Table 8.2, along with one in which a coloring test was utilized. As with the dolls tests, these nine studies have been scored in terms of the percentage of each group preferring own race, as photographed, as drawn or sketched, or in terms of color of crayon selected for the skin.

Evelyn Helgerson (1943) attempted to determine the relative importance of race, sex, and facial expression on Negroid and Caucasoid children's choice of playmate. The 108 Caucasoid and 27 Negroid subjects, $2\frac{1}{2}$–$6\frac{1}{2}$ years of age, attending Minneapolis nursery schools or kin-

[4] To *bad doll,* 34% of the Negroids and 29% of the Caucasoids chose own-race doll; for this item, 17% of the Negroids and 9% of the Caucasoids did not choose any doll. For the four positive or favorable items, from 1 to 3% of all subjects made no recognizable choice.

[5] $t = 2.12$, $p < .05$, calculated by reviewer from authors' Table V.

TABLE 8.2

Photographs, Drawings, Sketches

Author and date	Location	Instrument	Subjects			
			N			Age
Helgerson, E. (1943)	Minneapolis, Minn.	Two sets of photographs[a]		Negroid	Caucasoid	
			A		24	$2\frac{1}{2}$–$4\frac{1}{2}$
			B	10		3 –$6\frac{1}{2}$
			C		28	3 –5
			D	17	56	3 –$6\frac{1}{2}$
				27	108	
Morland, J. K. (1962)	Lynchburg, Va.	4 black-white 8 × 10 pictures of Negroid and Caucasoid children of same age of subjects.		Negroid	Caucasoid	
				42	74	3
				44	110	4
				34	89	5
				6	8	6
				126	281	

Subjects: Method of selection	Results	Comments of author
A-High ses Caucasoid subjects from nursery school Univ Minn; B-Black of low economic status. C-Nursery schools; Caucasoid of low economic status. D-Nursery school and kindergarten integrated and low economic status.	Choice of own-race picture as playmate by: (see tables below)	In segregated settings, Caucasoids indicated no preference, but Negroids slightly greater preference for own race. In integrated schools, both racial groups chose Caucasoid playmates more frequently.
Two day nurseries and 3 nursery schools for Caucasoids; 1 nursery school for Negroids. All present on testing days interviewed except for a few refusals.	Racial preference according to age (see table below)	Preferring the members of one race does not necessarily mean that members of the other race will be unacceptable.

Choice of own-race picture as playmate by:

Race of subjects[b]

Negroid	Caucasoid	t	p
47.2%	57.1%	.93	—

Racial composition of schools[c]

	Segregated	Integrated	t	p
Negroid	52.5%	44.1%	—	—
Caucasoid	50.5%	63.2%	1.4	—

Racial preference according to age

Subjects	Own race	Other race	?[d]
		Negroid	
3	11.9%	64.3%	23.8%
4	25.0%	59.1%	15.9%
5	14.7%	50.0%	35.3%
Total	17.5%	57.9%	24.6%
		Caucasoid	
3	59.5%	10.8%	29.7%
4	78.2%	10.9%	10.9%
5	75.3%	7.9%	16.9%
Total	72.6%	10.0%	17.4%[e]

TABLE 8.2 *(Continued)*

Author and date	Location	Instrument	Subjects			
				N		Age
				Negroid	Caucasoids	
Morland, J. K. (1966)	Lynchburg, Va., and Boston, Mass.	A different set of 6 black–white 8 × 10 pictures of Negroids and Caucasoids[f]	Lynchburg	41	41	3–6
			Boston	41	41	3–6

<table>
<tr><th>Subjects
Method of selection</th><th>Results</th><th>Comments of author</th></tr>
<tr>
<td>Lynchburg subjects from segregated nursery schools and day-care centers. Boston subjects mostly from integrated playground groups. Subjects matched by age and by sex placed in each of 4 groups: 6 3-year-olds, 16 4-year-olds, 7 5-year-olds, and 12 6-year-olds.</td>
<td>
Rather play with?
<table>
<tr><th>Subjects</th><th>Own race</th><th>Other race[g]
or ?</th></tr>
<tr><td>Northern Negroid</td><td>46.3%</td><td>53.7%</td></tr>
<tr><td>Southern Negroid</td><td>22.0%</td><td>78.0%</td></tr>
<tr><td>Northern Caucasoid</td><td>68.3%</td><td>31.7%</td></tr>
<tr><td>Southern Caucasoid</td><td>80.5%</td><td>19.5%</td></tr>
</table>
Rather be?
<table>
<tr><th>Subjects</th><th>Own race</th><th>Other race
or ?</th></tr>
<tr><td>Northern Negroid</td><td>56.1%</td><td>43.9%</td></tr>
<tr><td>Southern Negroid</td><td>43.9%</td><td>56.1%</td></tr>
<tr><td>Northern Caucasoid</td><td>78.0%</td><td>22.0%</td></tr>
<tr><td>Southern Caucasoid</td><td>78.0%</td><td>22.0%</td></tr>
</table>
</td>
<td>Scientific studies indicate that race in and of itself is not related to the intelligence, character, or creativity of the individual.</td>
</tr>
</table>

TABLE 8.2 *(Continued)*

Author and date	Location	Instrument	Subjects N			Age
				Negroid	Caucasoid	
Rohrer, G. K. (1972)	Several Southern California communities	Color photographs of 3 girls and 3 boys representative of the 3 racial/ethnic groups.	Boys	27	32	
			Girls	24	27	
				51	59	4–5
			School	Negroid	Caucasoid	
Radke, M., Trager, H. G., and Davis, H. (1949)	Philadelphia, Pa.	Two sketches of children at play from Social Episodes Test	1	51		
			2		45	
			3	6	15	
			4	9	41	
			5		50	
			6	29	4	
				95	155	5–8

Subjects: Method of selection	Results	Comments of author
170 Negroid, Mexican–American, and Caucasoid children in 16 Head Start classes in poverty areas, 8 segregated, 8 integrated. All present interviewed.	Preference for own race[h] (see tables below)	Present study did not confirm the Caucasoid-over-Negroid preference found elsewhere. Large Mexican–American population in Southern California may explain this.

Preference for own race[h] — Photograph

Subjects: Negroid	Subjects: Caucasoid	t	p
29.4%	44%	1.59	—

Subjects	Boys	Girls	t	p
Negroid	37%	21%	1.25	—
Caucasoid	50%	37%	1.00	—

Subjects	Integrated	Segregated	t	p
Negroid	41%	21%	1.55	—
Caucasoid	39%	52.2%	.99	—

Subjects: Method of selection	Results	Comments of author
Six PSs in sch district with varied ethnic, racial, & religious composition. Every third name in kindergarten, first, and second grades taken.	Perceived desirability of being Negroid or Caucasoid in terms of *yes* to questions[j] (see table below)	Self-hatred most extensive among Negroid children. Seldom refer to selves as Negroid; 33% (as compared to 8% of Caucasoids) never mention race.

Perceived desirability of being Negroid or Caucasoid in terms of *yes* to questions[j]

Subjects	1	2	3	4
Negroid				
Kindergarten	87%	70%	96%	61%
1	73%	79%	91%	70%
2	63%	74%	84%	58%
Caucasoid				
Kindergarten	46%	73%	78%	34%
1	24%	87%	91%	33%
2	25%	79%	94%	25%

TABLE 8.2 *(Continued)*

Author and date	Location	Instrument	Subjects N		Age
			Negroid	Caucasoid	
Stevenson, H. W., and Stewart, E. C. (1958)	Austin, Tex.	Seven drawings of Negroid and Caucasoid children at play	23	25	3
			13	25	4
			22	25	5
			17	25	6
			20	25	7
			95	125	

Author and date	Location	Instrument		Negroid	Caucasoid
Koslin, S. C., Amarel, M., and Ames, N. (1970)	Eastern city	Three pairs of sketches depicting classroom scenes, alike except for racial composition.	Segregated	35	48
			Integrated	17	19
				52	67

<table>
<tr><th>Subjects
Method of selection</th><th>Results</th><th>Comments of author</th></tr>
<tr>
<td>Ss in 10 private nursery and elementary schools. All in the nursery schools age 3 and over included. In elementary schools subjects were chosen at random, approximately evenly divided according to sex.</td>
<td>
Own-race choices according to age[k]
<table>
<tr><th></th><th>Card 4</th><th></th><th>Card 5</th><th></th></tr>
<tr><th>Age</th><th>Negroid</th><th>Caucasoid</th><th>Negroid</th><th>Caucasoid</th></tr>
<tr><td>3</td><td>70%</td><td>44%</td><td>49%</td><td>46%</td></tr>
<tr><td>4</td><td>33%</td><td>56%</td><td>52%</td><td>45%</td></tr>
<tr><td>5</td><td>23%</td><td>92%</td><td>48%</td><td>65%</td></tr>
<tr><td>6</td><td>35%</td><td>88%</td><td>41%</td><td>66%</td></tr>
<tr><td>7</td><td>40%</td><td>88%</td><td>31%</td><td>85%</td></tr>
<tr><td>Total</td><td>41%</td><td>74%</td><td>44%</td><td>61%</td></tr>
</table>
Difference between total own-race choices significant: Card 4 $p < .001$; Card 5 $p < .02$.
</td>
<td>By ages of 4, 5, and 6 these subjects were responding in a manner that indicated use of stereotyped roles and awareness of racial differences.</td>
</tr>
<tr>
<td>First and second graders in 3 elementary schools in a middle-sized Eastern city, one all Negroid, one all Caucasoid, and one about equally divided. Sample included approximately equal numbers by race, sex, grade, and school.</td>
<td>
Choice of own-race picture to 8 questions, according to:
<table>
<tr><th colspan="4">Race of subjects</th></tr>
<tr><th>Negroid</th><th>Caucasoid</th><th>t</th><th>p</th></tr>
<tr><td>48.7%</td><td>80.1%</td><td>3.59</td><td><.01</td></tr>
</table>
<table>
<tr><th colspan="5">Racial composition of school</th></tr>
<tr><th></th><th>Segregated</th><th>Integrated</th><th>t</th><th>p</th></tr>
<tr><td>Negroid</td><td>48.2%</td><td>49.9%</td><td>—</td><td>—</td></tr>
<tr><td>Caucasoid</td><td>81.4%</td><td>77.1%</td><td>—</td><td>—</td></tr>
</table>
Choice of own-race picture to "nicest children" question, according to race of subject:
<table>
<tr><th>Negroid</th><th>Caucasoid</th><th>t</th><th>p</th></tr>
<tr><td>44%</td><td>91%</td><td>5.60</td><td><.001</td></tr>
</table>
</td>
<td>Overwhelming tendency for Caucasoid subjects to choose Caucasoid classrooms. Negroids showed no such consistent attitudes, some preferring Caucasoid classes, some Negroid classes, and some no consistent preference.</td>
</tr>
</table>

TABLE 8.2 *(Continued)*

Author and date	Location	Instrument	Subjects				
			N				Age
Kircher, M., and Furby, L. (1971)	Calif.	64 colored drawings of children's faces varying in eye color, hair type (form), hair color, and skin color.[f]	Negroid	Caucasoid			
			15	15			3–5
Clark, K. B., and Clark, M. P. (1950)[g]	Hot Springs, Pine Bluff, and Little Rock, Ark., and Springfield, Mass.	Coloring test	Negroid				
						Age	
			Ark.	*N*	5	6	7
			Light	4	—	3	1
			Medium	36	3	14	19
			Dark	26	3	13	10
			Mass.	*N*		Age	
			Light	25	11	7	7
			Medium	46	15	18	13
			Dark	23	6	8	9
			Total	160	38	63	59

[a] Each set contained a photograph of: a laughing Negroid boy, a laughing Caucasoid boy, a serious Negroid boy, a serious Caucasoid boy, a laughing Negroid girl, a laughing Caucasoid girl, a serious Negroid girl, and a serious Caucasoid girl.

[b] Calculations by reviewer.

[c] Helgerson classified subjects also according to *younger* and *older* but did not define these groups by age nor report the number of Negroids and Caucasoids comprising them.

[d] Preference not clear. All responses categorized by examiner according to subject's most frequent choice.

[e] All *totals* include 6-year-olds.

[f] Two photos of adults excluded by reviewer.

[g] Morland combines Preference not Clear with Preference for Other Race.

[h] Reviewer has omitted all choices of the Mexican–American subjects as well as self-identification data; 57% of Negroid and 73% of the Caucasoid subjects identified themselves correctly. All *t*s calculated by reviewer from author's Table 6.

[i] The two sketches identified throughout monograph as Race Barrier and Non-Barrier pictures.

[j] The four questions were asked of each subject when shown the Race Barrier picture (first session) and a month later when shown the Non-Barrier picture: (1) Is the colored

<table>
<tr><th>Subjects
Method of selection</th><th>Results</th><th>Comments of author</th></tr>
<tr><td>Subjects in a racially mixed preschool; 5 of each race at each age. (No other information).</td><td>
<table>
<tr><th colspan="5">Preferences for Negroid features[m]</th></tr>
<tr><th>Subjects</th><th>Eye color</th><th>Hair color</th><th>Hair type</th><th>Skin color</th></tr>
<tr><td>Negroid</td><td>45.8%</td><td>54.2%</td><td>34.2%*</td><td>42.5%</td></tr>
<tr><td>Caucasoid</td><td>45.8%</td><td>51.7%</td><td>34.2%*</td><td>41.7%</td></tr>
<tr><td>Total</td><td>45.8%</td><td>52.9%</td><td>34.2%*</td><td>42.1%*</td></tr>
</table>
* Significant difference from chance (50%), $p < .01$.
</td><td>Development of preferences seems to be complex and affected by prevailing cultural values.</td></tr>
<tr><td>Southern subjects in segregated nursery schools and public schools. Northern in racially mixed nursery and public schools. 160 subjects used whose coloring responses "were stable enough to analyze."</td><td>
<table>
<tr><th colspan="5">Crayon choices of 6–7 year-old medium and dark subjects[o]</th></tr>
<tr><th></th><th colspan="4">Crayon</th></tr>
<tr><th></th><th colspan="2">Brown</th><th colspan="2">Other</th></tr>
<tr><th>Subjects</th><th>N</th><th>%</th><th>N</th><th>%</th></tr>
<tr><td>Ark.</td><td>41</td><td>73</td><td>15</td><td>27</td></tr>
<tr><td>Mass.</td><td>17</td><td>35</td><td>31</td><td>65</td></tr>
</table>
</td><td>The tendency to reject the brown color expressed not only in coloring their preferences white but also in irrelevant or escapist responses.</td></tr>
</table>

boy glad he is colored? (2) Would he sometimes like to be white? (3) Is the white boy glad he is white? and (4) Would he sometimes like to be colored? Each question was followed by *Why?*

[k] *Card 4*. A boy in foreground (seen from back and racially ambiguous); in background are a Caucasoid boy and a Negroid boy. Subject was told: This little boy is about ready to go home from school and wants a friend to go with him. Which boy do you think he will choose to go home with him?

Card 5. A girl is seated in left foreground (seem from back and racially ambiguous) with a birthday cake. Behind and to the rear are three Negroid and three Caucasoid girls, alternating according to race. Subject is told: This is Ann. See her birthday cake? Her mother said that she could invite three children to her birthday party. Which three girls do you think that she is going to invite?

[l] Eye color (blue and brown), hair color (brown and black), hair type (straight and curly), skin color (white and dark-brown).

[m] Brown eyes, black hair, curly hair, dark-brown skin.

[n] Or see Grossack (1963, pp. 53–63), for reprint of same.

[o] Difference between Northern and Southern subjects choosing brown crayon significant at .01 level. Authors include the few choices of the black crayon with the brown.

dergartens, were shown individually a booklet of photographs—presented in pairs—and asked to choose a playmate. Two sets of eight photographs were paired in three different ways: *(1)* sex as a variable, with race and facial expression as constants, *(2)* race as the variable, with sex and facial expression as constants, and *(3)* facial expression as the variable, with sex and race as constants. These two sets of pictures were shown in the first interview; weeks later, the procedure was repeated, the photographs having been transposed from left to right and vice versa.

It is evident from Table 8.2 that the Caucasoid more often than the Negroid children selected a picture of a child of *their own race* for a playmate and that the children in the segregated schools showed a negligible tendency to favor their own race, whether Negroid or Caucasoid, whereas subjects in the integrated school, Negroid and Caucasoid, tended to choose the pictured Caucasoids. The results of this research are not without ambiguity since the integrated schools making up *D* and the all-Negroid school *B* included 6-year-olds; whereas the Caucasoid school *A* included subjects below 3 years and none as old as 5 (Table 8.2). None of the differences noted proved to be significant at the .05 level of confidence.

Kenneth Morland (1962) endeavored to determine whether Negroid and Caucasoid nursery school children in a racially segregated community were willing to accept members of the other race as playmates and whether they had a preference for playmates of one race or the other. Accordingly, Morland and a number of Caucasoid senior sociology majors (women) examined 126 Negroid and 281 Caucasoid children between the ages of 3 and 6 years. The interviewer questioned each child tween the ages of 3 and 6 years. The interviewer questioned each child alone, using four photographs particularly selected for the project. They consisted of: *(1)* four Caucasoid children, two boys and two girls, sitting at a table looking at picture books, *(2)* four Negroid children, two boys and two girls at a table, also looking at picture books, *(3)* a Negroid boy and five Caucasoid boys and girls eating together at a table, and *(4)* a Caucasoid girl and a Negroid girl at play, with five Negroid boys and girls in the background. Every subject was given three opportunities to express a preference for Caucasoid or Negroid children in the pictures, the examiner never mentioning the words *race* or *color*. All responses were categorized by the examiner as: *prefer Negro, prefer white,* or *preference not clear,* according to the subject's most frequent choice.

The majority of both racial groups (about 73% of the Caucasoids and 58% of the Negroids) preferred to play with Caucasoid children as pictured. When grouped according to *age*, approximately 60% of the 3-year-old, 78% of the 4-year-old, and 75% of the 5-year-old Caucasoids chose their racial counterparts, whereas only about 12% of the 3-year-old, 25% of the 4-year-old, and 15% of the 5-year-old Negroids preferred *their race* as represented. These comparisons indicate rejection of the black children pictured as playmates, by Negroid as well as Caucasoid subjects. However, the interviewers were all Caucasoid and above the socioeconomic level of many of the children examined. Moreover, many of the Negroid children apparently did not understand what was required of them since 35% of the 5-year-olds could only be classified "preference not clear."

In a second study, Morland (1966) used other photographs to compare the responses of Negroid and Caucasoid children from Lynchburg nursery schools and day-care centers with those of a Boston sample. The Negroid and Caucasoid Boston subjects were examined during the summer of 1961, also by Caucasoid college women under the supervision of the author, and were drawn mainly from public interracial playgrounds. The primary goal of this work was to compare differences in racial recognition ability, racial preference, and racial self-identification of the Negroid and Caucasoid children living under conditions of segregation (Southern city) and less segregation (Northern city) where there was official disapproval of racial discrimination. Morland was able to match Northern Negroids, Southern Negroids, Northern Caucasoids, and Southern Caucasoids for age and sex, with 41 in each group. The groups were not matched for intelligence or socioeconomic level. The Caucasoid examiners showed pictures previously judged to be readily identifiable as to race and to be reasonably comparable in expression and dress. They comprised groups of Caucasoid, Negroid and Caucasoid, and Negroid children. Without mentioning race or color, the examiner pointed and asked: *Would you rather play with this child (these children) or with that one (those)?* As in the 1962 study, preferences were categorized according to the subject's most frequent choice as: *preference for own race, preference for other race,* or *preference not clear.* Morland (1966) combined *preference not clear* and *preference for other race,* but in referring to his table, Morland treats the combination as if it were simply preference for other race. (Referring to his Table 3, p. 26, Morland said that it revealed that a majority of the subjects in each of the four groups *preferred Caucasoids.* Actually, according to the table, 53.7% of the Northern Negroids and 78% of the Southern Negroids *preferred other race* [Caucasoid] *or preference not clear.*)

Another question, *Which child would you rather be?*, was designed to measure racial and self-identification, but as it seems to belong as appropriately to *racial preference,* we have included it in Table 8.2. This question was only asked once of each child—with the fifth picture if the subject was a girl or with the sixth if the subject was a boy. Seventy-eight percent of the Caucasoid subjects of each city preferred to be the Caucasoid child pictured, whereas more than half of the Northern Negroids and somewhat less than half of the Southern Negroids preferred to be the *Negroid child.*

To both *play with* and *rather be,* the differences between the combined Caucasoid groups and the combined Negroid groups were highly significant, the Caucasoids more often preferring the pictured child of their own race.[6] To the "play with" item, the Northern Negroids preferred their own-race photographs significantly more often than did the Southern Negroids. However, to "rather be," the difference between their preferences for own-race photographs was not significant.

Georgia Rohrer (1972) studied the influence of racial and ethnic group membership, sex, and segregation on racial–ethnic identification and preference among preschool children. Her measuring instrument consisted of three color photographs of young girls (a Negroid, a Mexican–American, a Caucasoid) to be shown together to the female subjects and three color photographs of the same size of young boys (a Negroid, a Mexican–American, a Caucasoid) to be shown together to the male subjects. These six photographs had been previously selected from a large supply of school pictures taken by a professional photography company. After several adult judges helped the author eliminate pictures of children with unpleasant expressions, extreme hair styles, unusual dress (and presumably all who looked older than 6 or 7 years) they selected 30 photographs, including five Negroid girls, five Negroid boys, five Mexican–American girls, and so on. From each of these five, one photograph was finally selected for the research by determining the picture that was most often preferred by samples of kindergarten and first-grade pupils in racially and ethnically mixed classrooms. A sample of 170 preschool children attending eight segregated and eight integrated Head Start classes in Southern California served as subjects. All were at least 4-years-old at the beginning of the school year, resided in designated poverty areas, and were assisted by government Aid to Families with Dependent Children. The children were approximately evenly divided into the three racial–ethnic groups and according to sex; 88 were labeled *integrated* and 82 *segregated.*

Rohrer, fluent in both Spanish and English, interviewed all children individually. Displaying the three photographs of the child's own sex, she asked: *Which one do you like best? Which one would you rather play with? Which one would you rather eat with?* Finding a high degree of

[6] The respective *t*s being 5.17 and 3.73, with corresponding *p*s of $<.0001$ and $<.001$, as calculated by reviewer.

response similarity to these three items by chi-square analysis, Rohrer restricted her analysis to the first item: *like best*. She found: *(1)* that the three racial–ethnic groups differed significantly in their own-group preferences, the Negroid own-group choices of 29% being the lowest and the Mexican–American choices of 53% being the highest. *The self-esteem of the Negroids and Caucasoids did not differ significantly from one another; (2)* integration did not affect own-group preferences in any predictable way; *(3)* own-group preference varied with the sex of the subject. Mexican–American girls and Negroid boys showed significantly higher own-group preference than the other groups.

The appearance of *high ethnocentrism* among the Mexican–American girls (69%) as against the slightly more than *chance own-group preference* among the Mexican–American boys (38%) and the *low ethnocentrism* obtained for the Negroid girls (21%) in contrast to the *moderate own-group preference* reported for the Negroid boys (35%) convinced us that we should consider the preferences for the Mexican–American photographs by race and sex. All girls combined preferred the photograph of the Mexican–American girl (59%) significantly more often than the total group of boys chose the Mexican–American boy (27.6%). Anticipating that a solution to this puzzle may lie in the specific photographs, we studied them and herewith offer a few pertinent observations. The Mexican–American girl appears to be the most outgoing of the six; she is laughing and seems ready to run, dance, or take part in any play another child might suggest; her hair is long and luxuriant, her bangs tousled, suggesting a carefree attitude. The Caucasoid and Negroid girls are pleasant looking, but the former looks shy; whereas the latter, the more serious of the two, has her hair done high on her head (not an Afro), giving her the appearance of a girl more sedate and older than the others. The three boys are comparable in appearance, all pleasing. However, the Negroid and Caucasoid boys look more friendly; the former is laughing, the latter smiling; whereas the Mexican–American boy is serious, face upturned and eyes wide open. We would like to see this provocative but inconclusive study replicated with other sets of photographs selected in much the same meticulous way.

Radke, Trager, and Davis (1949) have described in a relatively complex study the development of certain racial and religious attitudes among young schoolchildren. These attitudes appear to have culminated in self-rejection among the minority groups, the Negroid in particular giving evidence of self-hatred. The conclusions of the authors have been frequently cited by researchers in the area of self-esteem.

The subjects included 155 Caucasoids (35 Jewish, 58 Catholic, 61 Protestant) and 95 Negroids (Protestant), mainly of lower-income levels, with 2% of fathers in professions, 10% in small businesses, office and sales, 49% in skilled trades and factory work, and 16% in service trades. All subjects were attending public schools in one Philadelphia school district, two of the schools being predominantly Negroid, two mixed, and two predominantly Caucasoid. Each child was examined by a member of his own race, trained to accept uncritically and unemotionally any response of the subject. Test pictures were presented, the first being the *Race Barrier*, showing six Caucasoid children playing ball, with a Negroid boy watching. During the second session, four or five pictures were presented, the first always being the *Race Non-Barrier*, a city school playground where a Caucasoid girl, two Caucasoid boys, and one Negroid boy are playing ball together.

Presenting the Race Barrier scene, the examiner commented on it and questioned the subject in much the following manner: *Tell me about this little boy.* . . . He isn't playing. *Why isn't he playing?* . . . *Why don't they ask him to play?* . . . (if Negroid–Caucasoid has not been clearly specified in preceding answers) *This is a colored boy. These are white children* . . . *Is this little boy glad he is colored? Why?* . . . *Would he sometimes want to be a white boy? Why?* . . . *Is she* (the subject) *glad that she is a white girl? Why?* . . . *Would she sometimes want to be a colored girl?* . . .

Presenting the *Race Non-Barrier* scene, the examiner proceeded in this way: *Tell me about this picture.* . . . *These children are all playing together. This little boy* (pointing) *is colored. These aren't colored. Is this little boy glad he is colored?* . . . *Why?* . . .[7] Interspersed among selected comments and answers of the children to these and similar questions are the principal quantitative findings reported by the authors: *(1)* Seventy-four percent of the Negroids responded affirmatively to the questions *Is the colored boy glad he is colored?* and *Would he sometimes like to be white?* Eighty-eight percent of the Caucasoids responded positively to the question *Is the white boy glad he is white?* with 31% indicating that the Caucasoid boy would sometimes want to be colored.[8] *(2)* Negroid subjects were more likely to indicate that the colored boy is glad he is colored in *kindergarten* (87%) than in the *second grade* (63%), the corresponding figures for the Caucasoid subjects relative to the pictured Caucasoid boy being reversed—kindergarten (78%) and second grade (94%). *(3)* Thirty-three % of the Negroid subjects in contrast to 8%

[7] These questions (with answers) are given by authors as examples of their interviewing. However, the four questions listed in Footnote j, our Table 8.2, seem to have been addressed to all subjects. The Radke *et al.* results in Table 8.2 were adapted by reviewer from authors' Figure 4.

[8] These percentages only approximately accurate. Authors do not give number of cases. Sometimes the examiner directly addressed the subject with questions such as: *Are you glad that you are a white girl?*

of the Caucasoids did not at any time mention race, the authors attributing this difference to inhibition and avoidance rather than to lack of interest in or in awareness of race. *(4)* When asked to identify with the colored child and the white child in the pictures—and to tell why each wants to be or does not want to be colored or white—meaningless or irrelevant responses persisted among the Negroid children, 13% in kindergarten, 18% in first grade, and 13% in second grade. In contrast, the number of Caucasoid subjects giving such responses showed a sharp decrease with age, 20% in kindergarten, 13% in first grade, and none in second grade.

The majority of subjects, Negroid and Caucasoid, thought the child representing their own race was *glad* he was of that race. At the kindergarten level, the data indicate greater Negroid than Caucasoid ethnic identification, but in first and second grades the Caucasoids identified with a child of their race more than the Negroids with one of theirs. However, at *no grade level for either racial sample did the majority of the children report that the child of his own race was not glad he was of that race.* Depending in large part, presumably, upon the subjects' chronicled answers and comments in the interviews rather than tabulated data, the authors concluded that the Negroid children feel insecure, show ambivalence toward their own group, and develop self-hatred at an early age.

I cannot find sufficient support for these conclusions; on the contrary, the tabulated data would seem to have warranted a much more restrained statement. We will mention three areas in which specific criticisms seem appropriate; *(1) Experimental design:* Half the time (Barrier Picture) it is always the Negroid boy who is apart from the group of Caucasoids. In the other half (Non-Barrier Picture), only one of the four children playing is Negroid, the others being Caucasoids. In each of the two drawings pictured to bring out racial attitudes and feelings, the Negroid child is always in a minority of one and never a girl. *(2) Treatment of data;* The authors usually tabulated their data in percentages, omitting numbers of cases, thereby reducing the significance of the material presented. We are told that there were relatively more *Caucasoid* girls among the subjects than Negroid girls but not the numbers of each or whether the difference between the sex ratios varied with grade. *(3) Ambiguities:* Rather than drawing up a specific list of questions to be asked of each child in a prescribed order, the questions given were to be *used as a guide*, the examiner having the option of varying them and probing more or less deeply as the situation warranted. Sometimes the wording of the questions varied at critical points so that the reader is not certain just what is being tabulated. For example, indicating the Negroid boy pictured, the examiner asks: *Would he sometimes like to be white?* or *Would he sometimes want to be something else?* Similarly, referring to one of the Caucasoid boys pictured, she asks: *Would he sometimes like to be colored?* (For verification, see pp. 349, 350, 355, 377, and 424 of authors' monograph.) The word *sometimes* means *now and then, not always,* or *occasionally.* However, in the chapter on "Children's Social Perceptions of Negro and White," the authors typically omit the word *sometimes,* reporting instead: *The reasons for wanting to be white . . . The choice between remaining Negro or becoming white . . . Would the Negro boy like to be white? . . .* the questions and answers implying the presence of a relatively permanent and steadfast desire. (See pp. 377, 378, 380, 381, and 382 of monograph.)

In an effort to measure racial identification, preferences, and attitudes, Stevenson and Stewart (1958) constructed four tests for children. One of these—Incomplete Stories—consists of seven cards depicting different play situations, with one or two central figures in the foreground drawn with backs to viewer and ambiguous as to race. The secondary figures represent Negroid and Caucasoid children who are never grouped by race. In my opinion, it is only in reference to Cards 4 and 5 (Incomplete Stories 4 and 5) that the subject is required to make a preference for a Negroid or Caucasoid pictured child. The subjects cooperating in this research included 95 Negroid and 125 Caucasoid children, ages 3–7, from segregated private nursery and elementary schools in Austin, Texas. All children were examined individually, the Negroids by a Negroid examiner and the Caucasoids by a Caucasoid examiner.

The Negroid children chose a pictured member of their racial group significantly less often than the Caucasoids chose one of theirs in response to the questions *boy to go home with him* and *girls to invite to birthday party.* When separated by age, an insignificant *decrease* in the Negroid children's own-group preferences occurred from the 3–5 to the 6–7 level, whereas a significant *increase* in own-race choices occurred among the Caucasoid subjects.

Koslin, Amarel, and Ames (1970) studied the racial preferences of 119 Negroid and Caucasoid first and second graders in a middle-sized Eastern city from a Negroid, a Caucasoid and an integrated school. The children were individually shown successive pairs of sketches revealing three classroom scenes; one set included 10 Negroids and 2 Caucasoids, the other contained 10 Caucasoids and 2 Negroids. The pairs of sketches were rotated and nine questions were asked in random order of each subject. All children were retested after a 4-week interval.

Eightly percent of the Caucasoid children preferred the Caucasoid versions of the sketches, as opposed to about 49% of the Negroids who preferred the Negroid versions, the difference being significant at the .01 level of confidence. The Caucasoids showed the highest ethnocentrism when responding to the three questions relating to social desirability: *Which class do you think has the nicest children? Which class would you like to be in?* and *In which class do you think you would have the most friends?* with 91, 94, and 93%, respectively, choosing own-race version. The percentages of the Negroid children choosing own-race sketches for these items were 44, 48, and 58. In response to the five questions pertaining to academic achievement—*best work, neatest papers, learn most, read best, listen to teacher*—from 71–77% of the Caucasoids and from 42–54% of the Negroids preferred pictures in which their own race predominated.

An analysis of uncertainty was carried out for the Negroid subjects in an attempt to understand the distribution of their responses. Koslin *et al.* selected two items at random and pitted them against every other item in turn in a three-way contingency analysis. This showed that knowledge of what responses the subject had made to any two items considerably reduced the uncertainty concerning his other responses, suggesting that the black children were not responding in random fashion. The distribution of the blacks' scores could be accounted for by hypothesizing three distinct populations: a group (42%) that chose *white* classrooms two-thirds or more of the time; a group (35%) that chose *black* classrooms two-thirds or more of the time; and a group (23%) without consistent preferences. The data were transformed to adjust for the bimodality of the Negroid children's scores, and a four-way analysis of variance (race × grade × racial composition of school), adjusted for unequal cell sizes, was performed. The authors report a *highly significant main effect for race* ($F = 69$, $p < .001$) but no significant effects for grade, or sex, or racial composition of school and no significant interactions. (See Table 8.2 for our comparable findings on racial composition of school attended.)

Kircher and Furby (1971), in a brief report, describe an investigation of California children attending a racially mixed preschool to determine whether or not children between the ages of 3 and 5 showed preference for hair type and for eye, hair, or skin color. There were only five Negroid and five Caucasoid subjects at each age. Both Negroid and Caucasoid subjects chose the straight-haired child pictured to a significant degree, whereas preferences for Caucasoid skin were not significant at the .05 level for either sample. Eye and hair color choices did not differ significantly from chance expectancy. The authors point out the obvious fact that Caucasoid children have both blue and brown eyes and brown and black hair. It should be added that the authors give no information about the neighborhood in which the children lived, their socioeconomic status, their IQs, the method of selection, or the race of the interviewer.

Kenneth and Mamie Clark (1950) made use of a coloring device to study racial identification and preference in Negroid children. This relatively simple and easily administered test consists of a sheet of paper with printed outline drawings of a leaf, an apple, an orange, a mouse, a boy, and a girl. A box of crayons with "the usual assortment" was given the child. The subject was asked to color the first four items to determine whether or not he had a stable color concept. If the child "passed," he was told: *Color this little boy (girl) the color that you are* and then *the color you like little girls (or boys) to be*. Responses were categorized as *reality* if reasonably related to his own skin color, *fantasy* if the color markedly differed, or *irrelevant or escape* if bizarre, such as red, green, or purple. Subjects were divided by region of the country, age in years, and skin color (light, medium, or dark).

Eighty-eight percent of the 160 subjects colored the boy or girl realistically, with these responses increasing from 80% at age 5 to 97% at age 7. No significant differences were found between Northern and Southern or between medium- and dark-skinned children. Seventy percent of the *Southern* Negroid subjects preferred to color the drawing *brown* as against 36% of the *Northern* subjects. (Black color is always included under the term brown in this study.) However, the two regional groups differed in skin color and age. Thus, only four Arkansas children (6%) were reported to have light skin versus 25 Massachusetts subjects (27%); only six of the former group (9%) were in the youngest age group in contrast to 32 (34%) of the latter.

Fortunately, we can remove age and skin-color differences by comparing Southern with Northern Negroid 6-year-olds of medium-brown or dark skin. Sixty-seven percent of the Southern versus 15% of the Northern subjects preferred to use the brown (or black) color. Correcting for age and skin color, therefore, did not remove the regional difference in color preference. Referring to the greater tendency of Northern subjects to prefer Caucasoid or to give an irrelevant or escapist response, the Clarks find "further indication of a greater degree of emotional conflict centering around racial or skin color preference in the Northern children. (Grossack, 1963, p. 59)."

SUMMARY AND DISCUSSION OF OWN-RACE PREFERENCES AS DETERMINED BY DOLLS STUDIES, PHOTOGRAPHS, DRAWINGS, SKETCHES

When dolls, including puppets and form boards, were utilized, the questions or requests typically involved the key words of *play with, nice doll, nice color,* and sometimes *like best* or *good doll*. Usually, a control request for a *bad doll* was included among the positive items, many children responding meaningfully and harmoniously to this negative item relative to their other responses. However, often a significantly large percentage of the Negroid subjects failed to respond, making "bad doll" untrustworthy as a measure of racial preference. When photographs, drawings, and sketches were employed, the critical preference question was usually "play with" or, less often, "rather be" and "like best." While there is considerable variation in the key questions asked, particularly relative to the pictures presented, it is clear that all investigators have indicated serious attempts to measure the relative amount of racial preference among their subjects.

The results of each of the 19 studies have been tabulated to show own-race preference for each sex, race group, and age group, wherever possible. These proportions, of course, were multiplied by the number of cases reported for the specific groups before they were combined with comparable findings from other investigations.

The most impressive and consistent finding is the large difference separating the racial samples on the combined tests. Over a period of approximately 30 years, some 1681 Negroid and 1661 Caucasoid children—without duplication of cases—between 3 and 8 years of age, living in 10 states and Nova Scotia, were interviewed with dolls, photographs, drawings, or sketches.[9] The mean percentage of Negroid children preferring their own-race doll or picture was 45.4, in contrast to 72.0% of the Caucasoids. This large difference is highly significant (t = 15.59).

Few researchers have reported their data according to sex. Among those who have done so, it seems that 45.2% of the 352 Negroid boys and 50.4% of the 357 Negroid girls preferred their own-race doll or picture (t = 1.4). Comparable percentages for the 348 Caucasoid boys and the 332 Caucasoid girls were 74.4 and 67.8 (t = 1.92). Of course, neither of these differences is significant.

Own-race preference for dolls and pictures increased significantly from the 3–5 to the 6–8 age group in both races.[10] The average own-race preference of the 600 younger Negroids was 39.2%, that of the 691 older Negroids, 55.4 (t = 5.83). The own-race preference of the 694 younger Caucasoids was 68.2, that of the 543 older Caucasoids, 78.6% (t = 4.11).

All children described as living in California, Massachusetts, Minnesota, Nebraska, New Jersey, New York, as well as an Eastern city and Nova Scotia have been identified as *Northern;* all others who were living in Arkansas, the Carolinas, Florida, Texas, Virginia, or the Deep South we have classified as *Southern.* The mean own-race preferences of the

[9] Throughout summaries, we have avoided duplication of subjects in the event they were examined on two measures. We avoided duplication of the Clarks' subjects by eliminating the results on the not completely satisfactory coloring test; in two other duplications we have simply averaged the two percentages of own-race preference. See Radke and Trager (dolls) and Radke, Trager, and Davis (pictures), and the Stevenson and Stewart (dolls and pictures).

Goodman's research has been omitted from the summary since her figures are not exact when applicable to a specific racial group.

[10] In a few instances, subjects were older than 7 years. Where grade rather than age was recorded, subjects in Grades 1–3 have been grouped by reviewer with the 6–8-year-olds; kindergarten and nursery school subjects with the 3–5-year-olds.

1131 Northern Caucasoids and the 530 Southern Caucasoids were markedly similar, that is 71.2 and 73.3%, respectively, the difference being insignificant. The mean own-race choice of the 1193 Northern Negroid subjects proved to be 49.4%, that of the 488 Southern Negroids, 37.8, the difference being highly significant ($t = 4.33$, $p < .001$). It seemed possible that the North–South difference among the Negroid children might be due to age differences, the Northern subjects being older. To control the age variable, we matched 488 Southern Negroids with 491 Northern Negroids of the same age ranges (3–5 or 6–8). The own-race preference of the Southern Negroid subjects remained unchanged at 37.8%, while that of the matched *Northern* Negroids changed slightly, becoming 48.7 instead of 49.4%. The resulting difference between the Northern and Southern Negroid children matched for age range became 10.9% ($t = 3.41$, $p < .01$). Therefore, the North–South difference among the Negroid subjects cannot be attributed to age differences from these data.

Fourteen studies using dolls or pictures included 742 Negroid subjects in segregated and 619 in integrated schools. The segregated Negroids preferred their own-race designate 42.4% of the time, whereas the integrated Negroids chose their own-race object 51.2% of the time ($t = 3.24$, $p < .01$). In 12 studies covering 833 Caucasoid children in segregated and 482 in integrated schools, the segregated subjects preferred their own race more often, 72.9 versus 66.3% ($t = 2.56$, $p < .02$). Aware that the integration–segregation factor might have been contaminated by regional differences, we reexamined the data and found that all the integrated subjects lived in the North, whereas the segregated subjects lived in both regions. Eliminating the Southern subjects, we compared the 254 Negroid children in *Northern segregated schools* with the 619 in *Northern integrated schools* and found that their mean own-race preferences were practically identical, the respective percentages being 51.20 and 51.15. Caucasoid differences in own-race choices when corrected for region were also too small to be significant, the mean own-race preferences of the 303 Northern segregated subjects and the 482 Northern integrated subjects being 71.89 and 66.27 ($t = 1.65$). We conclude from the dolls-pictures studies that there is no evidence of significant differences in own-race preference attributable to racial composition of the school.

In general, the authors have avoided comparison between the intelligence levels of the groups interviewed.

With few exceptions, the socioeconomic provenance of the Negroid and Caucasoid children was stated. In some studies, middle and lower classes formed the background of both racial groups; frequently, however, the Caucasoid children seem to have been in a somewhat favorable position relative to the Negroid. On the whole, it appears that the various investigators made genuine efforts to select racial samples from the same schools or neighborhoods.

Three hundred thirty-three Negroid children are reported to have been interviewed on these tests by Caucasoid examiners, in contrast to 1141 Negroid subjects who were examined by Negroid examiners. The respective own-race preferences under the Caucasoid and Negroid interviewers were 33.3 and 48.1%, the difference being highly significant ($t = 4.80$). These results suggest that white examiners may exert a detrimental influence on own-race preference of Negroid children in the intimate one-to-one, face-to-face situation. In studying the seven reports in which examiners were Negroid and comparing them with the five in which examiners were Caucasoid, we found that those interviewed by a Negroid person averaged a year older than those interviewed by a Caucasoid (5.77 versus 4.75 years). Fortunately, there were 193 Negroid subjects in the 3–5 age group tested by Caucasoid examiners who could be matched with 316 Negroid subjects of this age group tested by Negroid examiners. The mean own-race preference of the latter is 48.5%; of the former, 25.6% ($t = 5.09$). Therefore, there is support for the view that young Negroid children between the ages of 3 and 5 are more likely to show preference for their own race when examined by a Negroid than by a Caucasoid adult.

CHILDREN'S SELF-SOCIAL CONSTRUCTS TESTS, SELF-SOCIAL SYMBOLS TASKS, CHILDREN'S SELF-CONCEPT INDEX, ILLINOIS INDEX OF SELF-DEROGATION, AND WHERE ARE YOU GAME

We will now examine six researches that have utilized various constructs or symbols to represent the self and other children, such as circles in rows or columns, stick figures of flag and balloon children, or simply stick figures. Summaries of these studies of Negroid and Caucasoid children in kindergarten, first, second, or third grade have been included in Table 8.3. The means reported from study to study cannot be directly compared since unit size and zero points vary, and variance has frequently been omitted. In the summary, the means have been transformed into percentages to combine them and make general comparisons.

In the Children's Self-Social Constructs Tests (CS-SCT), the subject is presented with symbolic arrays in which circles or other figures represent the self and others. The subject responds to each task by arranging the symbols in specific ways, by selecting a symbol to represent the self, or by drawing a symbol. The authors assume that the relations seen in symbols represent relations in the subject's life and that these arrangements are readily interpretable with translatable common meanings.

TABLE 8.3

Children's Self-Social Constructs Tests

Author and date	Location	Instrument	Subjects			
				N		Grade
Long, B. H., and Henderson, E. H. (1968)	A Southern rural community	Children's Self-Social Constructs—preschool form, vertical scale		Negroid	Caucasoid	1
			Boys	36	36	
			Girls	36	36	
Long, B. H., and Henderson, E. H. (1970)	Two Southern rural communities	Children's Self-Social Constructs—preschool form, vertical scale		Negroid	Caucasoid	1
			Boys	48	48	
			Girls	48	48	
Cornwell, H. G. (1970)	City in SE Pa.	S-SST, primary form, vertical and horizontal scales		Negroid	Caucasoid	
				138	387	Kindergarten
				122	353	3

<table>
<tr><th>Subjects
Method of selection</th><th>Results</th><th>Comments of author</th></tr>
<tr><td>72 Negroid boys and girls about to begin first grade; all in a Head Start program. Control of 72 Caucasoids beginning school, 45% of Caucasoids and none of Negroids had attended kindergarten.</td><td>Self-esteem means and significance of difference<table>
<tr><th colspan="2">Means</th><th></th><th></th></tr>
<tr><th>Negroid</th><th>Caucasoid</th><th>t</th><th>p</th></tr>
<tr><td>5.4</td><td>6.5</td><td>2.75</td><td>.01</td></tr>
</table></td><td>Maximum score for a single item was 5: there were two items of esteem. Scores were summed for a total score, making a possible range of scores from 2–10.*</td></tr>
<tr><td>From all pupils entering County A schools, random selection of equal numbers of subjects within each of 8 cells: black-male-middle; black-male-low; black-female-middle; black-female-low; white-male-middle; white-male-low; white-female-middle; white-female-low.</td><td>Self-esteem means[a]<table>
<tr><th>Class</th><th>Boys</th><th>Girls</th><th>Total</th></tr>
<tr><td colspan="4">Negroid subjects</td></tr>
<tr><td>Lower</td><td>10.4</td><td>10.6</td><td>10.5</td></tr>
<tr><td>Middle</td><td>10.4</td><td>11.1</td><td>10.8</td></tr>
<tr><td colspan="4">Caucasoid subjects</td></tr>
<tr><td>Lower</td><td>12.9</td><td>11.0</td><td>12.0</td></tr>
<tr><td>Middle</td><td>11.7</td><td>12.0</td><td>11.8</td></tr>
</table></td><td>Maximum score for single item was 5 with 4 items of esteem, making possible range of scores 4–20.</td></tr>
<tr><td>All pupils in kindergarten and third grade of a public school system if in attendance on testing days; 84% were present and had scorable responses.</td><td>Grade differences in esteem<table>
<tr><th></th><th colspan="2">Kindergarten</th><th colspan="2">3</th><th></th><th></th></tr>
<tr><th></th><th>Male</th><th>s</th><th>Male</th><th>s</th><th>t</th><th>p</th></tr>
<tr><td colspan="7">Vertical scale</td></tr>
<tr><td>Negroid</td><td>25.4</td><td>6.1</td><td>21.4</td><td>6.4</td><td>5.1</td><td><.001</td></tr>
<tr><td>Caucasoid</td><td>25.4</td><td>6.5</td><td>21.4</td><td>6.5</td><td>8.3</td><td><.001</td></tr>
<tr><td colspan="7">Horizontal scale</td></tr>
<tr><td>Negroid</td><td>21.3</td><td>7.1</td><td>21.3</td><td>6.0</td><td>—</td><td>—</td></tr>
<tr><td>Caucasoid</td><td>23.0</td><td>6.0</td><td>21.7</td><td>5.7</td><td>3.0</td><td><.01[b]</td></tr>
</table></td><td>Schools completely integrated, 26% Negroid at kindergarten and third grade. Preponderance of subjects in lower socioeconomic category.</td></tr>
</table>

TABLE 8.3 *(Continued)*

Author and date	Location	Instrument	Subjects N				Grade
Granger, R. L., Cicirelli, V. G., Cooper, W. H., Rhode, W. E. and Maxey, E. J. (1969)	United States, principally E and SE.	Children's Self-Concept Index[c]	Negroid (90–100%)		Caucasoid (90–100%)		
			HS	C	HS	C	
			88	88	152	152	1
			72	72	152	152	2
			88	88	104	104	3

Author and date	Location	Instrument	Subjects N				Grade
Posner, C. A. (1969)	Several Catholic schools of unidentified city and suburbs[d]	Meyerowitz' Illinois Index of Self-Derogation	Upper Middle Class				
			IQ range	Negroid	Puerto Rican	Caucasoid	1
			120–	20		20	
			90–110	20		20	
			50–75	20		20	
			Lower Class				
			IQ range	Negroid	Puerto Rican	Caucasoid	1
			120–	20	20	20	
			90–110	20	20	20	
			50–75	20	20	20	

<table>
<tr><th>Subjects
Method of selection</th><th>Results</th><th>Comments of author</th></tr>
<tr><td>From random sample of 300 Head Start centers, 104 selected. Control subjects eligible for Head Start but did not attend. Final Head Start and C samples equivalent on age, grade, sex, race, socioeconomic status, and residence in area.</td><td>Effect of summer program on CSCI means
<table>
<tr><th>Subjects</th><th>Head start</th><th>C</th><th>p</th></tr>
<tr><td colspan="4">Grade 1</td></tr>
<tr><td>Negroid</td><td>44.42</td><td>41.13</td><td>.01</td></tr>
<tr><td>Caucasoid</td><td>45.60</td><td>46.29</td><td>—</td></tr>
<tr><td colspan="4">Grade 2</td></tr>
<tr><td>Negroid</td><td>45.03</td><td>45.16</td><td>—</td></tr>
<tr><td>Caucasoid</td><td>47.45</td><td>47.64</td><td>—</td></tr>
<tr><td colspan="4">Grade 3</td></tr>
<tr><td>Negroid</td><td>46.59</td><td>46.81</td><td>—</td></tr>
<tr><td>Caucasoid</td><td>48.91</td><td>48.07</td><td>—</td></tr>
</table></td><td>Range of scores from 26–52, the lowest score indicating all self-derogatory responses.</td></tr>
<tr><td>300 subjects selected from first grade of "several" Catholic schools: 120 Negroids, 120 Caucasoids, 60 Puerto Ricans. Divided according to socioeconomic class and IQ level. Half of each subgroup were girls.[e]</td><td>Self-derogations by race[f]
<table>
<tr><th colspan="2">Means</th><th></th><th></th></tr>
<tr><th>Negroid</th><th>Caucasoid</th><th>F ratio</th><th>p</th></tr>
<tr><td>16.3</td><td>15.5</td><td>0.50</td><td>—</td></tr>
</table>
Self-derogations by IQ
<table>
<tr><th colspan="3">Means</th><th></th><th></th></tr>
<tr><th>High</th><th>Med</th><th>Low</th><th>F ratio</th><th>p</th></tr>
<tr><td>14.6</td><td>14.8</td><td>17.8</td><td>23.26</td><td>.01</td></tr>
</table></td><td>Self-image perceptions of the Caucasoid children in sample much less affected by social class membership than are perceptions of Negroid children.</td></tr>
</table>

TABLE 8.3 *(Continued)*

Author and date	Location	Instrument	Subjects N Negroid	Subjects N Caucasoid	Grade
Carpenter, T. R., and Busse, T. V. (1969)	Eastern City of medium size	Engel and Raine's Where Are You Game	20	20	1

* *Source:* Personal communication from B. H. Long.

Note: Item purporting to measure *realism* and *identification* omitted by reviewer. The Head Start teachers rated the classroom behavior of their pupils in the last week of the program; however, since the Caucasoid subjects were not in the HS program and were not rated, the ratings on the Negroid children have been deleted from this review.

[a] Significant race difference, $F = 4.27$, $p < .05$. Differences between classes and between sexes not significant. Dearth of middle-class Negroids and lower-class Caucasoids entering County A schools required filling these cells from adjoining County B.

[b] All of above calculations made by present writer from Cornwell's data as supplied in Table 1-A. The primary form of S-SST is usually administered as a group test. Instead of 5 circles presented two times (Long & Henderson, 1968) or four times (Long & Henderson, 1970), allowing for respective ranges of scores to be from 2 to 10 and 4 to 20, six circles are offered six times (Cornwell, 1970) permitting a possible range of scores from 6 to 36.

[c] Additional measures employed were: Metropolitan Readiness Tests, Stanford Achievement Tests, Illinois Test of Psycholinguistic Abilities, Classroom Behavior Inventory, Children's Attitudinal Range Indicator, and Parent Interview Questionnaire.

Basic to this is the view that self-esteem is a person's perception of his worth, derived from an accumulation of self–other comparisons.

Vertical esteem is measured by a column of five or six circles of identical size, representing children. In the Long and Henderson studies (1968, 1970), the subject is advised: *These circles stand for children. You pick one to be you.* Scores range from *one* for the lowest circle to *five* for the highest. The theory that high self-esteem is associated with a high—rather than with a low—position is supported by the work of De Soto, London, and Handel (1965), who found evidence that evaluative relations are tied to a vertical axis in most people's thinking.

The *horizontal esteem* task, considered by Long, Henderson, and Ziller to be appropriate for children *once they have learned to read,* involves a similar procedure. A page containing a row of six identical cir-

Subjects: Method of selection	Results					Comments of author
80 subjects from father-absent welfare families, equally divided by race, sex, and grade (1 or 5).[g]	Mean self-concept scores[h]					Influence of Negroid mother may contribute to highly negative self-concepts of Negroid girls. All lived in matriarchal homes.[i]
		Means		Mann–Whitney		
		Boys	Girls	U	p	
	Negroid	9.6	15.0	22.0	.05	
	Caucasoid	8.2	9.3	48.5	—	

[d] Probably Chicago area. Doctorate obtained from Illinois Institute of Technology.

[e] Note that the three levels of IQ selected are not continuous, nor are the two socio-economic classes.

[f] On the IIS–D, the higher mean indicates *more unfavorable evaluations* of the self.

[g] Results on fifth-grade children omitted by reviewer from table. Subjects, both grades, examined in own homes by one of 12 caseworkers, 7 of whom were Caucasoid males, 4 Caucasoid females, and 1 a Negroid female (personal communication, T. V. Busse).

[h] Lower mean scores indicate a more positive self-concept. Mann–Whitney U tests used due to extreme skewness of data.

[i] Note that mean of 15 is within range of means of the four groups examined in Grade 5 (14.3, 15.6, 18.2, and 16.6); 7 and 35 are the extremes of this scale, the midpoint being at 21.

cles is placed before the child, who is to select one to be himself. The highest score of *six* is given for choice of the extreme left circle, the lowest score of *one* for choice of the extreme right circle. Long, Henderson, and Ziller (1970–1971) link this association with the cultural norm of reading and writing from the left. As might have been surmised, Israelis—whose reading and writing begin on the right—show the opposite trend. (Authors cite work of Lila G. Braine: Asymmetries of pattern observed in Israelis. *Neuropsychologia,* 1968, 6, 73–88.)

Using a preschool form of the CS-SCT, Long and Henderson (1968) measured the *disadvantaged* Negroid child's concept of self in a rural Southern community. Thirty-six Negroid subjects were tested during the last week of a 7-week Head Start Program. None had attended kindergarten; their mean Otis IQ was 90.4; 80% of the chief earners in their families were in the two lowest categories of Hollingshead's Occupational Scale; 42% had been separated from their natural fathers; their mean number of siblings was 3.7. A control group of 36 Caucasoid subjects beginning school in the same community had not attended Head Start;

had mean Otis IQs of 110.8; had parents of higher occupational status; only 7% had been separated from their natural fathers; and had, on the average, 1.7 siblings. Consequently, the Negroid and Caucasoid groups, though equated for sex, grade, and community, differed significantly in IQ, occupational level of chief wage earner, number of siblings, separations from father, and kindergarten experience.

The examiner administered the vertical esteem test twice, then summed the values of the two items for a total score that had a possible range of from 2 to 10. Long and Henderson (1968) found the self-esteem means of the Negroid and Caucasoid subjects to be 5.4 and 6.5, respectively, the difference being significant. Low self-esteem of the Negroid children was reported to be associated with immature classroom behavior as judged by their teachers, which in itself was associated with a decrease in IQ.

These authors (Long and Henderson, 1970) gave the preschool form of the CS-SCT to children entering the first grade in 13 schools in two rural Southern counties. The final sample consisted of 192 subjects, half Negroid and half Caucasoid, half male and half female, and half lower class and half middle class. However, the Negroid children had significantly lower IQs, less preschool education, and more father absences than the Caucasoids. The esteem means of the Negroid and Caucasoids were 10.6 and 11.9, respectively, the difference being significant at the .05 level of confidence.

As a part of an HEW project, Henry Cornwell (1970) compared the self-esteem of Negroid and Caucasoid children from the kindergarten level through high school. His complete roster of subjects included all pupils enrolled in kindergartens and in Grades 3, 6, 9, and 12 of a Pennsylvania public school system who were present on testing days and gave scorable responses. The pupils were examined by their teachers, who administered all tests in at least two separate sessions. The children in the kindergarten and third grade (whose responses we are evaluating in this chapter) were given only the primary form of the Self-Social Symbols Tasks (S-SST), which includes two scales purporting to measure self-esteem, the *vertical* and the *horizontal* arrangements.

The vertical scales of the primary form of the S-SST and the preschool CS-SCT seem to be identical. The booklet is open in front of the child, the figures (circles) are the same, as are the directions that are read aloud by the examiner (the child pointing to the preferred circle or marking it in a specified way). The same procedure holds for the horizontal scale, the difference, of course, being that the circles are in a row on the page instead of a column. The scoring is simple and follows the same principle both in the S-SST and the CS-SCT in that one point is awarded for the position lowest on the vertical scale or that on the extreme right on the horizontal, with an additional point allowed for each progression from lowest to highest or from right to left. The preschool CS-SCT, however, is administered individually, whereas the primary form of the S-SST was developed for *group* testing. Further, instead of five circles presented two times (Long & Henderson, 1968) or four times (Long & Henderson, 1970), allowing for respective ranges of scores to be from 2 to 10 and 4 to 20, *six circles* are offered *six times* in the primary form of S-SST (Cornwell, 1970), permitting a possible range of scores from 6 to 36. It is obvious that the group means obtained in the three researches are not directly comparable.

From Table 8.3, the reader can readily ascertain that Cornwell's various means are 21 or above, 21 being the midpoint of the theoretical distribution, suggesting that self-esteem as measured by the S-SST appears to have been comfortably "normal" for the Negroids and Caucasoids in the kindergartens and third grades of this Pennsylvania city. The kindergarten children of both races earned significantly higher mean scores on the vertical than on the horizontal scale. In fact, neither race nor sex but *only grade differences* proved to be significant when measured by the vertical scale, the means of both racial samples declining equally. Significant differences were obtained on the *horizontal scale,* however, between the kindergarten and third-grade Caucasoids, between male and female Caucasoids, and between the total group of Caucasoids and Negroids. In these three comparisons, the third-grade Caucasoids, the male Caucasoids, and the total group of Negroids obtained the lower averages.

Granger, Cicirelli, Cooper, Rhode, & Maxey (1969) reviewed and assessed the effectiveness of summer and full-year Head Start programs, relative to cognitive and affective development of children in poverty areas. As a tool to serve in this project, the Children's Self-Concept Index (CS-CI) was developed by these authors from Myerowitz's Illinois Index of Self-Derogation (IIS-D). Some of the Myerowitz items were dropped, some modified, and a few additional ones were included. The major areas of emphasis of both the CS-CI and the IIS-D relate to the child's perception of self with respect to peer acceptance and positive reinforcements in the home and at school. The authors report the internal consistency reliability for the CS-CI to be .80, the test–retest reliability after two weeks, .66.

The CS-CI consists of 26 pairs of stick figures, one holding a balloon and one a square flag. Below each pair are two statements, one favorable, the other unfavorable. For example: *The balloon-child is learning a lot in school. The flag-child isn't learning very much.* Or: *Some children hate the balloon child. Children like the flag child.* The balloon child is always on the left side; the socially desirable is located half the time on each side. The CS-CI is designed to be administered by the classroom teacher in group testing. The examiner reads aloud all instructions, including two preliminary examples, noting that all subjects understand how they are to respond, that is, marking the balloon or flag child that is more like themselves. In scoring, a self-derogatory or undesirable response is given a weight of 1; a neutral or desirable response is given a weight of 2. The total score equals the sum of the weights; the larger the score, the more positive the self-concept.

A random sample of 300 Head Start centers from a total of 12,927 existing in continental United States in 1966–1967 was identified and listed in order of selection through the use of a coding procedure and a standard computer sampling program. The final sample comprised 104 centers, 70% having *summer programs* and 30% *full-year programs.*[11] Each of the 104 target areas included: (1) children who had completed Head Start programs and who were to enter the first, second, or third grade in September 1968 and (2) a control population of all remaining children in the first three grades of the same elementary schools who had been eligible for Head Start but had not participated. From each of the target areas, the researchers selected eight names at random, with an additional two as an oversample; these were matched with 8–10 controls, also selected at random, for ethnicity, race, sex, grade in school, and kindergarten experience. The two groups were later equated for socioeconomic status obtained from parental interviews, using Hollingshead's Two-Factor Index of Social Position by means of covariance analysis. "In the overall analysis for the Children's Self-Concept Index (CS-CI), a projective measure of the degree to which the child has a *positive self-concept,* the Head Start children from both the summer and the full-year programs did not score significantly higher than the controls at any of the three grade levels (pp. 4–5)."

It is evident from the tabulated data that the several differences between the means of the *summer* Head Start and their respective control groups were in general small, the only difference of significance being that between the two Negroid groups (90–100% Negroid) at Grade 1. The research staff report, in addition, that the self-concept of the total group of children attending the 68 summer Head Start centers was significantly lower than the controls at Grade 2 and that no significant differences were obtained between any of the summer Head Start groups and their controls at Grade 3.

All six of the primarily Caucasoid groups earned higher CS-CI means than those of the predominantly Negroid groups with whom they were compared, as will be noted in our Table 8.3. The average of the 496 predominantly Negroids proved to be 44.84; that of the 816 predomi-

[11] No comparisons were made by the authors between the 90–100% Negroid and the 90–100% Caucasoid groups in the full-year programs since there were too few of the latter in the year-centers to make valid judgments. Hence our racial comparisons are based on the analysis of summer programs only.

nantly Caucasoids was 47.19. With a difference of 2.25 between the means of the "Negroid" and "Caucasoid" groups and a sigma of the difference of 0.27 (approximate), the critical ratio is approximately 8.3. Even allowing for a small error in the size of the sigmas, it is evident that the difference between the racial means is highly significant.

One of the cognitive tests, the Illinois Test of Psycholinguistic Abilities (ITPA), used as a measure of language development, was administered to children in all three grades, and the means were tabulated for the primarily Negroid group and the primarily Caucasoid group at each grade level. The Negroid Head Start and control groups at each of the grade levels averaged, respectively: 187.25 and 177.87; 208.41 and 206.87; 244.98 and 238.59. The Caucasoid Head Start and control groups at each of the three grades averaged: 190.54 and 188.42; 230.59 and 230.65; 263.41 and 261.94. It is evident that the primarily Negroid and Caucasoid groups of children differed not only in self-esteem but in cognitive development as measured by these tests.

In 1969, Carmen Posner completed her doctorate, using the Meyerowitz Illinois Index of Self-Derogation in its unchanged form as well as the Children's Form of the Farnham-Diggory Self-Evaluation Scale. The former consisted of 30 pairs of stick figures with balloon and square flag, each pair followed by one favorable and one derogatory statement. The child's negative self-concept was measured by the number of derogatory statements ascribed to himself, the examiner having read aloud the statements to each group, reminding the children to *mark the child who is most like you.* As in the CS-CI adaptation, the balloon child is always on the left, the flag child on the right, the socially desirable figure located half the time on the left and half the time on the right.

Preliminary pretest studies with $2\frac{1}{2}$–$3\frac{1}{2}$-year-old children by Meyerowitz showed the pairs of stick figures to have been equally attractive when they were presented together with the request that the children label each child as *good* or *bad.* Posner's subjects had for the most part attended nursery schools in which they were tested for more than a year; the results of the *Index* agreed with the opinions of the staff about the children, indicating evidence of the validity of the items. Posner reports obtaining a product–moment correlation between the total IIS-D score in the test–retest presentations, the interval between them being one week. Analysis was carried out on 26 educable mentally handicapped children between 6 and 8 years of age, the r for the total number of derogatory choices being .75. As a result of these and other analyses, Posner concluded that the IIS-D was reasonably internally consistent and reliable.

A sample of 300 first-grade children, all from intact families and in Catholic schools, comprised 120 Negroids, 120 Caucasoids, and 60

Puerto Ricans. The Negroid and Caucasoid children were further divided into *upper-middle* and *lower* socioeconomic status and into three levels of intelligence: *below average* (50–75 IQ), *average* (90–110 IQ), and *superior* (120 IQ and above). Social class membership was determined merely by residence. Posner does not report the number of schools canvassed, the method of selection within school or classroom, nor mean IQ for each ethnic/racial group. Posner compared the mean number of self-derogations by race, sex, socioeconomic status, and intelligence, using analysis of variance. The test: (1) did not discriminate between Negroid and Caucasoid children, (2) did not discriminate between boys and girls, (3) did not discriminate between lower-class and middle-class children, but (4) did discriminate between IQ levels, *the low IQ group producing significantly more self-derogation than the medium and high IQ groups.*

Carpenter and Busse (1969) tested the hypothesis that Negroid children show increasingly negative self-concepts when compared with Caucasoid children of equivalent social status. The instrument, the "Where Are You Game," developed by Engel and Raine (1963), consists of seven bipolar dimensions thought to be important in self-concept: (1) seeing oneself as intellectually gifted versus ungifted; (2) as happy versus unhappy; (3) as well liked by peers versus unpopular; (4) as brave versus timid; (5) as physically attractive versus unattractive; (6) as strong versus weak physically; and (7) as obedient versus disobedient. Subjects score themselves on a five-point scale between each of these polar opposites. For each dimension, the subject is provided with a pen and a piece of paper with five horizontal lines vertically spaced between the two stick figures. The examiner points to the stick figures at the top and bottom of the paper and reports "a story" about each to the child. For example, he will say that this one (boy or girl, depending on the sex of the subject) is a very happy person, always smiling, full of fun. Then he points to the other figure, describing him as sad and unhappy. The examiner explains the meaning of the five steps and asks the subject to make a mark on the step on which he thinks he is between the two figures. Procedure is the same for all seven dimensions, each scale being presented on a separate sheet of paper. For the odd-numbered scales, the socially acceptable extreme is at the top of the paper; for the even numbered, it is at the bottom.

Carpenter and Busse randomly selected 40 Negroid and 40 Caucasoid children in Grades 1 and 5 from father-absent welfare families. The subjects were equally divided according to sex and grade and were interviewed by 11 Caucasoid and 1 Negroid social caseworkers. Differences between the means of the Caucasoid boys, Caucasoid girls, and Negroid boys in Grade 1 were insignificant, but the mean of the young Negroid girls (15.0) was significantly lower.

QUESTIONNAIRE AND RATING SCALES

Three primarily verbal studies of children in Grades 1–3 remain to be considered. The California Test of Personality, first published between 1939 and 1943, consists of five series, one of which is the *primary* for children in kindergarten to Grade 3. Each is designed to measure a number of self- or social adjustments. The Sense of Personal Worth in the Primary series is measured by responses to eight yes-or-no questions. (1) *Do the children think you can do things well?* (2) *Are the boys and girls mean to you?* (3) *Do you have less friends than other children?* (4) *Are most of the children smarter than you?* (5) *Do your folks think that you are bright?* (6) *Can you do things as well as other children?* (7) *Do people think that other children are better than you?* (8) *Do most of the boys and girls like you?* These questions are printed in relatively large type on one page of the booklet to be read aloud slowly and distinctly by the examiner, the children being directed to read the questions to themselves at the same time and to answer them by drawing a circle around *Yes* or *No*, which follow each question. In scoring, desirable responses are added, the maximum score being 8. L. F. Shaffer, in a review of the test (Buros, 1949, pp. 55–56), indicates that reliabilities of total scores on the tests range from .918 to .933, based on the split-half method, corrected by the Spearman–Brown formula, determined for *N*s of from 237 to 792 for the various forms. Subtest reliabilities are not given for the primary series.

T. L. Engle (1945) compared the personality adjustments of Indiana Negroid and Caucasoid public schoolchildren using the primary form of the CTP. All were enrolled in the second and third grades of 13 rural or three South Bend schools. The rural school pupils were Caucasoid, with 101 Amish and 76 non-Amish; the South Bend subjects comprised 107 Negroid and 92 Caucasoid (non-Amish) children. The reviewer combined the Amish and non-Amish Caucasoid averages and found that the observed Caucasoid–Negroid differences were insignificant at the .05 level of confidence ($t = 1.57$). See Table 8.4.

Carmen Posner (1969) examined her subjects with the Farnham–Diggory Self-Esteem Scale, which consisted of eight self-ratings covering such matters as: (1) competence at enjoyed tasks . . . and (4) ability to make others like you. The children score their own ability on these items by marking one number in a row of digits ranging between 0 and 10; then similarly score their ambitions in these areas; then score what they believe to be their parents' scoring of their capacities. Posner's main findings applicable to race are as follows: *(1)* Self-images of Caucasoids are significantly higher than those of Negroids. *(2)* The mean discrepancy between self-image and ideal self-image is significantly greater among Negroids. *(3)* The low IQ group ranks significantly lower in self-esteem

TABLE 8.4

Questionnaire and Rating Scales

Author and date	Location	Instrument	Subjects				Grade
				N			
Engle, T. L. (1945)	Rural NE Indiana and South Bend	CTP, primary form A		Negroid	Caucasoid		
					Con	Am	2–3
			Boys	51	90	51	
			Girls	56	78	50	
				107	168	101	
Posner, C. A. (1969)	Catholic schools of city and suburbs[b]	Farnham–Diggory S-E Scale, children's form	Upper middle class				
			IQ range	Negroid	Puerto Rican	Caucasoid	1
			120–	20		20	
			90–110	20		20	
			50–75	20		20	
			Lower class				
			IQ range	Negroid	Puerto Rican	Caucasoid	1
			120–	20	20	20	
			90–110	20	20	20	
			50–75	20	20	20	

Method of selection	Results	Comments of author
Second and third graders. Subjects in rural schools—*Controls* or *Amish;* subjects in city schools—Negroids and controls, no Amish.[a]	Sense of personal worth (see table below)	Scores of approximately 5.50 in sense of personal worth, feeling of belonging, and freedom from withdrawing tendencies at the 50th percentile of norms.
120 Negroids, 120 Caucasoids, and 60 Puerto Ricans from first grade of Catholic schools, divided equally according to 2 socioeconomic groups and 3 IQ levels. Half of each subgroup girls.	Self-images by race; Self-images by IQ level; Self-ideal discrepancy by race[c] (see tables below)	Negroid subjects characterized by negative self-images and less positive ideal self-perceptions than Caucasoid children.

Sense of personal worth

Subjects	Means: Negroid	Means: Caucasoid Con	Means: Caucasoid Am
Male	6.08	5.63	5.24
Fem	6.30	6.17	4.82

Self-images by race

Means: Negroid	Means: Caucasoid	*F* ratio	*p*
54.1	59.8	10.99	.01

Self-images by IQ level

Means: High	Means: Med	Means: Low	*F* ratio	*p*
59.9	57.6	52.4	4.95	.01

Self-ideal discrepancy by race[c]

Means: Negroids	Means: Caucasoids	*F* ratio	*p*
17.7	14.8	4.04	.05

TABLE 8.4 *(Continued)*

Author and date	Location	Instrument	Subjects: N	Subjects: Grade
			Negroid	
French, J. T. (1972)	Monticello and Tallahassee, Fla.	Gordon's How I See Myself, elementary form	225	3

[a] *Controls* in this study always signify non-Amish whites. Mean ages of Negroids: 8.7, controls: 8.4, and Amish: 9.3 years.
[b] See Table 8.3 for results obtained on these subjects with the IIS-D.

than the middle and high IQ groups, the two latter not differing meaningfully from each other. *(4)* Children of low SES have significantly more negative self-images than those of higher SES. *(5)* Both children of low socioeconomic status (SES) and of low IQ perceive their parents as ranking them significantly lower than do children of high SES or high IQ. *(6)* No sex differences were found among the children of either race.

Ira Gordon's "How I See Myself," elementary form, consists of 40 numbered items with parallel brief statements that are contradictory, though not necessarily bipolar. Evenly spaced below each pair of statements is a numbered sequence from 1 to 5.[12] After the examiner emphasizes that she alone will see their papers, she asks the children to think about each statement as she reads it aloud and to circle the one number that most accurately describes themselves. The test was reported by Yeatts (1967) to have a reliability coefficient of .78 for third-grade children obtained by retesting 34 pupils after nine days.

Jeana French examined the relation between self-concept of lower-class Negroid children as determined by How I See Myself and the racial composition of the school attended. Fifteen public elementary schools located in Monticello and Tallahassee, Florida, were classified as *(1)* with fewer than one-third Negroid, *(2)* one-third to two-thirds Negroid and *(3)* more than two-thirds Negroid. From each category, French selected by

[12] Examples of statements: *I wish I were smaller (taller) . . . I'm just the right height; Teachers like me . . . Teachers don't like me.* In scoring, 5 always indicates a positive view or favorable opinion of self.

<table>
<tr><th>Subjects
Method of selection</th><th>Results</th><th>Comments of author</th></tr>
<tr><td>75 lower-class Negroid third graders randomly selected from each of 3 groups of schools: I–fewer than one-third Negroid, II–one-third to two-thirds Negroid; III–two-thirds or more Negroid.[d]</td><td>Total self-concept scores
<table>
<tr><th>School category</th><th>Mean</th><th>Subject</th></tr>
<tr><td>I</td><td>275.6</td><td>44.0</td></tr>
<tr><td>II</td><td>258.3</td><td>46.6</td></tr>
<tr><td>III</td><td>278.2</td><td>40.3</td></tr>
</table>
F ratio = 4.57; p < .05; difference between I and III insignificant</td><td>Racially balanced schools tend to negatively affect self-concepts of Negroid lower-class subjects. Self-concepts of such children in mainly Caucasoid or Negroid schools are about the same.</td></tr>
</table>

[c] Self-Ideal discrepancy obtained by subtracting each self-image score from ideal-self score.

[d] Class having been determined by the short form of McGuire–White Index of Social Status.

random sampling 75 lower SES third-grade Negroid children. (See Table 8.4.)

French's findings may be summarized as follows: *(1)* Negroid third-grade children in mainly Caucasoid and in mainly Negroid schools do not differ significantly from one another in either total self-concept or in any of the eight factor subscores. *(2)* Negroid children attending racially balanced schools score significantly below children of this race in mainly Caucasoid or in mainly Negroid schools in total self-concept and in physical appearance and interpersonal adequacies. *(3)* Negroid children in the third grade of the racially balanced schools are significantly below the Negroid children in predominantly Caucasoid schools in physical adequacy, body build, and social adequacy among their peers. *(4)* The Negroid children in the racially balanced schools are likewise significantly below those in predominantly Negroid schools in teacher–school relationships, autonomy, and language adequacy. *(5)* Racial composition does not have a significant effect on these children's self-concepts of academic adequacy. As partial explanation for these findings, French observed that lower-class Negroid children in predominantly Negroid schools, being members of the majority group in their schools, may have self-acceptance without breaking down racial and social barriers. On the other hand, lower-class Negroids in predominantly Caucasoid schools may accommodate to the prevailing Caucasoid values and norms, increasing the probability of social acceptance (and self-approval); or their Caucasoid classmates may be willing to accept a few individuals whose race and norms differ from their own.

SUMMARY AND DISCUSSION OF SELF-ESTEEM AS MEASURED BY CONSTRUCTS AND VERBAL SCALES

In this summary, we will refer to the several instruments described in a previous section—Children's Self-Social Constructs, Self-Social Symbols Tasks, Children's Self-Concept Index, Illinois Index of Self-Derogation, and the Where Are You Game—simply as *constructs* and those included in the last section—Sense of Personal Worth, Farnham–Diggory SE Scale, and How I See Myself—as verbal scales. Approximately 1400 Negroid and 2133 Caucasoid children were examined on these instruments; the mean self-scores reported by the investigators were transformed by the reviewer into relatively equivalent units, that is, the proportion the mean self-score is to the maximum attainable. This frequently involved making use of specific zero points, such as 2, 4, 6, or more. (See footnote 13 for the formulas used by the reviewer in changing various raw scores into percentages.)

The Negroid and Caucasoid mean percentages of self-esteem are 64.3 and 68.0, respectively, a significant difference ($t = 2.63$). The self-esteem of the Negroid children determined by the constructs and verbal scales is significantly higher than that inferred from the dolls and pictures ($t = 10.44$), whereas the self-esteem of the Caucasoid children is

[13] By use of the following formulas appropriate to the several tests:

CS-SCT	$P = \left[\frac{\bar{X} - 2}{10 - 2}\right] \times 100$	(Long & Henderson, 1968)
CS-SCT	$P = \left[\frac{\bar{X} - 4}{20 - 4}\right] \times 100$	(Long & Henderson, 1970)
S-SST	$P = \left[\frac{\bar{X} - 6}{36 - 6}\right] \times 100$	(Cornwell, 1970)
CS-CI	$P = \left[\frac{\bar{X} - 26}{52 - 26}\right] \times 100$	(Granger *et al.*, 1969)
IIS-D	$P = 100 - \left[\frac{\bar{X}}{30} \times 100\right]$	(Posner, 1969)
WAY	$P = 100 - \left[\frac{\bar{X} - 7}{35 - 7} \times 100\right]$	(Carpenter & Busse, 1969)
SPW	$P = \left[\frac{\bar{X}}{8}\right] \times 100$	(Engle, 1945)
F-D Self-Esteem	$P = \left[\frac{\bar{X}}{80}\right] \times 100$	(Posner, 1969)
How I See Myself	$P = \left[\frac{\bar{X} - 70}{350 - 70}\right] \times 100$	(French, 1972)

significantly higher on the dolls and pictures ($t = 2.67$). Several explanations may account for this variance: (1) The children examined on the constructs and verbal scales were older. (2) Validity and/or reliability measures have generally been reported for the constructs–verbal tests, but not for the dolls–pictures. Test validity is sometimes assumed if the children identify with the doll or picture intended to represent their race. However, in six investigations that included questions as to *Which one looks like you?*, the Negroid children 3–5 years of age pointed to a "correct" doll less than 60% of the time.[14] (3) All of the subjects examined with constructs–verbal scales (with the exception of those tested by Long and Henderson) were examined in groups, whereas the subjects were interviewed individually when tested with dolls and pictures. The latter may have been disadvantageous to the child's own-race preference, particularly if the examiner was Caucasoid and the subject Negroid. (4) In the dolls–pictures situations, the choice is between *Negroid and Caucasoid;* whereas in the constructs–verbal tests, self-esteem or self–relative to other children—is rated. Two different attitudes are being measured: attitudes toward one's own race and attitudes toward self and others. Thus, in the opinion of the reviewer, the constructs and verbal tests, if the attitudes are stable, *measure self-esteem;* the dolls and pictures, *race esteem* and *only indirectly self-esteem.* In both, as we have seen, the esteem of the Caucasoid children is the higher.

The percentages of self-esteem attributed to the 246 Negroid boys and the 237 Negroid girls on the constructs–verbal measures are 57.2 and 58.1, respectively; to the 584 Caucasoid boys and 540 Caucasoid girls, 58.3 and 60.4. No significant sex difference was obtained on either the constructs–verbal or the dolls-pictures test employed.

On the constructs and verbal materials, the 138 younger and the 1258 older Negroid subjects (6–8 years) had mean self-esteem percentages of 57.8 and 65 ($t = 1.78$). However, on these instruments, the 387 younger Caucasoids scored significantly below the 1746 older subjects of their race, the respective percentages being 60.7 and 69.6 ($t = 3.42$).

The 507 Northern Negroid subjects examined on these instruments achieved a mean self-esteem score of 61.0%; the 393 Southern Negroids 59.0%, the difference being insignificant. No Southern Caucasoids were tested on verbal materials, but on constructs the 880 Northern and the 168 Southern subjects did not differ significantly, the respective mean percentages being 56.2 and 52.3.

[14] In the combined studies of the Clarks (1947); Stevenson & Stewart (1958); Gregor & McPherson (1966); Crooks (1970); Durrett & Davy (1970); and Fox & Jordan (1973) the 3–5-year-old Negroid subjects correctly identified themselves 56% of the time; the 6–7-year-old Negroids correctly identified themselves 82% of the time. In contrast, 80% of the 3–5-year-old Caucasoid and 92% of the 6–7-year-old Caucasoid subjects identified themselves correctly.

Young Negroid children of lower socioeconomic classes in the third grade of 15 Florida public schools were found to have significantly lower total self-concept scores if in racially balanced schools than those in mainly Caucasoid or in mainly Negroid schools. No Caucasoids served as subjects in this study.

An increase in the self-esteem of Negroid—but not of Caucasoid—children was reported at the end of the first grade following a summer session of Head Start. The increase disappeared by the end of the second grade.

In an investigation in which Negroid and Caucasoid children were matched according to three IQ levels (low, middle, high), the low groups were found to have significantly more self-derogations and to be significantly below the two higher-level groups on the self-image scale. Moderate retardation (50–75 IQ) seems related positively to low self-esteem in schoolchildren, a not surprising finding since the esteem scales employed include a number of academic situations in which the self is placed. In three studies, one or more cognitive test scores were reported for the Negroid and Caucasoid subjects, the means of the former being notably lower.

In the few studies relating self-esteem to class, there are enough inconsistencies to make generalizations inappropriate. One student found that self-images of Caucasoid children were less affected by social class than were those of Negroid children.

Precise information as to race of examiner was usually omitted.

CONCLUDING STATEMENT

Preference for one's own race as determined with the aid of dolls and pictures and self-esteem as measured by constructs and verbal tests have been studied extensively among American Negroid and Caucasoid children. Although *own-race preference* and *self-esteem* results are not identical, they are closely parallel and may even measure the same trait. Both own-race preference and self-esteem have been found to be significantly higher among Caucasoid children, the lower scores of the Negroids being particularly evident where dolls, puppets, photographs, and drawings were the instruments of measurement. No significant differences were found between the sexes of either race. Own-race preference and self-esteem increase in both races from 3 to 5 to 6 to 8 years. Only among the Negroid children were own-race preferences found to be significantly higher in the North than in the South. Neither racial group showed regional differences in self-esteem. Three- to 5-year-old Negroid children, when interviewed by a Negroid examiner, appear to show a higher percentage of own-race preferences than those examined by a Caucasoid examiner. We cannot be confident that the interviewer's race affects either own race preference, or self-esteem of Caucasoid children, or the self-esteem of older Negroids. It appears doubtful that the *racial composition* of the school attended—when considered apart from neighborhood, community, or section of the country—influences either own-race preference or self-esteem. The differences are not all clearly established. Probably some of them may be attributed to the instruments used, to the socioeconomic level of the children's families, and to the cognitive development of the children themselves.

REFERENCES

Asher, S. R., & Allen, V. L. Racial preference and social comparison processes. *Journal of Social Issues*, 1969, *25*, 157–166.

Bronfenbrenner, U. The psychological costs of quality and equality in education. *Child Development*, 1967, *38*, 909–925.

Buros, O. K. (Ed.) *The third mental measurements yearbook*. New Brunswick, New Jersey: Rutgers Univ. Press, 1949.

Carpenter, T. R., & Busse, T. V. Development of self-concept in Negro and white welfare children. *Child Development*, 1969, *40*, 935–939.

Clark, K. B. *Prejudice and your child*. Boston: Beacon Press, 1955. Pp. 151.

Clark, K. B., & Clark, M. P. Racial identification and preference in Negro children. In T. M. Newcomb, E. L. Hartley *et al.* (Eds.), *Readings in social psychology*. New York: Holt, 1947. Pp. 169–178.

Clark, K. B., & Clark, M. P. Emotional factors in racial identification and preference in Negro children. *Journal of Negro Education*, 1950, *19*, 341–350.

Cornwell, H. G. Comparison of changes in self-image of Black and White students kindergarten through high school. Off. Educ., Bur. Research, U.S. Dept. Health, Education, and Welfare. Project No. 0B003, Grant No. OEG-2-700014 (509), 1970.

Crooks, R. C. The effects of an interracial preschool program upon racial preference, knowledge of racial differences, and racial identification. *Journal of Social Issues*, 1970, *26*, 137–144.

De Soto, C. B., London, M., & Handel, S. Social reasoning and spatial paralogic. *Journal of Personality and Social Psychology*, 1965, *2*, 513–521.

Durrett, M. E., & Davy, A. J. Racial awareness in young Mexican-American, Negro and Anglo children. *Young Children*, 1970, *26*, 16–24.

Engel, M., & Raine, W. J. A method for the measurement of the self-concept of children in the third grade. *Journal of Genetic Psychology*, 1963, *102*, 125–137.

Engle, T. L. Personality adjustments of children belonging to two minority groups. *Journal of Educational Psychology*, 1945, *36*, 543–560.

Fox, D. J., & Jordan, V. B. Racial preference and identification of black, American Chinese, and white children. *Genetic Psychology Monographs*, 1973, *88*, 229–286.

French, J. T. Educational desegregation and selected self-concept factors of lower-class black children. Unpublished doctoral dissertation, Florida State Univ., 1972.

Goodman, M. E. Evidence concerning the genesis of interracial attitudes. *American Anthropologist*, 1946, *48*, 624–630.

Goodman, M. E. *Race awareness in young children*. Reading, Massachusetts: Addison-Wesley, 1952. Pp. 280.

Granger, R. L., Cicirelli, V. G., Cooper, W. H., Rhode, W. E., & Maxey, E. J. (Project Staff) The impact of Head Start, an evaluation of the effects of Head Start on children's cognitive and affective development. 2 Vol. Westinghouse Learning Corporation and Ohio University. Office of Research, Planning, and Progress Evaluation of Office of Economic Opportunity, 1969.

Greenwald, H. J., & Oppenheim, D. B. Reported magnitude of self-misidentification among Negro children-Artifact? *Journal of Personality & Social Psychology*, 1968, *8*, 49–52.

Gregor, A. J., & McPherson, D. A. Racial attitudes among white and Negro children in a Deep-South standard metropolitan area. *Journal of Social Psychology*, 1966, *68*, 95–106.

Grossack, M. M. (Ed.) *Mental health and segregation*. New York: Springer, 1963. Pp. 247.

Harris, S., & Braun, J. R. Self-esteem and racial preference in black children. *Proceedings, 79th Annual Convention, American Psychological Association*, 1971, 6 (Pt. 1), 259–260.

Helgerson, E. The relative significance of race, sex, and facial expression in choice of playmate by the preschool child. *Journal of Negro Education*, 1943, *12*, 617–622.

Hraba, J. The doll technique: A measure of racial ethnocentrism? *Social Forces*, 1972, *50*, 522–527.

Hraba, J., & Grant, G. Black is beautiful: A reexamination of racial preference and identification. *Journal of Personality & Social Psychology*, 1970, *16*, 398–402.

Kardiner, A., & Ovesey, L. *The mark of oppression: A psychosocial study of the American Negro*. New York: Norton, 1951. Pp. 396.

Karon, B. P. *The Negro personality: A rigorous investigation of the effects of culture*. New York: Springer, 1958. Pp. 184.

Kircher, M., & Furby, L. Racial preferences in young children. *Child Development*, 1971, *42*, 2076–2078.

Koslin, S. C., Amarel, M., & Ames, N. The effect of race on peer evaluation and preference in primary grade children: An exploratory study. *Journal of Negro Education*, 1970, *39*, 346–350.

Long, B. H., & Henderson, E. H. Self-social concepts of disadvantaged school beginners. *Journal of Genetic Psychology*, 1968, *113*, 41–51.

Long, B. H., & Henderson, E. H. Social schemata of school beginners: Some demographic correlates. *Merrill-Palmer Quarterly of Behavior & Development*, 1970, *16*, 305–324.

Long, B. H., Henderson, E. H., & Ziller, R. C. Manual for the self-social symbols tasks and the children's self-social constructs tests. Unpublished mimeograph, *circa* 1970–71. Pp. 71.

Morland, J. K. Racial acceptance and preference of nursery school children in a southern city. *Merrill-Palmer Quarterly of Behavior & Development*, 1962, *8*, 271–280.

Morland, J. K. A comparison of race awareness in northern and southern children. *American Journal of Orthopsychiatry*, 1966, *36*, 22–31.

Piers, E. V. *Manual for the Piers-Harris Children's Self-Concept Scale (The way I feel about myself)*. Nashville, Tennessee: Counselor Recordings & Tests, 1969.

Posner, C. A. Some effects of genetic and cultural variables on self-evaluations of children. Unpublished doctoral dissertation, Illinois Institute of Technology, 1969.

Radke, M. J., & Trager, H. G. Children's perceptions of the social roles of Negroes and whites. *Journal of Psychology*, 1950, *29*, 3–33.

Radke, M., Trager, H. G., & Davis, H. Social perceptions and attitudes of children. *Genetic Psychology Monograph*, 1949, *40*, 327–447.

Rohrer, G. K. Racial and ethnic identification and preference in young children. Unpublished doctoral dissertation, Univ. of California, Los Angeles, 1972.

Stevenson, H. W., & Stewart, E. C. A developmental study of racial awareness in young children. *Child Development*, 1958, *29*, 399–409.

Thorpe, L. P., Clark, W. W., & Tiegs, E. W. *Manual of directions: California Test of Personality-Primary Series*. Los Angeles: California Test Bureau, 1940.

Vontress, C. E. Black studies-boon or bane? *Journal of Negro Education*, 1970, *39*, 192–201.

Yeatts, P. P. An analysis of developmental changes in the self-report of Negro and white children, grades 3–12. Unpublished doctoral dissertation, Univ. of Florida, 1967.

9

Ethnic and Racial Differences in Intelligence: International Comparisons

RICHARD LYNN

New University of Ulster, Northern Ireland

INTRODUCTION

Views about subspecies differences in human intelligence have undergone some pronounced swings over the course of the last century. Three phases can be distinguished. The first was inaugurated nearly 110 years ago when Sir Francis Galton published the book *Hereditary Genius* (1869). Galton hypothesized that the Caucasoid race was the most intelligent and that the other races were characterized by varying degrees of lesser ability, particularly the Australoid and Negroid. At this time, and for the next half century or so, there were few who did not think his conclusions amply justified by common sense and observation. Later, the scholarly views of psychologists with the stature of R. S. Woodworth (1910), E. L. Thorndike (1914), and S. D. Porteus (1917, 1937) supported Galton with data suggesting a key role for heredity.

The second phase began to develop between the two world wars. An example is the antihereditarian book *Race Psychology* by T. R. Garth (1931). Opinion started toward a contradiction of Galton's position, and it was now commonly asserted that all the races had about equal mental endowment. By 1951, this view had become a kind of official doctrine of "respectable" academicians when a panel of social and biological specialists at the United Nations, under the auspices of the United Nations

The research for this chapter was supported by the Esmée Fairbairn Charitable Trust.

Educational, Scientific, and Cultural Organization (UNESCO), issued the following statement: "According to present knowledge there is no proof that the groups of mankind differ in their innate mental characteristics, whether in respect of intelligence or temperament. The scientific evidence indicates that the range of mental capacities in all ethnic groups is much the same (1951)." The reader may also consult Comas (1961) and Jensen (1972) for similar resolutions by other groups; a counter-resolution signed by 50 scientists was published in the *American Psychologist* (1972, *7*, 660–661), then another one in *Homo* (1973, *24*, 52–55).

UNESCO stated the prevailing view in the years following the Second World War. A few continued to entertain Galton's hypothesis (e.g., Garrett, 1961a, 1961b, 1962; Ingle, 1964; Porteus, 1961, 1967; Shuey, 1966; Weyl & Possony, 1963), but their voices were in the minority. Whatever some psychologists may have thought privately in this period, few asserted in public that there were important native differences in intelligence among the major divisions of mankind. What brought this consensus to an abrupt end was the publication by Arthur R. Jensen of an invited paper in the *Harvard Educational Review* in which he argued *inter alia* that the low mean IQs of American Negroids were probably substantially influenced by genetic factors (Jensen, 1969). Jensen's hypothesis aroused instant and vehement opposition, but there were also eminent scholars who supported him, notably the American psychologists S. S. Stevens and R. B. Cattell (Jensen, 1972, pp. 1–67) and the British psychologist H. J. Eysenck (1971). Coming to phase three, it can be said today that well-informed persons in this field are roughly equally divided on the issue. However, this is a rather specific problem that is treated elsewhere (e.g., Jensen, 1972, 1973; Loehlin, Lindzey, & Spuhler, 1975), hence will not greatly concern us in this chapter. Our subject matter is the much wider one of ethnic and racial differences in intelligence considered as a worldwide phenomenon.

GALTON'S *HEREDITARY GENIUS*

Sir Francis Galton (1869) took the view that it was possible to estimate the mean intellectual level of a population from the number of outstanding persons it produced. He argued that the higher the overall genetic endowment of a population, the greater would be the proportion of literate and proficient individuals in it. On this basis, Galton estimated the average hereditary brightness of a number of different ethnic groups. He concluded that the most intelligent population the world had yet seen was that of Attica from 530 to 430 B.C., which produced over 230 times the rate of British geniuses. Another highly intelligent ethnic group was the Lowland Scottish. Below them came the southern English. Much further down Galton put African Negroids, and below them he placed the Australoids.[1]

The only major recent use of this method is that of Weyl (1966), who has calculated achievement quotients for different ethnic groups in the United States. However, when the method is used to compare different societies, there are two difficulties. One is the assumption that the standard deviation or the range of intelligence is much the same from one population to another, which is not necessarily the case. A further weakness is that in backward societies there may be many potential geniuses who are unable to realize their talents because of inadequate education and the unsophisticated culture. These difficulties have meant that Galton's argument has not been taken very seriously by later investigators (e.g., Loehlin *et al.*, 1975).

S. D. PORTEUS AND THE MAZE TEST

Galton was considerably handicapped in his work on intelligence by the fact that despite much effort he did not succeed in inventing the intelligence test. This had to await the genius of Alfred Binet in France in the first decade of the present century. His test was soon translated into English by Lewis M. Terman at Stanford University, and it became, and remained for many years, the foremost intelligence test in the English-speaking world. Once Binet had shown how an intelligence test could be constructed, a number of other psychologists began to devise their own tests. Among them was a young psychologist in Australia named Stanley Porteus, who began work during World War I. His idea was to test intelligence by means of a series of paper and pencil mazes that required accurate tracing from the starting point through to the exit. He made up a series of mazes of different levels of difficulty, ranging

[1] Galton (1869) attempted to estimate quantitative differences in average intelligence among several ethnic groups by comparing the proportions of people who historically had reached outstanding levels of attainment in the various groups. He tried to supplement this by personal observations of work habits and by foreign travelers' reports. Using a 16-category, equal-interval scale, ranging from idiots to geniuses, Galton concluded that Australoids were "at least one grade below" Negroids, who were themselves "not less than two grades" below typical English Caucasoids, who were in turn "a fraction of a grade" below Scottish and northern English Caucasoids. A "grade" on Galton's scale corresponds to 0.695σ, or 10.425 points on an IQ scale with $M = 100$, $\sigma = 15$ (Jensen, 1973, p. 70). Assuming Galton's "ordinary Englishmen" were about average, this would imply for the ethnic groups of his day mean IQs of approximately 68.72 for Australoids, 79.15 for African Negroids, 100.00 for Englishmen, and perhaps 102.6 to 105.2 for Scotsmen (i.e., one fourth to one-half a grade). Although Galton's work was necessarily crude and liable to many sources of error, his estimates turn out to be tolerably close to modern measurements, as reported herein by Professor Lynn. [the Editors.]

from easy ones that could be done by 5- and 6-year-olds and increasing in difficulty up to tests for 14-year-olds. The Maze Tests were standardized on different age samples of children so that it was possible to determine which tasks could be done correctly by the average child of any particular age. This gave a mental age (MA) for each maze, as in the Binet test.

Porteus himself has been chary of calling his test a measure of intelligence and has preferred to regard it as a measure of planning ability. However, planning ability is an important component of intelligence. Numerous studies have shown that the Porteus Maze Test is an excellent measure of general intelligence, correlating about .60 with Binet tests, so the Maze is now widely accepted for this purpose (Porteus, 1961, 1965, 1967). Incidentally, Porteus (1967) has long criticized the term Intelligence Quotient (IQ), denoting the useful ratio of mental to chronological age (MA/CA), because no single test is an adequate measure of all-round behavioral adaptability. Rather, he insists, the proper term for the relationship is Test Quotient (TQ).

Quite early in his career, Porteus was interested in the possibility of using his Maze Test to measure the mental abilities of the Australoids. His first study was conducted in 1915, and he published the results a couple of years later (Porteus, 1917). His sample consisted of 28 children of mixed Australoid and Caucasoid stock. He found that they scored a little lower than the Caucasoid standardization samples, noting that this poor performance was more pronounced among the older children. The number of children tested in this study was hardly sufficient to yield any definitive conclusions. But in 1929 Porteus carried out another study, this time on 56 adult pure Australoids. He asserted that the Australoids were manifestly interested in the test and were perfectly able to grasp the nature of the problems, but they had genuine difficulty in solving the complex mazes. The group achieved a mean MA of 10.48 years on the tests. Two subsequent investigations on Australoids have been made using the Porteus Mazes. The first was carried out on a sample of 24 Australoids by M. Piddington and R. Piddington (1932) and showed a mean MA of 10.52. The second was done some 30 years later by Gregor on a group of 50 adult male Australoids; it revealed a mean MA of 10.4 (Porteus & Gregor, 1963). An account of the whole set of studies is given by Porteus (1965). In all three samples, the numbers of cases are small. On the other hand, the MAs obtained are quite similar in central tendency, so they are probably fairly reliable. The mean MA is very low and corresponds to an IQ in the range of 50–70.

Porteus's studies of 1915 and 1929 were among the first to find that primitive peoples living outside advanced Western societies do poorly on intelligence tests, a finding that has been confirmed many times on different groups. The question that such results raise is what precisely they mean. This, in turn, raises the question of what is meant by "intelligence" and IQ. For Porteus, the Maze Test taps hereditary talent.

A considerable source of confusion in this field is that there are several different meanings of the term "intelligence." We must now pause to clarify these. Three major usages have to be distinguished. First, there is intelligence as "innate, general, cognitive ability," as Burt (1972) defined it. This has also been called Intelligence A by Hebb (1949), or genotypic intelligence in biological terminology. Second, there is the set of mental skills actually exhibited by an individual in performance, Hebb's Intelligence B, that is, the psychological phenotype. An individual's behavioral skills are not entirely a reflection of his innate intelligence since they are partly determined by his environment. Third, there is the score that an individual obtains on an intelligence test, his IQ, which Vernon has designated Intelligence C (Vernon, 1969). An IQ estimates, with varying degrees of accuracy, an individual's psychological aptitudes or innate abilities.

Of these three meanings of intelligence, the only one measured directly is the third one, the IQ or intelligence test score. To what degree is it legitimate to argue from one's IQ (Intelligence C) to one's practical mental skills (Intelligence B), thence to the genotype (innate intelligence or Intelligence A)? Some experts argue that an IQ provides a reasonably good index of both actual mental skills and innate aptitudes. Others disagree. Thus, to take the racial difference in intelligence that has received most attention, everyone agrees that American Negroids have a mean IQ of approximately 85 as compared with an American Caucasoid mean of 100. But there is dispute about whether intelligence tests provide a valid measure of Negroid mental abilities, some authorities arguing that the Negroid is handicapped in his performance on the test because it is couched in a Caucasoid middle-class idiom, administered by a Caucasoid tester, and so on. There is even more dispute about the degree to which the test results reflect innate differences. Quite a number of authorities are willing to concede that the average American Negroid does indeed operate at a significantly lower level of mental ability than the average American Caucasoid, as suggested by the intelligence tests, but they are not prepared to allow that this is due to a genetic difference. This problem of the validity of the intelligence test arises whenever comparisons are made between the test results of Western populations and groups outside Western culture, such as Indo-Dravidians, Negroids, Australoids, Eskimos, and Negritos. The reason that the validity of the tests is questioned is that different cultures foster different kinds of intellectual skills. For instance, in Western society, people need to be literate and numerate, but in Australoid society they need to be good at throwing boomerangs. Hence, it is reasonable to measure the mental abilities of Western populations by giving them tests of language and number but not to apply these tests to people in other cultures who have developed their mental abilities in other directions.

While these distinctions in the meaning of intelligence may be confusing, most psychologists who are well informed in this area would probably agree that as long as we are working in a Western culture, intelligence tests give a fairly accurate measure of mental ability (Intelligence B, phenotypic intelligence) and a reasonably accurate, though certainly less adequate, measure of innate aptitude (Intelligence A, genotypic intelligence). But this does not hold outside Western culture. At the borderline of this culture there is some difficulty, which we shall be taking up later when we consider the IQ results for different ethnic populations.

To return now to Porteus's finding that the Australoids have mean IQs in the range of 50–70, we probably ought to conclude that the Australoids tested by Porteus were too far removed from Western culture for the test to give a valid index of their level of mental ability. The reason for this lies in the lack of familiarity of Australoids with the task presented in the test. The essential concept required for the Maze Test is that of the cul-de-sac, as contrasted with the through road. These are familiar enough concepts to people brought up in houses and in towns but not to people like the Australoids reared in the Australian bush. There are no cul-de-sacs in the deserts of central Australia.

R. B. CATTELL AND THE CULTURE-FAIR INTELLIGENCE TEST

By the 1930s there was a growing realization that most intelligence tests were heavily loaded with material that reflected the intellectual skills developed in advanced societies. How, then, might it be possible to measure the intelligence levels of primitive peoples where these intellectual skills were not acquired? An attempt to solve this problem was made by a British–American psychologist, R. B. Cattell, who devised a so-called "culture-free intelligence test," later more generally known as the Culture-Fair Test (Cattell, 1940). His idea was to produce a test whose problems were universally familiar, such as those involving the detection of an odd item among a set, the principle underlying a series, and so on. In this way, Cattell maintained, his Culture-Fair Test would measure hereditary mental aptitude, uncontaminated by the particular intellectual skills acquired in a given culture.

Cattell's Culture-Fair Test needs to be understood in the context of his theory of intelligence. For Cattell, there is not one general intelligence, as maintained in the classical theory of intelligence derived from Spearman, but two. The first kind Cattell calls *fluid* intelligence, which is pure mental ability, the second *crystallized* intelligence, which consists of learned cognitive skills. The form of crystallized intelligence depends on the cognitive skills that are taught and practiced in a particular culture. Fluid intelligence is free from these cultural conditions, although not from broader environmental effects such as severe malnutrition, birth injury, and so on. Thus, fluid intelligence is not innate ability, or Intelligence A, but pure mental aptitude independent of cultural artifacts. Fluid intelligence is measured by the Culture-Fair Tests, whereas many conventional intelligence tests measure crystallized intelligence.

Cattell's claim that his test measures fluid intelligence unaffected by cultural influences has met wide acceptance. His Culture-Fair IQs are not as highly correlated with socioeconomic status (SES) as are conventional intelligence tests (Cattell, 1971), an indication that they are less affected by the cultural environment. Operationally, one distinction between fluid and crystallized intelligence is that provided by the difference between culture-fair and culture-loaded tests, respectively.

Another argument for regarding the Culture-Fair Test as independent of social and economic conditions comes from its correlation with critical flicker frequency (Barratt, Clark, & Lipton, 1962). The critical flicker frequency (CFF) threshold is the speed at which a flashing light can no longer be seen as flickering and is instead perceived as a steady light. This CFF threshold may be the point at which the brain fails to process information accurately. Barratt, Clark, and Lipton carried out two studies, in both of which they found that the CFF threshold was significantly correlated with IQs measured by the Culture-Fair Test but not with those of more culture-loaded intelligence tests. This result seems to indicate that the Culture-Fair Test gives a measure of the efficiency of the brain on a simple perceptual task that should be independent of the particular skills acquired in different cultures since in all cultures people need to be able to perceive accurately.

Apart from Cattell's Culture-Fair Tests, there are others that have frequently been given to people in a wide range of different societies on the assumption that such tests are largely culture fair. The most commonly used is Raven's Progressive Matrices, a test of perceptual reasoning involving geometric designs and the completion of two-dimensional sequences. Another test that has been used in a number of studies is Goodenough's Draw-a-Man (DAM) Test, in which the subject's drawings of a man and a woman are scored for completeness of structure and detail. Not all critics are willing to concede that such tests are culture fair. But the problem of the interpretation of test results will be set aside for the moment and taken up later in this chapter.

ETHNIC AND NATIONAL DIFFERENCES IN INTELLIGENCE

The major results of studies on ethnic and national intelligence levels will now be reviewed swiftly and summarily. Most of the intelligence tests that have been given in different parts of the world are American or British, standardized on American or British Caucasoid populations yielding IQs with $M = 100$ and $\sigma = 15$ or 16, and it is convenient to use this scale for considering the performances of people in other parts of the world.

Caucasoids

It has generally been found that populations of northern European extraction have mean IQs of approximately 100. Thus, well-drawn samples of Scottish children tested in 1932 and 1947 on the American Stanford–Binet and Terman–Merrill tests obtained mean IQs of approximately 100 (Scottish Council for Research in Education, 1933, 1949). Garth's (1931) review of American Army Alpha Test data from World War I immigrants showed Scots, English, and northern Europeans at the top of the scale. In New Zealand, 26,000 children of European extraction obtained a mean IQ of 98.5 on the American Otis Test (Redmond & Davies, 1940). On the same test, an Australian sample of 35,000 children had a mean IQ of approximately 95 (McIntyre, 1938). In Belgium, a standardization of Cattell's Culture-Fair Test gave Belgian children a mean IQ of approximately 104 (Goosens, 1952). In France, children drawn as a representative sample and given Raven's Colored Progressive Matrices also obtained a mean IQ of approximately 104 (Bourdier, 1964).[2] In the city of Rostock, East Germany, the mean IQ of children on Raven's Colored Progressive Matrices is approximately 100 (Kurth, 1969). In Denmark, representative samples of children tested on Raven's Progressive Matrices obtained mean scores that are virtually the same as the original standardization samples in Britain (Vejleskov, 1968).

In the countries of southern Europe, the mean IQ appears to be somewhat lower (Hirsch, 1926), except perhaps in Italy in which a recent sample of adolescents in Florence had a mean IQ around 100 on Raven's Test (Tesi & Young, 1962).[3] However, in Spain, the mean IQ of 113,749

[2] All Progressive Matrices data in this chapter have been transformed to IQs with a British $M = 100$ and $\sigma = 15$.

[3] These results conflict with Garth's (1931) review of the IQs of Italian immigrants to the USA between the two world wars. They averaged about 1σ lower on individual Binet tests ($M = 83.97$, $N = 500$), as well as on nonlanguage group tests ($M = 84.80$, $N = 446$). By contrast Swedish immigrants performed slightly above the American norms ($M = 102$, $N = 419$) during the same period (computed as weighted means from Garth, 1931, p. 80). Lowest of all Europeans were the Portuguese ($M = 82.7, N = 671$). [the Editors].

army conscripts tested on Raven's Matrices in 1965 was roughly 87 (Nieto-Alegre, Navarro, Santa Cruz, & Domínguez, 1967). In Zagreb, Yugoslavia, a sample of schoolchildren obtained a mean IQ of 89 on the same test (Sorokin, 1954). In Greece, children in the city of Thessaloniki obtained a mean IQ of approximately 89 on the Wechsler Performance Scale (Fatouros, 1972). The Caucasoid race extends eastward from Europe into the Near East and India. Samples from Baghdad in Iraq tested on the Goodenough DAM Test have mean IQs around 80 (Al-zobaie, 1965). In Iran, the mean IQ of children in the city of Shiraz is in the low 80s (Mehryar, Shapurian, & Bassiri, 1972). Some authorities identify the local race here as Iranian.

In India, there is a considerable literature on intelligence testing. Fifty years ago, the Stanford–Binet was given to a sample of students at the University of Calcutta; the mean IQ was found to be 95 (Maity, 1926). A more recent investigation, using a small sample of 25 postgraduate students at the University of Calcutta, who took Raven's Test, produced an incredibly low mean IQ of 75 (Sinha, 1968). Several Indo-Dravidian studies have employed Raven's Test. Sinha's review provides data for 17 groups of children aged between 9 and 15 years drawn from a variety of Indian states and numbering in excess of 5000 cases. All the mean IQs lay in the range from 81 to 94, the overall mean being about 86.

If these results from Caucasoid, Iranian, and Indo-Dravidian nations are considered in the light of the racial composition of the populations, it is apparent that where the people are predominantly of northern European stock, as in Britain, northwestern Europe, the United States, Australia, and New Zealand, their mean IQs are approximately 100. The other Caucasoid peoples inhabiting the more southerly latitudes from Spain through the Middle East to India score substantially lower. This also holds true for immigrants to the United States (Garth, 1931; Goodenough, 1949) and for their offspring (Hirsch, 1926, Pintner, 1931).[4]

[4] Based on U.S. Army data from World War I recruits, the Combined Scale (8 Alpha subtests, 4 Beta subtests, and Stanford–Binet) gave the following data for 11,446 foreign-born Caucasoid immigrants.

Northern Europeans and British:	$M = 13.09$,	$N = 6{,}442$
Southern and Eastern Europeans:	$M = 11.64$,	$N = 5{,}004$

As for the 3,627 children of Caucasoid immigrants, examined mostly with nonlanguage tests during the same decade, the following IQs were reported and compared here with 1,030 native-born American Caucasoid children.

Native Americans:	$M = 98.3$,	$N = 1{,}030$
Northern Europeans and British:	$M = 97.4$,	$N = 2{,}336$
Southern and Eastern Europeans:	$M = 85.4$,	$N = 1{,}291$

(Footnote continues on p. 270)

Negroids

The Negroid or Congoid race originally inhabited west-central Africa (the Congo region), and it is only during the last 1500 years that Negroids have spread over most of Africa south of the Sahara (Coon, 1965; Darlington, 1969). Further, from the sixteenth century onward, groups have migrated voluntarily or otherwise to various parts of the world. Substantial numbers now reside in the United States, the West Indies, and Great Britain. There are few satisfactory studies of the intelligence of Negroids in Africa, largely because it is difficult to obtain accurate information on their ages, a necessity for the calculation of children's IQs. There are additional difficulties in obtaining representative samples of the population.

By far the greatest amount of research on Negroid intelligence has been conducted in the United States. The standard work is that of Shuey (1966), who reviewed 382 studies that employed 81 different tests of mental ability. The results taken as a whole indicate that the mean IQ of American Negroids is approximately 85 and that they score somewhat higher on verbal tests than on nonverbal tests. Of course, Negroids in the United States are not necessarily representative of all Negroids, and there are two reasons in particular for regarding them as atypical. One is that they are descended from slaves who were either purchased or captured by tribal chiefs and were probably selected for nonintellectual traits and genetic docility (Darlington, 1969, pp. 650–668). Another reason is that many Afro-Americans have some Caucasoid ancestry, the average admixture being in the region of 20–30% (Reed, 1969). In both these respects, they are unrepresentative Congoids. That the average IQ of American Negroids is around 85 is not a subject of much dispute, but the explanation of the population mean is a matter of considerable controversy. (See Jensen, 1969, 1972, 1973.) Negroids from the Southeastern United States, who tend to be less Caucasoid than in the North and West, have both lower IQs and smaller dispersions. Based on 1800 representative elementary schoolchildren tested by Kennedy *et al.* (1963), M = 80.7 and σ = 12.4 (see also Jensen, 1973).

A few studies of the intelligence of Negroids are available from other

---

It is obvious that sizable and consistent average differences appeared among the test scores of various ethnic and linguistic groups of Caucasoids comprising the European migrations to the United States around the time of World War I. Mean differences among these groups, however, may be attributed to various factors: climatology, hereditary aptitudes, biased selection, differential socioeconomic status, training, and education, language barriers, and diverse customs and attitudes. There are also intriguing test-score differences associated with religious affiliation, Jews tending statistically to outrank Protestants as the latter outrank Catholics (Weyl & Possony, 1963).

parts of the world. From Uganda, Silvey (1972) reported an investigation in which Raven's Test was given to approximately 470 children who had successfully passed the examination for entrance into secondary schools. The mean IQ of this group was about 88, but the sampling suggests that this value may be higher than the mean of the general population. Another Uganda study was conducted by Vernon (1969). His sample was composed of 50 boys aged 12 years taken from two schools in Kampala, the capital city. The average SES of the boys' families was described as much higher than that of the general population in Uganda. Vernon gave a number of cognitive tests, including the Terman vocabulary scale and Kohs Blocks, and the nine best measures of general intelligence produced a sample mean IQ of approximately 80.

The level of intelligence in Jamaica is discussed by Vernon (1969). Large numbers of the children there take British intelligence tests for selection at the age of 11 for secondary schools, and the results indicate that the mean IQ as assessed by these tests is approximately 75. Vernon has also carried out a study of his own using individual tests on a sample of 50 boys. He gave a variety of cognitive tests, including the Terman vocabulary scale and Kohs Blocks. The mean IQ of the Jamaican children was in the low 80s. These children tended to do better on verbal and educational tests than on nonverbal and spatial tests, as in the United States.

A study of the intelligence of 2959 Negroid children in Tanzania has been made by Klingelhofer (1967). He administered Raven's Progressive Matrices to a sample consisting of approximately one-fifth of all adolescents in the first three forms of secondary schools in the country; the mean IQ was approximately 88. (Also tested in the same study were 727 Asian children drawn as part of the same sample. The mean IQ of the Asians was approximately 98.) In Ghana, Jahoda (1956) has reported a study of 317 boys attending schools in Accra. They were given Raven's Progressive Matrices. The mean score was not cited, but it is apparent from the data reported that the mean IQ was approximately 75.

A number of studies of the intelligence of Negroids have been conducted in South Africa. A typical result is that of Lloyd and Pidgeon's (1961) investigation of 275 Negroid children in Natal. The children were at two schools, one urban and one rural, and were considered a fairly representative sample. The test used was one of the British National Foundation for Educational Research nonverbal tests; the Negroid children obtained a mean IQ of 87. A standardization of Raven's Progressive Matrices on the Zulu population of South Africa was made by Notcutt (1950). The mean IQ of this sample of 1220 Zulu school children aged from 8 to 16 was approximately 81. Also tested were 703 adults, who obtained a mean IQ of approximately 75.

Mongoloids

Little is known about the intelligence levels of Mongoloids in their homelands. The majority of studies have been made on Chinese and Japanese emigrants to the United States. One of the first of these studies was by Yeung (1921) on 109 Chinese immigrant children aged 5–14. The children were tested on the Stanford–Binet and obtained a mean IQ of 97. The study was a small one, but the essential finding that Mongoloid groups score close to the means of Caucasoids has now been confirmed by several later investigations. One of the best studies (Coleman *et al.*, 1966) drew more than half a million children from all over the United States. The Coleman Report noted that Mongoloid children obtained approximately the same mean nonverbal IQs as Caucasoid children. They placed somewhat lower on verbal tests, but this is probably because in many cases they did not speak English at home. The nonverbal test should be taken as the best index of their intelligence.

An investigation using approximately 10,000 California children enrolled in kindergarten and the first four grades, employing the Gesell Institute's Figure Copying Test, revealed that Mongoloid youngsters generally exceeded the mean scores of Caucasoid as well as Mexican and Negroid children (Jensen, 1973, pp. 304–305). For each of the four ethnic groups, the test was found to be heavily loaded with a general intelligence factor (Spearman's *g*), somewhat like that of Raven's Matrices, and the test appeared to be quite status fair based on a composite index of socioeconomic status (SES). Mexican–American children, for example, placed in the lowest of these categories (SES = 6) but scored quite near two groups of Anglo-American children (SES = 1,3) in figure-copying skill. The Chicanos were consistently more proficient than two groups of Negroids (SES = 4,5) at all school grades.

A further study of 1703 Caucasoid, Mexican–American, and Negroid schoolchildren in California, using the culture-fair Raven's Matrices and the culture-biased Peabody Picture Vocabulary Test, gave more complex results. To quote Jensen (1973), "California Orientals bear a similar relationship to whites as the Mexicans bear to the Negroes, that is, a higher average genotype and lower average environmental advantages (p. 312)." The shortcoming of these studies of Mongoloids in the United States is that they are not necessarily representative either of China or Japan, whence the subjects' ancestors mainly originated, or of Mongoloids as a whole.[5]

One of the relatively few studies of the intelligence of a Mongoloid

[5] For Chinese, Garth (1931) cites IQ means from 513 subjects in Hawaii ($M = 99.3$) and from 224 subjects in Vancouver ($M = 107.2$) that equal or exceed the Caucasian norms. For Japanese, the IQ medians from 536 subjects in Tokyo (median = 99.0) and from 276 subjects in Vancouver (median = 114.2) are similar, if not higher. It may be noted that Porteus (1967) found Mongoloids from the island of Saipan and Amerinds from North America also surpassed average Caucasoids on his Maze Test.

population in Asia is that of Rodd (1958) on a sample of children in Taiwan. He used Cattell's Culture-Fair Test and found that Taiwanese children obtained approximately the same mean IQ as American children. Unfortunately this study has not been published. An estimation of the intelligence of native Japanese can be made from the normative data of the Wechsler tests. The Wechsler Intelligence Scales for Children (WISC) and for Adults (WAIS) were standardized on 1070 children and 1682 adults in Japan in the early 1950s. Many of the performance subtests were retained unaltered in the Japanese versions of the WISC and WAIS, so it is possible to use the Japanese means on these to estimate mean IQs for the Japanese normative samples. In addition, the Wechsler Preschool and Primary Scale of Intelligence (WPPSI) was standardized on 600 children in the late 1960s. Taking a weighted average of these three studies, based on 3352 Japanese cases, the resulting mean intelligence in terms of an IQ of 100 for American or British Caucasoids, is a remarkable 106.6 (Lynn, 1977). This appears to be the highest mean IQ ever recorded for a national population. A small but apparently representative sample of Chinese boys in Singapore scored significantly higher on Raven's Progressive Matrices than the British standardization sample (Phua, 1976).

Mongoloids extend into Indonesia. A study of children's intelligence was conducted in Bandung, Java, a city of about one million inhabitants (Thomas & Sjah, 1961). The test used was Goodenough's Draw-a-Man (DAM) Test. Schools were selected by random sampling and the test was given to every child who was present on a particular day. Since not all children in Bandung attend school, the sample is probably biased in favor of middle and upper SES categories. Unfortunately, the authors did not adjust the sample to make it representative of the city. Results are reported for ages from 5 to 12; the overall mean IQ was approximately 96. This is probably an overestimate because of the defective sampling. Furthermore the DAM Test is not a particularly good one. Nevertheless, the mean is a high one for an economically undeveloped country.

The Eskimos are a Mongoloid sub-race, living above the Arctic Circle. There has been some interest in Eskimo intelligence following a study by Berry (1966) purporting to show that a group of 14–15-year-old Eskimos had approximately the same mean IQ as Scots in Scotland. However, the claim cannot be accepted. Berry's study consisted of two samples of Eskimos drawn from different localities. The test on which they scored almost as highly as the Scots was Raven's Colored Progressive Matrices. The Eskimos achieved a mean score of about 28. Reference to Raven's norms shows that 28 is approximately the mean score obtained by British 11-year-olds, so the Eskimos had a mean MA around 11; this suggests an IQ in the range 70–80. Berry's Scots scored about 10 IQ points higher than the Eskimos. A few years later, Vernon (1969) published a study of the intelligence of 50 Eskimo children.

He administered a number of tests, and on the nine best measures of general intelligence, the Eskimos obtained a mean IQ of approximately 85. On the other hand, MacArthur (1969) found that Arctic Eskimos placed at or above Caucasoid norms for Canada using Raven's Test. On Piagetian clinical tests, the Eskimos outperformed Caucasoid Canadian children from urban areas, and Amerinds in Canada were similar to the Eskimos (Jensen, 1973). Let us examine the Amerind studies more closely.

Amerinds

Quite early in the century, a number of studies were made of the cognitive abilities of Amerinds living on reservations in the United States. Evidence up to 1930 was summarized by Garth (1931) and Pintner (1931). All the results showed that the mean intelligence levels were lower than those of American Caucasoids, but there was considerable variability, the mean IQs ranging between 69 and 97. Some of these early studies also revealed that Amerinds tended to do relatively better on tests of nonverbal and spatial ability than on verbal tests. These findings have several times been substantiated by later research (Tyler, 1965).

The most extensive investigation of the cognitive abilities of Amerinds is that made by Coleman (1966) as part of his study of a sample of over half a million children drawn from all parts of the United States and including the major ethnic minorities. A number of tests were given, including what are described as verbal and nonverbal achievement tests. Coleman states that "these tests do not measure intelligence," but this can hardly be the case since it has usually been found that cognitive tests serve as good measures of general intelligence *(g)*. If Coleman has indeed found some tests of verbal and nonverbal achievement, mathematics, and general information that do not have *g* loadings, then he has made a remarkable discovery. But probably he simply wished to avoid the controversy aroused in the United States by the word "intelligence." No doubt this is also why he gave the test results a mean of 50 instead of the 100 associated with IQs. If his results are transformed into IQs, they show that in relation to American Caucasoid IQs of 100, Amerinds have a verbal IQ of approximately 91 and a nonverbal IQ of approximately 96. It will be noticed that this study confirms the earlier results, indicating that Amerinds have relatively higher nonverbal than verbal abilities. The Coleman Report also revealed that, in spite of Amerinds having a lower average SES than American Negroids, their mean aptitude and achievement scores are higher; indeed, by about the same margin as that by which the Caucasoids exceed the Amerinds.

Forty Amerind children living in Canada were studied by Vernon (1969). On the nine best measures of general intelligence, they obtained a mean IQ of approximately 79. Once again, this group showed better spatial than verbal ability.

Back in the 1920s, Garth and others had found positive correlations in the neighborhood of .41–.42 between IQ and degree of Caucasoid admixture among Amerinds on government reservations and in Amerind schools. The subjects were administered the Otis and the National Intelligence tests. Garth (1931) preferred an interpretation that was primarily environmental, suggesting that language handicap was a particularly detrimental factor among the full-blooded Amerinds of the American Southwest.

Australoids

Variously referred to as the Australasid, Australid, or Australoid race (Baker, 1974; Coon, 1965), the full-sized aborigines of Australia, New Zealand, and India, the hybrids of Indonesia, Micronesia, and Polynesia, along with the dwarfed Negritos of the Philippine Islands, the Malay Peninsula, and the Andaman Islands doubtless comprise some of the most ancient-appearing yet handsome of all living peoples. Comprising only 0.4% of the world's population, the Australoids represent one of the endangered subspecies of mankind.

Groups of Australoids were given intelligence tests early in the century by Porteus, as we saw earlier, but these groups were probably living in conditions too remote from European culture to give valid results. But in recent decades many Australoids have been brought up in Western society and now attend ordinary schools with European children. In these conditions, intelligence tests probably give reasonably valid measures of mental ability. A number of studies have been conducted. They show that Australoids invariably obtain mean scores below Australian Caucasoids (Kearney, de Lacey, & Davidson, 1973). A typical study is that of Bruce, Hengeveld, and Radford (1971). They attempted to find all the Australoid children aged 5–13 attending primary schools in the state of Victoria. Obtaining a group of 83, they administered a number of tests, including the Peabody Picture Vocabulary Test and the Illinois Test of Psycholinguistic Abilities. The mean IQ of the Australoid children was approximately 80.

A similar result was published by McElwain and Kearney (1973), who devised a nonverbal intelligence test called the Queensland Test. It resembles the performance scale of the Wechsler and was specifically designed for the native peoples of Australasia, but many Europeans have also taken it. When the Queensland Test was given to over 1000 Australoids, the mean IQs of the samples varied between approximately 78 for those who live in isolation to approximately 85 for those in close contact with Caucasoid Australian society. The second mean of 85 can probably be taken as a reasonably valid index of the present level of Australoid mental ability because these youngsters go to school with European Australians and are brought up in an advanced Western culture.

A number of research studies have been conducted on the intelligence of the Maoris in New Zealand. St. George (1972) concludes in a recent review that they do not perform as well as Caucasoids on a variety of tests, including the Otis, Wechsler, Raven, and Thurstone Primary Mental Abilities. One of the better investigations is that of Du Chateau (1967). He tested 236 Maori and 719 Caucasoid adolescents on the Otis, which is a well-standardized test in New Zealand on which New Zealand Caucasoids score approximately at the same level as American Caucasoids. The mean IQ obtained by the Maori sample was 84. Probably the best study using a nonverbal test is that of R. and A. St. George (1975) using the Queensland Test. The mean IQ of these Maoris was approximately 94.

The Micronesians are Australoid–Mongoloid hybrids of small stature (M = 5 feet 4 inches) who inhabit a group of the Pacific islands to the north of New Guinea. Largest of these islands are the Carolines, the Marianas, and the Marshalls. Many of the children attend village schools administered by the United States government in which they are taught basic subjects. A sample of about 400 Micronesian youngsters aged 12–18 were given Cattell's Culture-Fair test by Jordheim and Olsen (1963). The mean IQ was roughly 88, but this may be an overestimate because the test was given without time limits.

The Polynesians, who are also hybrids but generally taller (M = 5 feet 7 inches) and less variable in appearance than the Micronesians, are found in the Pacific islands lying broadly northeast of New Zealand. Among them are the Cook Islands, where a survey of intelligence has been carried out by A. St. George (1974). She used the Pacific Infants Performance Scale, a nonverbal intelligence test, and reported norms for both New Zealand and Cook Island children in the age range from $4\frac{1}{2}$ to $7\frac{1}{2}$. The mean IQ of the Polynesian children was about 88, like that of the Micronesians to the north.

Other Pacific islanders of mixed Mongoloid–Australoid origin are the Melanesians and Papuans inhabiting New Guinea and the islands east of Australia, but they have received little psychometric attention.

Capoids

The Capoids, or Khoisans, inhabit the southern tip of Africa and consist mainly of the hybrid Hottentots of Capetown and the relatively purer Bushmen of the Kalahari Desert. Baker (1974) calls them Khoisanids, whereas Coon (1965) calls them Capoids (after the Cape of Good Hope).

Numbering only 126,000 souls, less than .01% of the world's population, very little of a quantitative nature is known of their psychometric intelligence. Porteus visited the Kalahari in 1934 and administered his Maze Test to 25 adult Bushmen (Porteus, 1937). These small (5 feet ± 2 inches) Capoids achieved a mean MA of 7.6 years, which, considering their CAs, corresponds to an IQ range of 50–60. It is doubtful whether this can be regarded as a valid measure of their intelligence because of the lack of familiarity of desert-living Bushmen with some of the procedures and "civilized" aspects (e.g., cul-de-sac) of the Maze Test. Furthermore, it strains one's credulity that a population could long survive the rigors of the Kalahari with a true mean IQ around 55.

Bushmen display an impressive level of practical intelligence in their daily lives. They hunt skillfully with bow and arrow, treating the arrowheads with a lethal substance concocted from snake venom, plant juices, or local beetle poisons. They shrewdly gather nuts all year round, eating meats and vegetables whenever possible, and they move their camps systematically to take advantage of the water holes' variable contents. Bushmen also exploit the occasional desert thunderstorms by filling ostrich eggs with water and burying them in the sand for future use. As hunter–gatherers, Bushmen may be the world's finest. According to Coon, the superb hunting technique of these remarkable people "sets a premium on litheness, endurance, economy in consumption of water, and a defiance of the elements (1965, p. 111)." More recently, Coon (1971) has argued that hunting is fully as complex an industry as crop growing or cattle raising; hence, one may conclude that the intelligence of neither the African Bushmen nor the Australoids should be underestimated. (See also Porteus, 1967.)

CAUSES OF NATIONAL AND ETHNIC DIFFERENCES IN IQ

We come now to the problem of the causes of the different mean IQs obtained by various ethnic, racial, and national populations. That such differences do exist is about the only thing in this field on which there is agreement. When it comes to the variables responsible for these divergences, there are wide differences of opinion. Broadly, five positions are taken, which we will consider in turn.

The first is that despite possible appearances to the contrary, all ethnic groups have much the same innate intelligence. It is not easy to find evidence to support this position. It has been argued that all the races of mankind evolved from a common stock and that there is no particular reason to suppose that they would have developed different average levels of cognitive ability (Comas, 1961; Garth, 1931). Alternatively, it has been argued that some races have been exposed to more difficult conditions, such as a more severe climate for those in northern latitudes, where there would have been stronger selective pressures for high intelligence (Weyl & Possony, 1963). Some writers have attempted to establish the similarity of the innate intelligence of underdeveloped peoples and those in technologically advanced societies by comparing the mean scores of the less developed peoples with the scores of lower SES samples drawn from advanced nations. This argument is advanced among others by Berry (1966) in comparing the intelligence test scores of Eskimos with those of lower SES Scotsmen living in Scotland. He argues that this group is most appropriate for comparative purposes because the members are reared in a relatively unsophisticated and impoverished environment like that of the Eskimos. The lower SES classes in Scotland have mean IQs in the range of 85–100, which is only a bit higher than the IQ range of Eskimos. Thus, the argument runs, if we take northern European Caucasoids brought up in a similar environment to Eskimos, the IQ difference is reduced. Therefore, the innate intelligence must be approximately the same.

The critic of this argument will not admit that lower SES groups in Britain can be taken as a random sample of the total British population with respect to innate intelligence. In Britain, there has been social mobility for a number of centuries during which more intelligent persons have tended to rise in the SES hierarchy. The result is that now there are genetic differences in intelligence between these classes (Burt, 1961). If this argument is accepted, it is not legitimate to take lower SES groups in Britain as representative of the total population.

A second point of view is that intelligence tests do not give valid measures of mental ability beyond the culture for which they were designed; hence, many of these mean IQs are invalid and may even be meaningless. Critics of this position reply that many nations and populations share the same culture sufficiently for the tests to give approximately valid measures of mental aptitude. It is noted that American and British intelligence tests apparently give valid results in Belgium, Denmark, France, East Germany, New Zealand, and Australia since all these populations obtain mean IQs around 100. Why, then, should not the tests give valid results in Spain, Yugoslavia, or Greece? And if tests give approximately valid results here, why not also for children attending schools elsewhere in the world? The mean IQ differences, a critic will assert, are too great to explain in terms of the unfairness of the tests as measures of mental ability. The position that the intelligence test results are invalid because they are not culture fair implies that the populations that obtain low mean IQs have the same *fluid intelligence* as those of advanced societies. If this is so, it should be possible to devise tests on which these populations perform as well as the populations of northern Europe, the United States, Australia, and New Zealand. The critic will say that those who take this position ought to produce such tests. In 70 years of intelligence testing, no such test has yet been found.

One of the reasons sometimes advanced for the supposed lack of validity of intelligence test results on populations outside advanced societies is that these populations lack practice or test sophistication. One of the best studies of this issue is that of Lloyd and Pidgeon (1961) on European, Negroid, and Indian children in South Africa. They found that after coaching on the principles of the test, European children gained 10.6 points, Negroid children 14.6 points, and Indian children 6.1 points. The authors state that these differences show that the test is not culture fair and that valid conclusions cannot be drawn from test results on different racial groups. But this inference does not seem to follow. It is not part of the theory of culture fair tests that the principles involved in the problems cannot be acquired by coaching. In this study, on first testing, the European children obtained a mean IQ of 103, and the Negroid and Indian children mean IQs of 87. The point here is that the European children were better at seeing the principles of the test for themselves when the test was first given. And even after the principles of the test had been explained, the European children still scored substantially higher on the test, and the gap was hardly diminished. One conclusion indicated by this study is that neither practice nor coaching do much to reduce the superiority of children of northern European origin on intelligence tests.

A third position on mean IQ differences among different ethnic groups is to concede that the mean IQs are valid as approximate measures of the average levels of mental ability among the various nations and populations but to maintain that these differences can be readily explained in terms of known environmental factors. The four most important of these are probably the standard of living, the quality of education, the general intellectual sophistication of the culture, and the level of nutrition. Taking the standard of living first, there is certainly a striking association between the per capita incomes of populations and their mean IQs. The affluent nations of northern Europe, the United States, New Zealand, Australia, and Japan have mean IQs around 100, whereas the underdeveloped nations have mean IQs around 75–90. The same association between income and mean IQ has often been found within nations, the higher SES groups tending to have both higher incomes and higher mean IQs. This is notably true in the United Kingdom, the United States, and the USSR.

It is often asserted that income, or the various socioeconomic advantages associated with income, determines intelligence, but it is by no means certain that this is the case. It can be argued equally that intelligence is an important determinant of income. Intelligent people are able to acquire the complex cognitive and psychomotor skills for which others are prepared to pay high prices, for example, the skills of physicians, engineers, inventors, administrators, and so forth. Less intelligent people find it hard or impossible to acquire these skills; most of them are only able to work in simpler occupations for which there is less demand and for which others are not prepared to pay so highly. This principle would be expected to operate both within nations and among nations, so that an intelligent population should be able to discern where the best opportunities lie in the world markets, produce for those markets, and prosper accordingly. On the other hand, it seems probable that the causal connection between national per capita income and mean population IQ is really quite complex. A nation's per capita income is related to a variety of other social phenomena, such as education, literacy, technology, and nutrition, all of which may have some effect on the mean level of intelligence. In light of the available data, however, what some call the "culture hypothesis" (i.e., psychological differences among races are basically social rather than genetic) is not corroborated by research done in the United States on Negroid–Caucasoid IQ comparisons since the acceleration of equal opportunities.[6]

Turning now to educational differences as one of the main factors responsible for the national and ethnic differences in intelligence, it is not easy to estimate the magnitude of the effect. In advanced countries like the United States and Great Britain, additional education, such as nursery schooling or remaining at school in late adolescence, has virtually no effect on IQ (Jencks *et al.*, 1972). But this is probably a diminishing-returns effect and does not imply that better education in underdeveloped nations might not raise the measured intelligence of the populations. It may also be that the quality of education in some underdeveloped nations is lower than it appears on paper because the schools put much emphasis on learning and memorization and comparatively little emphasis on training in thinking and the analysis of problems. This would probably have some tendency to reduce mean IQs. Critics often make three points. First, the poor quality of education cannot bear the burden of explaining anything like the full range of IQ disparities. Second, it is not clear why poor education should impair all facets of intelligence, including, for instance, vocabulary, which requires little

[6] A recent analysis by McGurk (1975) of 80 relevant articles comparing the IQs of American Negroids and Caucasoids that were published between 1951 and 1970 was directed toward the evaluation of two hypotheses: (1) that improvement in the SES of Negroids in the United States since the time of World War II has resulted in a relative gain in their mean IQ compared with that of Caucasoids and (2) that test items with verbal loadings are more disadvantageous to Negroids than are items with nonverbal loadings. The author's conclusion was that neither hypothesis was supported by the studies he examined: "Intellectually, the Negro of today bears the same relationship to the contemporary white as did the Negro of the World War I era to the white of that time. Socioeconomic changes have not resulted in a higher relative intellectual status for the Negro (McGurk, 1975, p. 235)."

problem-solving ability and is mainly a matter of efficient incidental learning. Third, a critic may question why some societies have evolved educational systems that put comparatively little emphasis on thinking, and he may suggest that the chief reason is that thinking is a trait for which few people in such societies have much talent.

A third environmental variable that may be invoked to account for the low mean IQs in a number of nations and ethnic groups is the general level of intellectual sophistication of the culture. This is the position taken by Vernon (1969) in his discussion of the low average IQs in Jamaica and Uganda. He proposes that the chief factors responsible for low scores in those countries lie in the emphasis their societies place on conformity and the tradition of magical belief. However, critics will argue that it is not clear why such factors as these should impair mental ability on so many different tests, including vocabulary items, since there seems no good reason why a conformist and superstitious culture should reduce a child's capacity to learn the meanings of words.

A somewhat similar explanation is put forward by Mehryar *et al.* (1972) to account for the low mean IQ of their sample of middle-class children in Iran. They regard the chief responsible factor as a deficiency in the amount of intellectual stimulation given in childhood. To this it can be objected, first, that it would be difficult to establish that median SES Iranian parents provide their children with less intellectual stimulation than parents in the economically advanced nations; and, second, that the evidence that a low level of intellectual stimulation in childhood can reduce the IQ is itself shaky. In the case of Iran, in which the mean IQ of predominantly middle-class schoolchildren is some 17 IQ points lower than in the United States and the United Kingdom, it is doubtful whether any supposed deficiency of intellectual stimulation in childhood can bear the weight of explanation placed on it.

The last major environmental variable frequently advanced to explain the low mean IQ generally present in undeveloped nations is inadequate nutrition. It is reasonably well established that severe protein deficiency during prenatal or early postnatal life can impair brain development and intelligence (Jensen, 1973), but it has not been easy to demonstrate how important this is in practice in underdeveloped countries. Severely undernourished children almost invariably come from the poorest families in which there are a whole host of both environmental and genetic factors that could be responsible for depressing the IQ. The problem of controlling for these is very difficult. Furthermore, considerable resistance of the human embryo to malnutrition was shown in the study of World War II data by Stein *et al.* (1972), in which it was found that the children of Dutch mothers who were reduced to a diet of around 730 calories a day during 1944–1945 suffered no impairment of intelligence (see also Loehlin *et al.*, 1975, Appendix N). This result throws considerable doubt on malnutrition as a significant factor in the low mean IQs characteristic of underdeveloped and semideveloped countries.

We come now to a fourth position that can reasonably be taken in regard to national and ethnic differences in mean IQs. Proponents of this position will admit that the IQs reflect differences in average scholastic aptitude, but they will say that not enough is known about the determinants of intelligence to reach any conclusion about the factors responsible. They will concede that it is doubtful whether the usual set of environmental explanations (e.g., low per capita income, educational deficiencies, poor nutrition, impoverished intellectual and cultural conditions) can plausibly explain the low mean IQs obtained by many populations throughout the world. Nevertheless, they will say, these mean differences in IQ may be brought about by unknown environmental factors such as subtle nutritional deficiencies, arcane differences in methods of bringing up children, and climatic factors. Therefore it is best to regard the matter of population differences in mean IQ as a completely open question that cannot be solved by existing techniques. This is the view taken by Bodmer (1972).

Finally, there is a fifth position. Its advocates believe there is no convincing environmentalistic explanation for the national and ethnic differences in mean IQ, and it is beginning to seem unlikely that any will be found. In view of the absence of adequate environmental explanations, it is understandable that interest is directed toward genetic factors. This hypothesis is supported by the fact that *within* economically advanced populations differences in intelligence are substantially determined by heredity, about which there is no serious dispute among research specialists. This makes it not improbable that mean differences *between* populations may also be substantially determined by heredity. Some will take the view that this position is strengthened by the fairly close association between a population's mean IQ and its racial composition. For it will be apparent from the results summarized earlier in this chapter that the national and ethnic populations which are predominantly Caucasoid and North European in origin usually obtain mean IQs around 100. The same has generally been found in Mongoloid populations in Japan, the United States, and Taiwan. The environmentalist, however, would argue that there may be cultural traditions of child rearing in Mongoloid and North European Caucasoid populations which are sufficient to account for the high mean IQs obtained by these groups living in different parts of the world.

The final conclusion must be that it is not at present possible to determine what factors are responsible for ethnic and racial variations in measured intelligence. Different readers may form different judgments on the evidence as it now stands, but definitive answers will have to await further advances in population genetics.

SUMMARY

Studies of intelligence quotients (IQ) in different parts of the world show that the highest mean IQs are generally obtained by Mongoloid populations originating in Japan and China, and by Caucasoid populations deriving from Northern Europe. On the average, the Japanese exhibit IQs significantly above the Americans and the British. Caucasoid peoples residing in the Middle and Near East and in Southern Europe (except Italy) obtain lower average scores, and the same is true of most other racial and ethnic populations. Amerinds and Mexican-Americans rank next, followed by Afro-Americans, African Negroids, Australoids, and Capoids in that order. Among and within nations mean IQs are closely related to per capita incomes. Five possible explanations for the average differences in population IQs are examined but none of them can be regarded as securely established at the present writing.

REFERENCES

Alzobaie, A. J. The validity of the Goodenough Draw-a-Man Test in Iraq. *Journal of Experimental Education*, 1965, *33*, 331–335.

Baker, J. R. *Race.* London: Oxford Univ. Press, 1974.

Barratt, E. S., Clark, M., & Lipton, J. Critical flicker frequency in relation to a culture fair measure of intelligence. *American Journal of Psychology*, 1962, *75*, 324–325.

Berry, J. W. Temne and Eskimo perceptual skills. *International Journal of Psychology*, 1966, *1*, 207–229.

Bodmer, W. F. Race and IQ: The genetic background. In K. Richardson & D. Spears (Eds.), *Race, culture, and intelligence.* Harmondsworth, England: Penguin Books, 1972.

Bourdier, G. Utilisation et nouvel étalonnage du PM47. *Bulletin de Psychologie*, 1964, *235*, 39–41.

Bruce, D. W., Hengeveld, M., & Radford, W. C. *Some cognitive skills in Aboriginal children in Victorian primary schools.* Victoria: Australian Council for Educational Research, 1971.

Burt, C. Intelligence and social mobility. *British Journal of Statistical Psychology*, 1961, *14*, 3–24.

Burt, C. Inheritance of general intelligence. *American Psychologist*, 1972, *27*, 175–190.

Cattell, R. B. A culture-free intelligence test. *Journal of Educational Psychology*, 1940, *31*, 161–179.

Cattell, R. B. *Abilities: Their structure, growth and action.* Boston: Houghton Mifflin, 1971.

Coleman, J. S. *et al. Equality of educational opportunity.* Washington, D.C.: U.S. Office of Education, 1966.

Comas, J. "Scientific" racism again? *Current Anthropology*, 1961, *2*, 303–314.

Coon, C. S. *The living races of man.* New York: Knopf, 1965.

Coon, C. S. *The hunting peoples.* Boston: Little, Brown, & Co., 1971.

Darlington, C. D. *The evolution of man and society.* New York: Simon & Schuster, 1969.

Dasen, P. R. Cross-cultural Piagetian research: A summary. *Journal of Cross-Cultural Psychology*, 1972, *3*, 23–29.

Du Chateau, P. Ten point gap in Maori aptitudes. *National Education*, 1967, *49*, 157–158.

Eysenck, H. J. *Race, intelligence and education.* London: Temple Smith, 1971.

Fatouros, M. The influence of maturation and education on the development of mental abilities. In L. J. Cronbach & P. J. D. Drenth (Eds.), *Mental tests and cultural adaptation.* The Hague: Mouton, 1972.

Galton, F. *Hereditary genius.* London: Macmillan, 1869.

Garrett, H. E. Comments. *Current Anthropology,* 1961, *2,* 319–320. (a)

Garrett, H. E. The equalitarian dogma. *Perspectives in Biology and Medicine,* 1961, *4,* 480–484. (b)

Garrett, H. E. The SPSSI and racial differences. *American Psychologist,* 1962, *7,* 260–263.

Garth, T. R. *Race psychology.* New York: McGraw-Hill, 1931.

Goodenough, F. *Mental testing.* New York: Holt, 1949.

Goosens, G. Une application du test d'intelligence de R. B. Cattell. (Échelle 2, Forme A.) *Revue Belge de Psychologie et de Pedagogie,* 1952, *14,* 115–124.

Hebb, D. O. *The organization of behavior.* New York: Wiley, 1949.

Hirsch, N. D. M. A study of natio-racial mental differences. *Genetic Psychology Monographs,* 1926, *1,* 231–406.

Ingle, D. J. Racial differences and the future. *Science,* 1964, *146,* 375–379.

Jahoda, G. Assessment of abstract behavior in a non-Western culture. *Journal of Abnormal and Social Psychology,* 1956, *53,* 237–243.

Jencks, C. *et al. Inequality: A reassessment of the effect of family and schooling in America.* New York: Basic Books, 1972.

Jensen, A. R. How much can we boost IQ and scholastic achievement? *Harvard Educational Review,* 1969, *39,* 1–123.

Jensen, A. R. *Genetics and education.* New York: Harper, 1972.

Jensen, A. R. *Educability and group differences.* New York: Harper, 1973.

Jordheim, G. D., & Olsen, I. A. The use of a non-verbal test of intelligence in the trust territory of the Pacific. *American Anthropologist,* 1963, *65,* 1122–1125.

Kearney, G. E., de Lacey, P. R., & Davidson, G. R. *The psychology of Aboriginal Australians.* New York: Wiley, 1973.

Kennedy, W. A., Van de Riet, V., & White, J. C., Jr. A normative sample of intelligence and achievement of Negro elementary school children in the southeastern United States. *Monographs of the Society for Research in Child Development,* 1963, *28,* No. 6.

Klingelhofer, E. L. Performance of Tanzanian secondary school pupils on the Raven Progressive Matrices Test. *Journal of Social Psychology,* 1967, *72,* 205–215.

Kurth, E. von. Erhöhung der Leistungsnormen bei den Farbigen Progressiven Matrizen. *Zeitschift für Psychologie* 1969, *177,* 86–90.

Lloyd, F., & Pidgeon, D. A. An investigation into the effects of coaching on non-verbal test material with European, Indian and African children. *British Journal of Educational Psychology,* 1961, *31,* 145–151.

Loehlin, J. C., Lindzey, G., & Spuhler, J. N. *Race differences in intelligence.* San Francisco: Freeman, 1975.

Lynn, R. The intelligence of the Japanese. *Bulletin of the British Psychological Society,* 1977, *30,* 69–72.

MacArthur, R. S. Some cognitive abilities of Eskimo, white, and Indian-Metis pupils aged 9 to 12 years. *Canadian Journal of Behavioral Sciences,* 1969, *1,* 50–59.

McElwain, D. W., & Kearney, G. E. Intellectual development. In G. E. Kearney *et al.* (Eds.), *The psychology of Aboriginal Australians.* New York: Wiley, 1973.

McGurk, F. C. J. Race differences-twenty years later. *Homo,* 1975, *26,* 219–239.

McIntyre, G. A. *The standardisation of intelligence tests in Australia.* Melbourne: Univ. Press, 1938.

Maity, H. A report on the application of the Stanford adult test to a group of college students. *Indian Journal of Psychology,* 1926, *1,* 214–222.

Mehryar, A. H., Shapurian, R., & Bassiri, T. A preliminary report on a Persian adaptation of Heim's AH4 test. *Journal of Psychology*, 1972, *80*, 167–180.

Nieto-Alegre, S., Navarro, L., Santa Cruz, G., & Domínguez, A. Diferencias regionales en la medida de la inteligencia con el test M.P. *Revista de Psicología General y Aplicado*, 1967, *22*, 699–707.

Notcutt, B. The measurement of Zulu intelligence. *Journal of Social Research*, 1950, *1*, 195–206.

Phua, S. L. Ability factors and familial psychosocial circumstances: Chinese and Malays in Singapore. Unpublished doctoral dissertation, Univ. of Alberta, 1976.

Piddington, M., Piddington, R. Report of field work in northwestern Australia. *Oceania*, 1932, *2*, 342–358.

Pintner, R. *Intelligence testing*. New York: Holt, 1931.

Porteus, S. D. Mental tests with delinquents and Australian aboriginal children. *Psychological Review*, 1917, *24*, 32–42.

Porteus, S. D. *Primitive intelligence and environment*. New York: Macmillan, 1937.

Porteus, S. D. Comments. *Current Anthropology*, 1961, *2*, 327–328.

Porteus, S. D. *Porteus Maze Test*. Palo Alto: Pacific Books, 1965.

Porteus, S. D. Ethnic groups and the Maze Test. In R. E. Kuttner (Ed.), *Race and modern science*. New York: Social Science Press, 1967.

Porteus, S. D., & Gregor, A. J. Studies in intercultural testing. *Perceptual and Motor Skills*, 1963, *16*, 705–724.

Redmond, M., & Davies, F. R. J. *The standardisation of two intelligence tests*. Wellington: New Zealand Council for Educational Research, 1940.

Reed, T. E. Caucasian genes in American Negroes. *Science*, 1969, *165*, 762–768.

Rodd, W. G. A cross-cultural study of Taiwan's schools. Unpublished doctoral dissertation, Western Reserve Univ., 1958.

St. George, A. Cross-cultural ability testing: The Pacific Infants Performance Scale. Unpublished manuscript, 1974.

St. George, R. Tests of general cognitive ability for use with Maori and European children in New Zealand. In L. J. Cronbach & P. J. D. Drenth (Eds.), *Mental tests and cultural adaptation*. The Hague: Mouton, 1972.

St. George, R., & St. George, A. The intellectual assessment of Maori and European school children. In P. D. K. Ramsey (Ed.), *The family and the school in New Zealand society*. London: Pitman, 1975.

Scottish Council for Research in Education. *The intelligence of Scottish children*. London: London Univ. Press, 1933.

Scottish Council for Research in Education. *The trend of Scottish intelligence*. London: London Univ. Press, 1949.

Shuey, A. M. *The testing of Negro intelligence*. (2nd ed.) New York: Social Science Press, 1966.

Silvey, J. Long range prediction of educability and its determinants in East Africa. In L. J. Cronbach & P. J. D. Drenth (Eds.), *Mental tests and cultural adaptation*. The Hague: Mouton, 1972.

Sinha, U. The use of Raven's Progressive Matrices in India. *Indian Educational Review*, 1968, *3*, 75–88.

Sorokin, B. Standardisation and analysis of Progressive Matrices Test by Penrose and Raven. Unpublished manuscript, 1954.

Stein, Z., Susser, M., Saenger, G., & Marolla, F. Nutrition and mental performance. *Science*, 1972, *178*, 708–713.

Tesi, G., & Young, H. B. A standardisation of Raven's Progressive Matrices. *Archivio Psicología Neurologica Psiquiatra*, 1962, *5*, 455–464.

Thomas, R. M., & Sjah, A. The Draw-a-Man test in Indonesia. *Journal of Educational Psychology*, 1961, *32*, 232–235.
Thorndike, E. L. *Mental work and fatigue and individual differences and their causes*. New York: Columbia Univ. Press, 1914.
Tyler, L. E. *The psychology of human differences*. (3rd ed.) New York: Appelton, 1965.
UNESCO. *Statement on the nature of race and race differences*. Paris: United Nations, 1951.
Vejleskov, H. An analysis of Raven matrix responses in fifth grade children. *Scandinavian Journal of Psychology*, 1968, *9*, 177–186.
Vernon, P. E. *Intelligence and cultural environment*. London: Methuen, 1969.
Weyl, N. *The creative elite in America*. Washington: Public Affairs Press, 1966.
Weyl, N., & Possony, S. T. *The geography of intellect*. Chicago: Regnery, 1963.
Woodworth, R. S. Race differences in mental traits. *Science*, 1910, *31*, 171–186.
Yeung, K. T. The intelligence of Chinese children in San Francisco and vicinity. *Journal of Applied Psychology*, 1921, *5*, 267–274.

10

Age, Race, and Sex in the Learning and Performance of Psychomotor Skills

CLYDE E. NOBLE
University of Georgia

INTRODUCTION

Skilled behavior on the part of *Homo sapiens* is a function of many variables. Paramount among these are the manifold conditions of practice and the three organismic factors lending title to the present volume. This chapter treats of their effects and interactions.

The Domain of Human Skills

Our interest centers on those nonverbal human abilities classified by learning and performance specialists as *psychomotor* (or *perceptual-motor*) *skills*. We shall be concerned with their acquisition and execution as a function of age, race, and sex. Fundamentally, psychomotor skills may be described as organized patterns of neuromuscular activities that are usually sequential and continuous in form and that operate under the guidance of changing feedback signals from the learner and the environment. Although verbal and linguistic processes are traditionally deemphasized in our definition, psychomotor skills abound in arts and crafts, business and industrial jobs, special education, games and sports, the music profession, scientific, medical, and engineering specialties, and in various military occupations. Overviews of the role of learning in this domain are given elsewhere (Noble, 1974a, 1977a).

Familiar examples of psychomotor behaviors include the fine coordinations of eyes, ears, hands, trunk, and feet involved in playing tennis, sewing, driving a racing car, trapshooting, painting a landscape, swimming, playing a piano, typing, riding a bicycle, skating, twirling a baton, dancing, or drilling a tooth. Notice the predominance of gerunds. My emphasis is upon dynamics rather than statics, upon doing rather than being. Action, process, and continuity are key ingredients in the analysis of those human skills called psychomotor. Now what are the physiological substrata? Molar abilities do not arise from empty organisms.

Psychomotor behaviors are mediated by electrochemical processes that, although widely distributed throughout the brain, are partially specialized. They occur in (*1*) the cortical projection areas for sensory and motor functions of the new forebrain; (*2*) the large cortical association regions of the prefrontal, temporal, and parietal lobes; (*3*) the basal ganglia, limbic system, and thalamus of the old forebrain; (*4*) the bundle of commissural fibers (e.g., corpus callosum) connecting the left (speech–language) and right (perceptual–motor) cerebral hemispheres; and (*5*) the subcortical integrative and communicative mechanisms of the midbrain (reticular activating system) and hindbrain (cerebellum, pons, medulla). Functioning as a whole yet selectively, the brain evaluates a multitude of stimulus inputs and directs the responses of specific effectors in an integrated plan of action (R. F. Thompson, 1967).

In general, the learning and performance of complex, intricately coordinated acts that are under voluntary control (e.g., flying an airplane, playing a trumpet) depend on the neurophysiology of the higher brain centers, whereas simple reflexes or automatic, routinized movements (e.g., exhibiting conditioned eyeblinks, maintaining static equilibrium) can be served by lower brain centers. For this reason, and also the fact of their acquisition and maintenance being correlated with certain intellectual aptitudes and capacities, *psychomotor skills* are regarded by some specialists as being more cognitive than ordinary *movement behaviors* (Bartlett, 1958; Fitts, 1964; Noble, 1972, 1974a, 1977a; Singer, 1975). I shall focus on psychomotor skills while trying not to slight the important related fields of perceptual skills and athletic skills. But first a bit of history.

Historical Perspective

The confluence of research on the learning and performance of psychomotor skills and the differential psychology of individuals and groups is a phenomenon of recent date, although its origins are traceable to the nineteenth century. James McK. Cattell, Charles Darwin, Hermann Ebbinghaus, Francis Galton, Karl Pearson, Adolphe Quetelet, W. H. R. Rivers, and Wilhelm M. Wundt are the remote ancestors of our modern synthetic viewpoint about the role of human variation in psychomotor behavior. Another six scientists, research psychologists all, have made signal contributions to the empirical and theoretical foundations of human skill during the twentieth century. These more recent antecedents include Clark L. Hull, Arthur W. Melton, Charles E. Spearman, Edward L. Thorndike, Herbert Woodrow, and Robert S. Woodworth. Let us now survey the main developments in this history.

In 1884, Galton opened his Anthropometric Laboratory in London, just 5 years after Wundt had defined psychology as a natural science with the establishment of his Psychological Institute in Leipzig. The two men provide a study in contrasts: the Englishman unsystematic, original, versatile, idiographic—the German systematic, erudite, methodical, nomothetic. Yet each founded a school, or at least a research tradition, that today complements the other. General–experimental psychology is descended from Wundt, differential–correlational psychology from Galton. Each man was a pioneer of a "new psychology" in his native land. Galton, the dilettante, believed that psychology ought to develop whatever methods lead to the solution of practical or theoretical human problems. Wundt, the professional, argued that a comprehensive approach to mankind might include not only experimental, general, and physiological psychology but also developmental, folk, and social psychology. Of course, Wundt preferred brass-instrument *experiments* whereas Galton was less doctrinaire with his mental *tests*. Their objectives were not the same; one was trying to understand structure, the other function. Wundt was looking for uniformities, Galton for variations. Such distinctive emphases appeared to be incompatible. Although there may have been some basis for harmonizing the two approaches in the 1880s, too many differences between British and German science were manifest in those early days. It remained for experimental and correlational students of the second and third generations to begin an integrative trend, one that is still in progress. Details of the history may be found in Boring (1950), Du Bois (1970), Hearnshaw (1964), and Irion (1966).

Working under the combined influence of Darwin's theory of evolution and Quetelet's application of probability theory, Galton collected a variety of anthropometric and psychometric data from 9337 British males and females ranging in age from 5 to 80. He not only measured standing and sitting height, arm span, and body weight but also recorded his subjects' vision, hearing, fingerprints, and hand steadiness; their grip, strike, and pull strength; their exhalation pressure, memory of form, and discrimination of lifted weights. Norms for age and sex were worked out, and he theorized about racial and nationality differences too, as mentioned by Lynn in the present volume (Chapter 9). His intent was to measure people as they are, with all their human strengths and weaknesses. Empiricist and evolutionist, Galton concentrated on human differences as the gateway to human betterment.

Cattell, Galton's assistant and a doctoral student of Wundt's, inaugurated programs of psychological testing at the University of Pennsylvania in 1888 and at Columbia University in 1891. His subjects were mostly freshmen, women as well as men, and like Galton he found psychological differences between the sexes. In addition to perceptual functions, Cattell studied rate of hand movement, dynamometer pressure, reaction time, association, and memory. He computed measures of central tendency (M), variability (σ), and correlation (r).[1] As it turned out, the laboratory tests showed low interrelationships and were poorly correlated with academic course grades. Meanwhile Pearson, another of Galton's assistants, invented the specialty of mathematical statistics and began to develop formulas for the coefficient of correlation, biserial correlation, partial and multiple correlation, tests of goodness of fit, and statistical corrections for restrictions in range. Pearson also did biometric research on the relationship between indices of intelligence and anthropometric measures, finding some positive but not very dramatic correlations. The correlations between siblings (2000 brothers and sisters) were higher for a number of psychological traits, which led Pearson to conclude (1904) that human behavior is as much a function of the laws of heredity as physical features are.

Between the dates that Wundt's and Galton's laboratories were founded Ebbinghaus was experimenting upon human learning and memory. Although taking verbal behavior as his dependent variable, Ebbinghaus set a methodological pattern for later associative research in psychomotor skills to follow. In quantitative terms, he demonstrated the power of practice to develop one's proficiency; he collected data representing acquisition and retention phases that were orderly as well as replicable; and he obtained preliminary evidence that distributed practice was more efficacious than massed practice. In 1896, Ebbinghaus invented the completion test of intelligence, anticipating Binet by nearly a decade; and he devised a procedure for group administration roughly 20 years ahead of the large-scale intelligence testing program of the U.S. Army psychologists in World War I.

During the 1890s, a number of biologically oriented scientists in both

[1] The raw-score equation for calculating r from any bivariate distribution of X and Y scores is:

$$r = \frac{\Sigma(X - M_x)(Y - M_y)}{N\sigma_x\sigma_y},$$

where M_x and M_y are the means of the two distributions, σ_x and σ_y are their respective standard deviations, and N is the number of paired scores. The standard deviation of the X distribution is defined as $\sigma_x = [\Sigma(X - M_x)^2/N]^{1/2}$. Its variance is σ_x^2. Comparable operations apply to σ_y and σ_y^2.

Britain and the United States began investigating phenotypic similarities and differences among various human populations in psychomotor traits having strong genotypic determinants. A team of anthropologists and psychologists at Cambridge University, the latter led by Rivers (1901, 1905), embarked upon field expeditions to the Torres Straits Islands and to India. There they studied the basic sensory, perceptual, and motor abilities of nonliterate Australoids and hybrid peoples of mixed Australoid–Caucasoid ancestry. Intrigued by his research on primitive cultures, Rivers made an effort to combine anthropology and psychology into what may be termed *ethnopsychology*.

The American pioneer in the ethnopsychology of psychomotor skill was Woodworth (1910), who administered perceptual and motor tests to about a thousand people attending the St. Louis World's Fair of 1904. The visitors included persons from several primitive societies. Later, I review that study, along with the data of Rivers. Woodworth (1899, 1901, 1903) had already reported seminal work on the accuracy of voluntary movement, the improvement of handwriting, and the perception of time, force, and extent of motion. For Woodworth, the spatiotemporal organization of much skilled behavior proceeds by way of a blending of diphasic motor units; e.g., the integration of backswing and forehand strokes in tennis or the upswing and strike in hammering nails. These cyclical patterns, he believed, are initiated as wholes and run their courses without further sensory control. More complex polyphasic behaviors (e.g., two-plate tapping or rotary pursuit tracking) derive their smoothness from practice in the timing and sequencing of the diphasic part-skills, although there may be periodic adjustments in extended serial-action tasks. Woodworth's analysis of skill has shown remarkable survival value over the years, as has his functional S–O–R formula (Noble, 1966b).

Between 1890 and 1910, other American psychologists conducted numerous experiments on the learning of telegraphy, fencing, mirror tracing, ball tossing, dart throwing, and typing. Transfer of training was examined by Thorndike and Woodworth (1901) in a landmark study of perceptual discrimination and clerical skills that led to a rejection of the traditional formal discipline hypothesis in favor of the hypothesis of common elements. Armchair speculation about "faculties" was supplanted by laboratory investigations based on stimulus–response analyses (Noble, 1977b). Within a decade, appropriate control groups to measure the extent and direction of effects became standard practice in experiments on transfer. By the heyday of functionalism, the psychology of skill acquisition had taken its rightful place as a scientific research specialty. Columbia University was one of its centers.

Under Thorndike, the author of the Law of Effect, research on the psychology of human learning and performance benefitted from vigorous experimentation and thought-provoking theorizing. He elaborated the principles of association discovered by Ebbinghaus, extending their application to a variety of other species, tasks, and classes of behavior. Implicitly accepting Woodworth's (1929) formulation that *stimuli* (S) do not directly cause *responses* (R) without the mediation of empirical *organismic* (O) variables, Thorndike developed a continuity theory of cumulative habit growth and diminishing returns with practice trials that nevertheless left room for anatomical, physiological, and genetic factors to operate in determining behavior. "Learning is connecting," he said; "the mind is man's connection system (1931: 122)."But he also said, "Human individuals differ by original nature as cats and dogs and tulips and roses do (p. 198)." This biopsychological orientation produced a more viable theory than the one that emerged from the brittle behaviorism of Watson (Noble, 1977a). In particular, the importance of *reinforcement* (e.g., reward, punishment, feedback, knowledge of results), demonstrated in numerous Thorndikean experiments, came to exert a profound influence upon Hull and other neobehaviorists—and thence upon a corps of psychomotor skill researchers during and after World War II.

In the third volume of his *Educational Psychology* (1914) Thorndike published an extensive review of research on human differences, and he devoted a chapter each to sex, race, and age in which form-board skill, reaction time, card sorting, rate of tapping, handwriting ability, illusory perceptions, and various sensory discriminations figured prominently. Speaking quantitatively, he thought males more variable in behavior than females (the hypothesis examined by Lehrke in Chapter 7); he considered race a source of multimodality, yet with much overlapping of distributions; and he regarded aging as one of the keys to differences in rates and limits of maturation. Although he found it impossible to estimate precisely the relative contributions of these organismic factors to the behavioral variations encountered on different tasks, Thorndike gave the term *original nature* new scientific respectability and insisted that the analogy between associative learning and natural selection be taken seriously.

Continuing in the Galton–Pearson line at the University of London, albeit with training under Wundt, Spearman began a program of correlational research in 1904 that led to the birth of factor analysis, the discovery of hierarchical intercorrelations, the law of tetrad differences, and the theory of general and specific factors in intelligence. Spearman made many notable contributions to psychometrics, including the concept of Gaussian oscillation, the rank-difference correlation method, and a statistical technique for computing the changes in reliability of a test resulting from doubling or tripling its length. Whereas Cattell had failed to uncover much evidence for the overlapping of human abilities, Spearman made appropriate corrections to offset random errors of measurement and canceled out the *specific (s)* factors. Consequently, he revealed a theoretical picture of strong intercorrelations centering around his hypothetical *general* (*g*) factor. In *The Abilities of Man* (1927) he argued that *g* is involved to some extent in all psychological tests, laboratory experiments, and practical activities—and further that a person talented in one or two areas is probably going to be talented in many others. Horn develops this point in Chapter 5. For Spearman correlation, not compensation, was the rule of nature.

A logical development of Spearman's 2-factor theory was the extension to learning phenomena undertaken by Woodrow at the University of Illinois in the 1930s and 1940s. Correlations among successive practice trials reveal a dual pattern of decreasing commonality with initial performance and increasing commonality with final performance as training progresses (Irion, 1966; Jones, 1966). An example of this bidirectional phenomenon is presented in Figure 10.1. The data come from an experiment of mine (Noble, 1970a) in which 500 adult Caucasoid subjects practiced a rotary pursuit task for 100 20-sec. trials spaced by 10-sec. rests. Confirming the expected superdiagonal form of the correlational matrix pattern, and the associated aspects of Spearman's empirical law of single tetrad differences, adjacent-trial correlations are generally higher than remote-trial correlations.

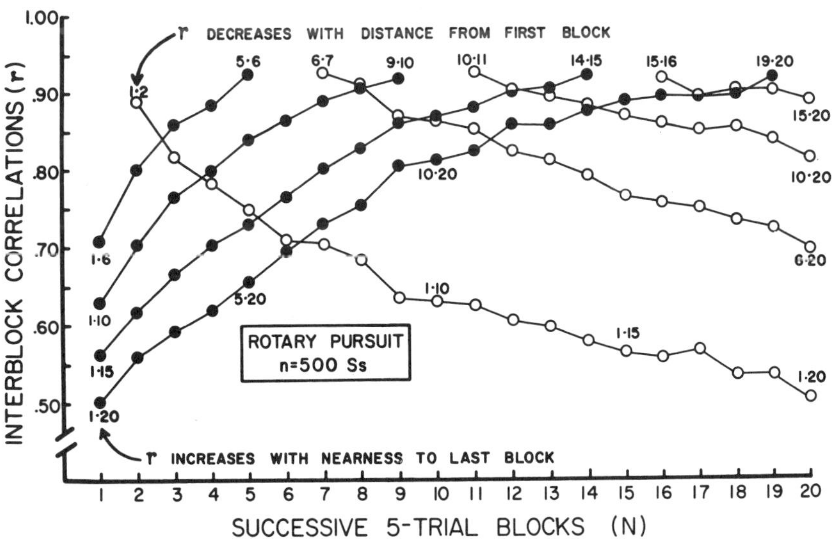

Figure 10.1. Interblock correlation coefficients (r) based on increasing distance from Block 1 (falling curves) and on increasing nearness to Block 20 (rising curves) as functions of successive practice blocks (N) for 500 subjects learning a pursuit tracking task (from Noble, Clyde E., "Acquisition of pursuit," *Journal of Experimental Psychology*, 1970, *86* (No. 2): 360–373, Fig. 4, p. 369. © 1970 by The American Psychological Association. Reprinted by permission.

Woodrow's explanation of such results was couched in terms of a systematically changing pattern of abilities as a function of practice. These alterations in the way subjects perform the task culminate in factorial complexity because gain scores correlate better with final proficiency than with initial proficiency. He concluded (1946) that there is no *general learning factor* (*g*), and others have concurred (Roff & Payne, 1956); but the question is still moot (Allison, 1960; Duncanson, 1964; Glaser, 1967; Stake, 1961). Although subsequent researchers have proposed alternative explanations for the principle illustrated in Figure 10.1, such as the *simplification* hypothesis (Jones, 1966, 1969, 1972; Noble, 1970a; Reynolds, 1952), Woodrow's *complication* hypothesis has a number of supporters (Fleishman, 1960, 1966, 1972; Fleishman & Hempel, 1955; Hinrichs, 1970; Irion, 1966). I return later to the question of *g* in psychomotor skills. Woodrow's impact on this field is further notable for his research program in the distribution of practice, which confirmed and extended Ebbinghaus' early studies and forged a link with the inhibition theory of Hull's system. However, in theorizing about the quantitative properties of acquisition functions he eschewed intervening variables.

Whereas Woodrow employed an ultrapositivistic, empirical-equation approach to determining the mathematical characteristics of learning and performance data, Hull preferred the hypothetico-deductive method. At Yale University, he integrated a biological view of psychology with experimentation and quantification in developing the reinforcement theory of behavior (Hull, 1943, 1952). It was a unique combination of four sets of concepts: *(1)* the empirical laws of British associationism; *(2)* the laboratory findings from Russian research on conditioned reflexes; *(3)* the data resulting from American research on trial-and-error learning and operant conditioning; and *(4)* the intervening-variable technique of theorizing by means of hypothetical mediating processes. Behaviorism's influence on Hull was reflected mainly in the objective orientation of his theory, while the influence of functionalism could be seen in his search for quantitative relationships among S–O–R variables, and in his penchant for explanatory principles consistent with biological evolution.

Hull patterned his neobehavioristic system on the hypothetico-deductive models of physics, which drew fire from the Gestaltists, but he clearly recognized that behavior is a statistical phenomenon and that it is the product of a continuous interaction between an organism and its environment. We are indebted to Hull for such hypothetical concepts as habit (*H*), drive (*D*), stimulus dynamism (*V*), incentive (*K*), inhibition (*I*), oscillation (*O*), and the reaction limen (*L*), along with the theory that response (R) is a multiplicative function of *H*, *D*, *V*, and *K*, from which positive tendency is subtracted the sum of the negative effects of *I* and *O*, the net result of which must exceed *L* in order for R to occur. The Hullian mathematical approach to learning phenomena set a brilliant example for later theoreticians to follow. As I commented elsewhere:

> Hull's explicit paradigm was to proceed inductively from simple to complex behavior, then to employ the more abstract generalizations deductively (in combination with local initial and boundary conditions) in order to derive the empirical laws and particular observations of the learning laboratory [Noble, 1977a: 17].

Thus, the often misunderstood hypothetical *intervening* variables (not to be confused with Woodworth's empirical *organismic* variables) are, in the last analysis, summary conceptualizations that tentatively coordinate a broad spectrum of operations, actions, and situations (Noble, 1966b, 1970b, 1976b; Spence, 1948, 1966).

Hull revised continually. Shortly after *Principles of Behavior* (1943) appeared, he began altering his theory to increase its flexibility and range of application. On a significant new tack, Hull (1945, 1952) suggested that individual and group differences might affect the numerical parameters of behavioral equations (e.g., origin, rate) rather than their mathematical types (e.g., exponential, hyperbolic). Furthermore, he assumed that general laws obtained inductively from statistical samples would probably include the entire natural range of innate aptitudes, capacities, traits, and abilities. Differences among individuals or groups (attributable to sex, age, genera, species, subspecies, or native talent) would presumably be reflected in parametric variations of the basic functions. While acknowledging the great power of environmental sources of differential behavior among organisms (e.g., prior training and reinforcement histories), Hull (1945) said:

> There is much reason to believe, however, that even if organisms could be subjected to identical environmental conditions from the moment of conception, great differences would still be displayed in the behavior of different species as a whole and in the behavior of the individual organisms of each species. Such differences must presumably be regarded as dependent upon, i.e., derived from, differences in the innate or original nature and constitution of the individual organism [p. 56].

In conclusion, he hypothesized

> that the *forms* of the equations representing the behavioral laws of both individuals and species are identical, and that the differences between individuals and species will be found in the empirical constants which are essential components of such equations [p. 60].

Several investigations have followed Hull's initial foray into this field (Adams, 1957; Noble, 1961, 1966a, 1969a, 1969b, 1970a; Noble, Baker, & Jones, 1964; Spence, 1956; Zeaman & Kaufman, 1955). Even though human subjects and a variety of learning tasks were used, the results were generally supportive of Hull's position.

My last historical figure, Melton, is a contemporary psychologist. Trained at Yale, he was largely responsible for two quantum leaps forward in research on psychomotor skills—the first an applied venture, the second rather basic. During World War II, Melton directed the U.S. Army Air Forces' psychomotor testing program, the most ambitious effort ever made to employ a battery of complex apparatus tests in order to select and classify hundreds of thousands of persons for specialized jobs. Detailed results are presented in the book *Apparatus Tests* (Melton, 1947), portions of which I review in the next section. The importance of this large-scale program in military psychology was that measurements of psychomotor skill were needed for optimally predicting the success of aircrew candidates training to become pilots, bombardiers, navigators, and aerial gunners. Fortunately, the psychologists were able to construct a variety of apparatus devices, to maintain a high degree of calibration constancy, and to administer them as psychomotor tests in a standardized fashion so that scores obtained by different candidates at far-flung installations between 1942 and 1945 would have uniform psychometric meanings. The program was eminently successful, thanks to inspired leadership and a gifted, industrious staff of psychologists and technicians. As we shall see, Melton's program produced a set of culture-fair psychomotor learning devices well suited to laboratory research on the interaction of organismic factors and practice variables.

Melton's second major contribution to the domain of human skills came during the Korean War when he founded and headed the U.S. Air Force's Human Resources Research Center (later the U.S.A.F. Personnel and Training Research Center) from 1949 to 1957. Again there was superb leadership within the aeronautical setting, plus another generation of highly competent research scientists, but this time Melton laid greater emphasis on fundamental or general-purpose research in several of his key laboratories. He created "an extraordinary, active program of in-service and contract research that in its first five years set the field ahead by more than 20 (Bilodeau & Bilodeau, 1961: 245)." Some of this work is cited in subsequent sections, so we can now bring our historical survey to a close.

Before concluding, however, I should mention that Melton's personal research skills facilitated his administrative achievements by providing a scientific model worthy of emulation and respect. The two-factor theory of retroactive inhibition, the research and integrative essays on memory, and the seminal contributions to methodology over the past 40 years are benchmarks in the experimental psychology of learning and performance. Ending this section with the career of a leading functionalist is not only historically correct, it is a strategic necessity if one is to appreciate the thrust of this chapter. To a considerable extent, Melton's legacy can be appreciated by studying the research on psychomotor skills cited in the *Annual Review of Psychology* (e.g., Adams, 1964; Bilodeau & Bilodeau, 1961; Noble, 1968;. Styled neobehavioristic, or functionalistic in the literal sense, this recent work reflects the *rapprochement* of the experimental and correlational psychologies of which I spoke earlier. As a functionalistic neobehaviorist himself, Melton (1967) has argued effectively in support of formulating hypotheses about human variation within the context of explicit theories of learning and performance.

ANALYSIS OF SKILLS

Psychomotor Skills

Laymen find it natural to think of psychomotor skills as reflecting some unitary aptitude or general capacity, much as they regard IQ; i.e., a global kind of human capability that is fully indexed by a single score. When one says that Tom is more "skillful_1" than Dick, or Harry more "skillful_2" than Jimmy, or Alice more "skillful_3" than John, one may be masking some vital information with an overly generic adjective. Perhaps Tom and Dick differed in *speed* (skill_1) because of *age*; Harry and Jimmy in *coordination* (skill_2) because of *race*; or Alice and John in *dexterity* (skill_3) because of *sex*. Or perhaps all three pairs simply differed in amount of *practice*. The term skill as such is unanalytical. Other questions come to mind that call for explicit research efforts.

Is there a general psychomotor factor that can be distinguished from specific factors? If not, in what way do the separate aptitudes combine to produce all-round proficiency? How many specific factors are there? Do perceptual and motor abilities overlap? Are psychomotor abilities correlated with athletic abilities? Answers to questions of this sort have been sought by means of the statistical technique of factor analysis (Spearman, 1927; Thurstone, 1947). This is a method of manipulating and interpreting clusters of test-score correlations obtained from the performance data of large samples of subjects. The basic procedure is to compute product–moment correlation coefficients among the various subtests that make up a battery of printed or apparatus tests (e.g., of psychomotor skill). Subtests overlapping with one another form a cluster of intercorrelations having a theoretical *common factor*. By analyzing the clustering and nonclustering subtests of the correlation matrix according to certain mathematical criteria (e.g., simple structure, oblique rotation), one can account for the total variance in the subjects' scores by a minimum number of these hypothetical concepts called *factors*. If the analyses are done properly, the investigator will be able to reproduce the original matrix with high fidelity. He may also *believe* that he has laid hold of an independent variable in the process. To digress a moment, I must sound a warning here about some inappropriate techniques and fallacious conclusions. The core of the problem seems to be a failure to recognize the difference between independent and dependent variables, or between empirical and hypothetical concepts, or both (see Noble, 1966b, 1975, 1976a, 1976b).

Thoughtful readers will carefully distinguish between empirical *abilities* or *skills*, which are operationally defined behavioral concepts, and hypothetical *aptitudes* or *capacities*, which are postulated intervening states or processes. The former tend to be specific, concrete, situational, and hedged about with qualifications; the latter tend to be general, abstract, transsituational, and unabashedly categorical. Because their mathematical solutions are not unique, their behavioral identifications are rather idiosyncratic, and their hypothesis-testing methodologies are sometimes less than kosher, certain factorial investigations of the learning and performance of psychomotor skills (e.g., those of Edwin A. Fleishman and his associates) have evoked serious technical criticisms (e.g., Adams, 1964; Bechtoldt, 1962, 1970; Jones, 1966; Noble, 1968, 1970a) to which pertinent rejoinders are overdue. The trenchant and altogether cogent analyses of Fleishman's work by Harold P. Bechtoldt provide an effective antidote to unbridled "cognitive" speculation about these factors (i.e., primary mental abilities in Spearman's or Thurstone's sense) having ideomotor efficacy as intervening variables of some unspecified information-processing type (see Fleishman, 1966:162). What Bechtoldt (1970) has recommended and applied is the multiple-regression technique (see also Adams, 1957). Briefly, when some of Fleishman's most salient data were recomputed by Bechtoldt, modeling his regression analyses upon the original factor analyses, Fleishman's oft repeated and generally accepted conclusions (e.g., Singer, 1975) about the changing factor structure of psychomotor tasks (e.g., Fleishman, 1966:159; 1972:99) failed to be supported by statistical reanalyses that properly treated combinations of reference-test data and practice-task data. I am inclined, therefore, to concur with Adams (1964), Bechtoldt (1962, 1970), and Jones (1966) that factor analysis is all right in its place (Thurstone, 1947), but not in learning experiments where distinctions between independent and dependent variables are theoretically crucial. Fleishman's hypotheses presuppose a causal nexus, hence Spearmanesque or Thurstonean methods are incorrect. Whether Fleishman's "abilities" (read *factors*) operate in the fashion he believes cannot be determined by the techniques he has employed. He is trying to unscrew the inscrutable. Correlation is not causation.

Freed of confusion, factor-analytic studies by psychometric psychologists have identified a number of purportedly basic aptitudes or capacities (factors) that are considered to be important for success on a wide variety of apparatus tests (Fleishman, 1964, 1966, 1972; Guilford, 1959; Michael, 1949; Roff, Payne, & Moore, 1954). Among the chief *psychomotor skill factors* are

1. arm-hand steadiness
2. control precision
3. finger dexterity
4. manual dexterity
5. multilimb coordination
6. rate control
7. reaction time
8. response orientation
9. speed of arm movement
10. wrist-finger speed

There seems to be no "general psychomotor" factor (Henry, 1968; Singer, 1975).

Psychomotor skills are widely distributed in musical, athletic, industrial, and military settings; but such complex, practical situations are rarely suitable for rigorous experimental research. Scientists find it more analytic to investigate psychomotor learning and performance with specially built devices employed under controlled laboratory conditions. Proficiency measures obtained in the laboratory exhibit increasing accuracy and decreasing variability of a learner's performance on such equipment as practice continues. Assuming the relevant genetic aptitudes are present, a person's mastery of a given psychomotor task will depend on the reliability of the learning apparatus, on motivation to improve, on receiving continuous feedback about the adequacy of performance, and on such variables as the reinforcing effects of corrections made during training (Noble, 1970b, 1974a, 1977a).

Ten examples of psychomotor apparatus used in basic and applied research of either the experimental or the psychometric variety are illustrated in Figure 10.2. On the Complex Coordinator (device A), the operator (subject) must rapidly move the airplane stick (aileron, elevator) and pedal (rudder) controls to change three green response lights in order to match the positions of three red stimulus lights (Melton, 1947). The Discrimination Reaction Timer (device B) requires the operator quickly to throw one of the four toggle switches in response to changing color–spatial light patterns in order to turn off the white light (Melton, 1947). With the Manual Lever (device C), the operator learns to make and recall a simple positioning response on the basis of numerical feedback provided after successive blindfolded movements of the lever through various degrees of arc (Bilodeau, 1969). Using the Mirror Tracer (device D), the operator has to trace quickly around the 6-pointed star pattern with the electrified stylus, relying on indirect mirror vision while avoiding contact with the sides of the path (Snoddy, 1935).

The Multidimensional Pursuitmeter (device E) requires the operator to scan the four display dials and restore to zero settings the frequent drifts of the indicators by making appropriate corrective movements on the four airplane controls (Payne & Hauty, 1955). On the Rotary Pursuitme-

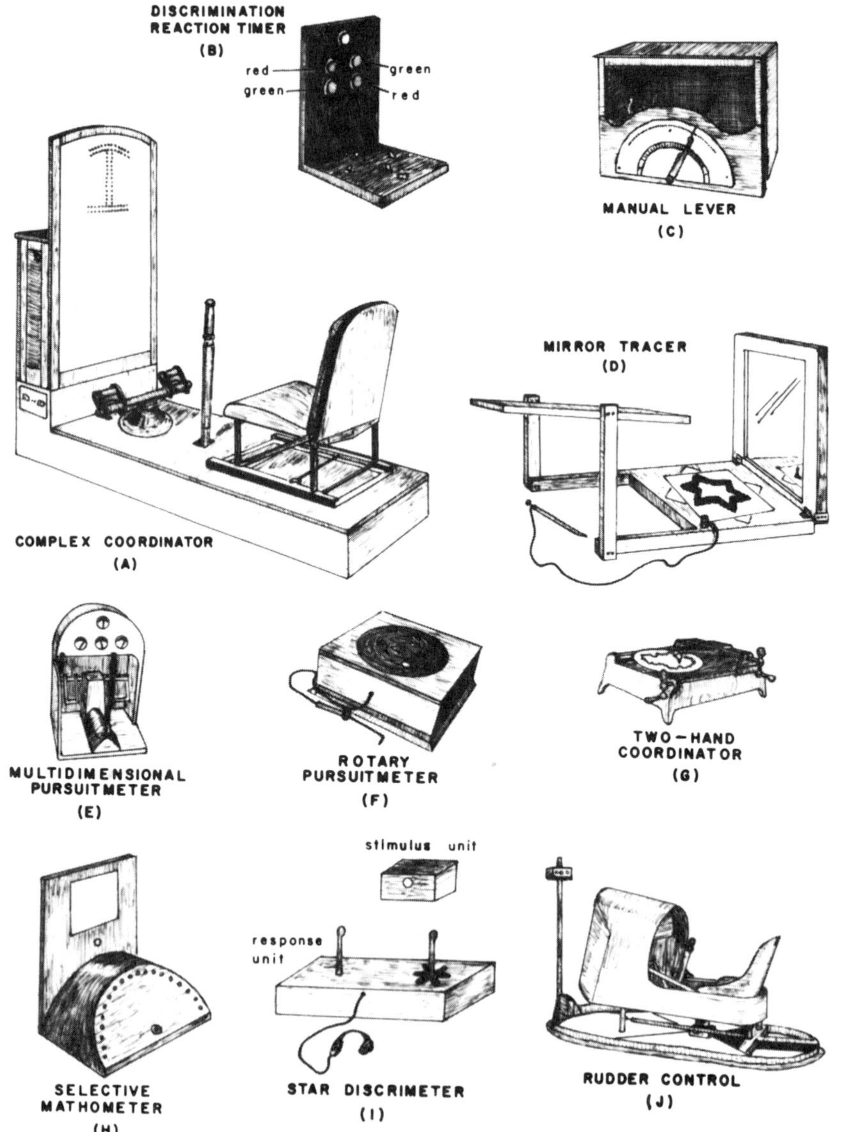

Figure 10.2. Psychomotor devices for studying learning and performance. See text for descriptions.

ter (device F), the operator attempts to keep the electrified, hinged stylus tip in continuous contact with the metal target disc as it revolves clockwise in a circular pattern at 60 rpm (Melton, 1947). Using the Two-Hand Coordinator (device G), the operator simultaneously manipulates two hand cranks, one of which moves the cursor right and left, the other fore and aft, in an effort to maintain electrical contact with the metal target disc as it moves clockwise in an irregular pattern (Melton, 1947). With the Selective Mathometer (device H), the operator's task is to discover, by means of visual or auditory feedback signals, which one of several reaction keys is to be pressed in response to each of a series of distinctive stimulus patterns appearing on the screen (Noble, 1969a). On the Star Discrimeter (device I), the operator must, on the basis of auditory feedback signals, learn to respond to each of six colored lights appearing in the stimulus unit by moving the right-hand lever into one of six corresponding slots while executing a hand-steadiness action with the left hand (Duncan, 1953). The Rudder Control (device J) requires the operator to coordinate a set of foot pedals inside the cockpit in order to keep the nose of the plane aligned with one of three target lights as they change in an irregular manner (Melton, 1947).

Five of these psychomotor devices (A, B, F, G, J) were found to be very useful in the selection of aircrew personnel by the U.S. Army Air Forces (AAF) during World War II. These tests were useful because of their correlations with the jobs required of pilots, bombardiers, navigators, and aerial gunners (Melton, 1947). By far, the main emphasis and greatest success of the AAF psychologists lay in the selection of potentially skilled pilots. The magnitudes of the biserial correlations or validity coefficients (r_b) for various groups of pilot candidates differing by stage of training, race, and sex are shown in Table 10.1[2] Quantitative test scores were correlated with the pass–fail criterion of graduation or elimination from elementary or advanced flying training, a subjective criterion that was much less reliable than the apparatus tests themselves.[3] Nevertheless, the proficiency of most groups of pilot candidates was significantly predictable by these devices, especially when they appeared in a sophisticated battery of 20 tests. Whereas all 5 of the

[2] The formula for calculating the biserial correlation coefficients (r_b) in Table 10.1 is

$$r_b = \frac{M_g - M_e}{\sigma} (pq/z),$$

where M_g is the mean test score of those graduated, M_e is the mean test score of those eliminated, σ is the standard deviation of all cadets' scores, p is the proportion of graduates, q is the proportion of eliminees, and z is the height of the ordinate at the point on a Gaussian frequency distribution that dichotomizes the area into proportions p and q.

[3] Intratest reliability coefficients, based on odd–even trial correlations for several thousand cases, ranged between .89 and .94 for the 5 psychomotor tests cited in Table 10.1. Intercorrelations among instructors' ratings ran only from .58 to .79, with a median of .65 (Melton, 1947). Such low reliability in the criterion places limitations upon the validity of even a perfectly reliable test. One of the most realistic of all the tests developed by the AAF psychologists, the Pedestal Sight Manipulation Test, was essentially a piece of equipment lifted directly out of the nose of a B-29 bomber. Despite high reliabilities (> .9) for azimuth tracking, elevation tracking, framing, and triggering scores by aerial flexible gunners on this device, its validity for predicting gun-camera records during air-to-air training missions was essentially zero. This situation posed a double problem of low criterion reliability and an excessively complex learning task (Melton, 1947).

TABLE 10.1

Predictive Validities and Test-Aptitude Correlations of Five Psychomotor Tests Used in Pilot Selection by the U.S. Army Air Forces during World War II[a]

Psychomotor test and pilot aptitude correlation[b]	Stage of training	Validity and sample size	Taxonomic group of pilot candidates[c]		
			Male Caucasoid	Female Caucasoid	Male Negroid
Complex coordinator ($r = +.68$)	Elementary	r_b	+.33		+.04
		N	10288		658
	Advanced	r_b	+.33	+.22	
		N	10143	191	
Two-hand coordinator ($r = +.54$)	Elementary	r_b	+.33		+.10
		N	6993		657
	Advanced	r_b	+.36	+.34	
		N	1017	191	
Discrimination reaction timer ($r = +.51$)	Elementary	r_b	+.24		+.14
		N	15078		644
	Advanced	r_b	+.42	+.38	
		N	1017	191	
Rotary pursuitmeter ($r = +.41$)	Elementary	r_b	+.20		+.11
		N	4875		643
	Advanced	r_b	+.18	+.21	
		N	5720	191	
Rudder control ($r = +.65$)	Elementary	r_b	+.41		+.14
		N	2007		657
	Advanced	r_b	+.40	+.31	
		N	1017	191	

[a] Adapted from Melton, 1947.

[b] Comparable versions of the standard tests and testing conditions were selected for the various groups. All aviation students were tested and trained between 1942 and 1944. The correlations shown in parentheses are product–moment coefficients (r) between each test and the aptitude index for male fighter pilots based on 5000 cases.

[c] Validity coefficients are biserial correlations (r_b) computed between psychomotor test scores and frequency of elimination from each stage of training as the criterion. Medians of r_b are given when data from two or more pilot classes were available. Validities have not been corrected for restriction of range due to selection on the basis of the pilot aptitude index. Usually this adjustment raised r_b by only a small amount (.04 to .08), but the curtailment of the distribution of aircrew talent actually sent to training that resulted from selection undoubtedly reduced the validity coefficients.

psychomotor tests were valid for predicting the performance in flying schools of male Caucasoids (both stages of training), few were valid for male Negroids (elementary stage). The validities for female Caucasoids (advanced stage) were lower than those of their male counterparts in four out of five cases, but the Rotary Pursuitmeter gave a somewhat better prediction for women than for men. Reasons for these differences are unknown (Melton, 1947), although they may be related to the phenomenon of range restriction (see Table 10.1, note *c*) because aviation students were initially selected in terms of their performance on the AAF Qualifying Examination (see also Michael, 1949).

Mainly because of the high reliability of the *pilot aptitude index*, a special standard score[4] derived from the combination of weighted scores on all printed and apparatus tests, the correlations (r) between that index and each of the psychomotor tests shown in Table 10.1 are higher than any of their respective validities (r_b). The quantitative pilot aptitude index is a more stable criterion than the qualitative pass–fail dichotomy imposed during flying training. As a matter of fact, based on 5000 cases, the mean test–aptitude correlation for 5 *apparatus* tests is +.56 whereas the comparable statistic for 15 *printed* tests is +.38, a value only slightly above that of the mean validity of the apparatus tests for predicting which Caucasoid male cadets will survive one of the two stages of training recorded in Table 10.1 (Melton, 1947).

What all these comparisons imply, at least for some taxonomic groups, is that well-designed psychomotor devices *(1)* provide very consistent learning and performance tasks, and *(2)* exhibit significant commonalities with standardized criteria of pilot aptitude as well as pilot training. Briefly, these apparatus tests are characterized by high reliability and high validity. It follows that such training devices are ideal for applied research in selection and classification or for basic research in learning and performance. Their value is not only practical but also theoretical.

Postwar research was directed toward finding out why some psychomotor tests were valid and others not, and trying to determine what basic aptitudes (factors) were being engaged by these devices (Fleishman, 1953, 1954). The data are voluminous, and some of the factors appear under various names or with different loadings, but a few examples may be cited. In summaries of this work (Fleishman, 1964, 1972), it has been reported that the Complex Coordinator (Figure 10.2) involves the factors of response orientation, control precision, and multilimb coordination; the Discrimination Reaction Timer (Figure 10.2) involves the factors of response orientation, speed of arm movement, and reaction time; the Multidimensional Pursuitmeter (Figure 10.2) involves the factors of control precision and response orientation; the Rotary Pursuitmeter (Figure 10.2) involves the factors of response orientation, control precision, and rate control; the Two-Hand Coordinator (Figure 10.2) involves the factors of multilimb coordination and rate control; and the

[4] The normalized *standard* score scale of *nine* units, called a "stanine" by the AAF psychologists, used $M = 5$ and $\sigma = 2$ for each aircrew specialty. Stanines 1 (low) and 9 (high) each included 4% of the cases, 2 and 8 had 7% each, 3 and 7 had 12% each, 4 and 6 had 17% each, and stanine 5 contained the middle 20% of the cases. All sum to 100%.

Rudder Control (Figure 10.2) involves the factor of multilimb coordination. Many of these psychomotor tests also exhibit factors peculiar to the apparatus itself. For instance, the Discrimination Reaction Timer and the Rotary Pursuitmeter both show evidence of strong specific factors whose loadings allegedly increase with continued practice (Fleishman, 1960, 1966, 1972). However, Bechtoldt's (1970) reanalysis of Fleishman's data via multiple regression indicates a rather steady degree of predictive usefulness over trials.

Varying with the practical job in question will be the unique combination of factors as well as the relative loadings of the 10 basic psychomotor aptitudes. For example, from what I have said, one would not be surprised to learn that multilimb coordination, reaction time, rate control, and response orientation are important factors in the selection of future fighter pilots. Moreover, one might expect finger dexterity and arm–hand steadiness to be valid indicators of those medical students who are destined to become capable surgeons. And that speed of arm movement and wrist–finger speed would be relevant to the selection of young musicians interested in the art of percussion.

Do flyers, surgeons, and drummers inherit their psychomotor predispositions to any appreciable extent? The answers are not clear yet, but it is a reasonable hypothesis that a person's genetic endowment plays some functional role, along with his or her environmental opportunities and experiences, in the origin and development of most of these psychomotor factors, and therefore in the determination of the complex behaviors represented by the aviation, medical, and musical specialties I cited. For instance, take rhythm—a key ingredient in the make-up of a competent percussionist. This is generally regarded as a musical aptitude of the perceptual variety, and it is presented thus in the next section, but research has shown that the discrimination of rhythm is significantly correlated with rhythmic performance (Geldard, 1962). In short, if one can tell the difference between two rhythmic patterns, one can probably reproduce those patterns, given adequate training. Now it happens that rhythm has a strong hereditary component (Vandenberg, 1971). Comparisons of test scores for rhythmic ability among identical or *monozygotic* (MZ) twins and fraternal or *dizygotic* (DZ) twins reveal that the intraclass correlation coefficients are higher for the MZ twins ($r_{mz} = .59$) than for the DZ twins ($r_{dz} = .28$). In other words, there is greater similarity in the test scores of pairs of identical twins than in those of fraternal twins. On the basis of Vandenberg's data, we may estimate the heritability index (h^2) of rhythmic aptitude by the simple equation:

$$h^2 = \frac{r_{mz} - r_{dz}}{1 - r_{dz}}. \qquad (1)$$

When the two twin correlations are substituted in this formula, the value of $h^2 = .43$. This statistic may be interpreted to signify that, for the sample of subjects and testing conditions used, the genetic contribution to individual differences in observed (phenotypic) rhythmic ability represents about 43% of the total variance (σ^2) in the scores. The nongenetic proportion is given by $1 - h^2 = .57$; i.e., 57% of the phenotypic variance in rhythm scores is attributable to environmental factors. Heritability in the broad sense may be thought of as the degree of genetic determination of a trait (structure, function, behavior), i.e., the proportion of variance attributable to all genetic components of the genotype.

Another way of grasping the meaning of a heritability index is to think of it as a ratio of two variances, one genotypic (σ_g^2) and the other phenotypic (σ_p^2). Thus:

$$h^2 = \frac{\sigma_g^2}{\sigma_p^2}. \tag{2}$$

All heritability methods are based on comparisons of similarities among relatives of different degrees of kinship, ranging from MZ and DZ twins down to cousins and nonrelatives. Heritability estimates can also be derived from correlations among parents, full siblings, and half siblings. Other techniques for computing h^2 employ variances and covariances. Twin and family studies have produced evidence of appreciable heritabilities for such diverse skills as auditory and visual acuity, reaction time, hand steadiness, tapping speed, finger dexterity, manual dexterity, card sorting, rotary pursuit, mirror drawing, certain musical aptitudes, spool packing, spatial relations, rowing, and other athletic skills (de Garay, Levine, & Carter, 1974; Jones, 1972; Loehlin, Lindzey, & Spuhler, 1975; W. R. Thompson, 1974; Vandenberg, 1966, 1971).

Confusions about heritability are commonplace, so it may be desirable to point out that h^2 is a statistic referring to a collection of scores made by many people rather than to a single measurement of any given individual in the population. The h^2 index is susceptible to errors of sampling, however the greater the number of cases the smaller the error; heritabilities based on large samples are more reliable than those based on smaller ones. As a statistic based on samples from the population, h^2 will naturally be influenced by population characteristics. That includes the task requirements, the subjects actually tested, and the nature of the environmental conditions. If the environment is fairly homogeneous relative to the trait we are interested in, then h^2 will be higher there than in a more heterogeneous environment. Moreover, when genetic characteristics are rather uniform but the pertinent environmental characteristics are highly variable or changing systematically, as in a learning experiment, then h^2 will be lower. For instance, the heritability of the Rotary Pursuitmeter is close to 90% at the beginning of a laboratory training session, but the effect of continuous practice and reinforcement is to reduce that value by nearly half within 45 min. or an hour (McNemar, 1933; Noble, 1970a)—so high heritability does not imply low learnability. In short, h^2 depends upon environmental variation as well as genetic variation. A continuous interaction takes place between environments and genotypes. Heritability is not a physical constant like gravitation or the velocity of sound, nor is it recalcitrant to learning variables.

The foregoing remarks may help dispel the widespread misconception that human skills having high heritabilities are those least amenable to practice. Even if the tracking ability demanded by the Rotary Pursuitmeter were associated with an h^2 of 1 (heritability of 100%), most people would still exhibit dramatic practice gains within half an hour, and there would be both individual and group differences in the acquisition of skill. What the high heritability of a training device does imply is that, despite the uniform conditions of learning and performance controlled in the laboratory, subjects will nevertheless differ quantitatively from one another in the origins, rates of gain, and final levels (asymptotes) of proficiency. Heritability is concerned with differences of genotypic provenance, and nature's theme is variation.

Turning to psychomotor skills of greater cultural interest, we find rather persuasive evidence for the heritability of musical talent in familial research on professional musicians. About 40 years ago Scheinfeld (1965) studied the families of 36 outstanding instrumental virtuosi (e.g., Jascha Heifetz, Guiomar Novaes), 36 leading Metropolitan Opera singers (e.g., Kirsten Flagstad, Ezio Pinza), and 50 conservatory graduate students at the Julliard School of Music. The mean age at which their talents appeared was 6.67 years. Precocity was greatest for the instrumentalists (4.75 years), and their professional debuts came at 13 years on the average. In nearly every case, then as now, top-flight instrumental soloists in the field of classical music first performed in public as child prodigies. Table 10.2 presents a summary of the familial data.

It is notable that the majority of the subjects reported that one or both of their parents (64% to 68%), their brothers or sisters (52%), and other relatives (54%) were also musically gifted. The families of the three groups of musicians were genetically unrelated, and they differed widely in socioeconomic status (SES), but the high incidence of musical talent was similar in all groups. Contrary to popular belief, SES was not an important factor. "Some of the greatest virtuosi came from the humblest and least musical homes . . . some of the lesser ones came from highly musical backgrounds . . . corresponding situations were found in our vocalist and Julliard groups and were consistent with the stories of many other musicians outside of our study (Scheinfeld, 1965:391)." Further analysis of the data revealed that when neither parent showed any talent for music only 15% of the siblings did; when only one parent was musi-

TABLE 10.2

Familial Evidence of Musical Talent among 122 Professional Musicians[a,b]

Talented kinfolk	Frequency of musical talent	Total number in sample	Percentage showing musical talent
Mothers	78	122	63.93%
Fathers	83	122	68.03%
Siblings	149	285	52.28%
Other relatives[c]	66	122	54.10%

[a] Adapted from Scheinfeld, 1965.

[b] The survey of 122 musicians included 36 instrumentalists, 36 singers, and 50 graduate students.

[c] Having one or more musically talented kinfolk, other than parents or siblings, counted as a maximum of one per subject (n = 122).

cally gifted, about 60% of the siblings were; but when both parents exhibited musical talent, over 70% of the siblings did so too. The rising probability of gifted musical offspring with single-talent and double-talent matings among the parents lends support to the hypothesis of a polygenic mechanism of inheritance for predispositions to great musicianship (Scheinfeld, 1965), as is the case for general intelligence (Jensen, 1972, 1973); but other quantitative genetic hypotheses have also been considered (W. R. Thompson, 1974).

In terms of race, most of the professionals were Caucasoid, with the Jewish subpopulation contributing a remarkable portion of the supreme talent.[5] As for sex, 91.7% of the instrumental artists were men and 8.3% women, but the opera stars were equally representative of the two sexes. Scheinfeld found no female composers or conductors of the first rank prior to World War II.[6] If such a study were repeated today, it would

[5] In another musical survey of the 1930s (Sward, 1933), it was established that Jews were far more prominent in the field of classical music than their numbers in the U.S. population (then about 3.6%) would lead one to expect. Jews constituted between 9% and 10% of the woodwind and brass sections of the 12 major symphony orchestras; they made up about 15% of the American composers listed in reference volumes; 24% of the American composers selected for the concert programs; 34% of the string players of the major orchestras; 46% of the leading symphonic conductors; 51% of the first violins in the major orchestras; and 70% of the violin soloists in greatest demand (those featured four times or more) with symphonic organizations. The correlation coefficient between the proportion of Jewish musicians and the symphony orchestras' rated excellence was +.41. Jews were even more prominent in popular musical organizations of the period. In 23 of these orchestras, they comprised over a third of the total personnel and nearly three-fourths of the violinists. For the period examined by Sward, Jews are therefore seen to be members of a gifted musical elite who (in the best sense of the word) were overrepresented by factors that, depending on the specialty, ranged from nearly 3 to about 21 times their population percentage.

[6] He might have selected Nadia Boulanger and Antonia Brico.

undoubtedly produce a few different conclusions about race and sex. The recent emergence of superb instrumental, operatic, and conducting talent among Negroids (Grace Bumbry, Mattiwilda Dobbs, James De Priest, Leontyne Price, Andre Watts) and among Mongoloids (e.g., Yin Cheng-chung, Kyung-Wha Chung, Hiroyuki Iwaki, Seiji Ozawa, Yoshio Unno) is noteworthy. Some impressive female conductors (e.g., Sarah Caldwell, Margaret Hillis) and composers (e.g., Jean E. Ivey, Thea Musgrave) may be added to the list. Similar familial investigations should also be made of musicians in the various fields of popular music: jazz, for instance, a medium in which Negroids have so successfully held the spotlight during this century in the roles of instrumentalists, singers, conductors, and composers (e.g., Louis Armstrong, Duke Ellington, Ella Fitzgerald, Scott Joplin).

Pedigree studies of famous musical families (e.g., those of Bach and Strauss) corroborate Scheinfeld's observations, although they cannot be regarded as conclusive of hereditary transmission. More definite is the evidence from twin studies mentioned earlier; for example, musical aptitude test scores show greater similarity in several specific traits for MZ twins than for DZ twins (W. R. Thompson, 1974; Vandenberg, 1971). Furthermore, tendencies toward musical stardom usually emerge early in life and improve steadily in a variety of environmental contexts, often without manifest practice and sometimes in spite of firm parental opposition. Phenomena of this sort suggest the maturation of time-released genes in the steady flowering of the musical phenotypes. But the foregoing remarks are primarily concerned with rare professional talents—with profound musicianship—not with the garden variety of amateurs. Although a few gifted musicians may be able to get by with a minimum number of rehearsals, most singers and instrumentalists have to spend many hours in the studio in order to maintain their proficiency levels and achieve a satisfactory degree of artistic expression. To quote concert pianist Jan Smeterlin, "Heredity without training would not go far, but training without heredity would not go anywhere (Scheinfeld, 1965:407–408)." In his textbook on the psychology of music, Lundin (1967) takes a more proenvironmentalistic position. While not denying the inheritance of favorable structures, he prefers to emphasize musical surroundings and individual effort.

Perceptual Skills

The domain of psychomotor behavior is bordered by perceptual abilities at one extreme and by athletic skills at another. Perceptual performance is conventionally studied under the headings of stimulus detection, recognition, discrimination, identification, and judgment. The role of association is powerful, even in simple sensory tasks such as learning to reduce the psychophysical threshold of discriminating two points of pressure stimulation on the skin. Operationally, the threshold is the median distance associated with judgments of "one" and "two" points. Practice decreases this two-point threshold, enabling one to detect progressively shorter cutaneous gaps between stimuli. More complicated tasks require subjects to learn a unique response to each member of a set of similar configurations (e.g., aircraft recognition and identification); to adjust themselves to stimulus transformations (e.g., altered or contradictory visual–spatial relationships); or to apply what is learned in one sensory modality to another (e.g., cross-modal transfer of training from vision to touch). There is not only a process of discovery but also one of enrichment (Epstein, 1974).

Reading is a crucial intellectual ability that—once the biological readinesses to speak and read have matured—appears to proceed through several psychomotor stages. Initially, there is increased specificity of response to stimulation; this is accompanied by the detection of distinctive stimulus features; then later by the identification of patterns and properties of stimulation. The response learning of speech phases into the discrimination of letters, which in turn is followed by the acquisition of letter–sound combinations, and eventually by the mastery of words, phrases, and sentences. Evidence exists that many other practical skills entail these stages. Perceptual learning, in short, is a systematic change in a person's ability to obtain information from the environment (Gibson, 1969).

Perceiving is highly contingent upon modification by action and by training, but not always in the direction of improvements in people's ability to sense their environments. Learned inferences about perceptual relationships sometimes lead unsophisticated observers into error (Epstein, 1974). In the main, however, one who engages in continuous tactual, manual, and locomotor activities acquires the skill of perceiving because of the actual doing (Noble, 1977a). Experimental psychologists are confirming G. H. Mead's insight of the 1930s that percepts are essentially collapsed acts.

The methods of factor analysis have revealed no more than a dozen basic perceptual aptitudes of interest to us (Hakstian & Cattell, 1974; Horn, 1976; Thurstone & Thurstone, 1941; Tyler, 1965). The six chief *perceptual skill factors* are

1. aiming
2. clerical perception
3. flexibility of closure
4. perceptual speed
5. spatial orientation
6. speed of closure

There seems to be no "general perceptual" factor.

The factor of aiming involves the speeded execution of precise movements requiring eye–hand coordination as a subject draws pencil lines between narrow printed boxes or places dots in a series of small printed circles as fast as possible. Clerical perception indicates the ability to grasp verbal and numerical details in print and to spot similarities and differences between sequences of words and numbers. Flexibility of closure requires a subject to disregard or see through irrelevant visual stimuli in order to discover embedded stimulus figures. Perceptual speed reflects the capacity to scan geometrical figures quickly and detect similarities and differences among them. Spatial orientation taps one's aptitude for visualizing and remembering spatial relationships and for recognizing the same form presented in various orientations. Speed of closure measures how well a subject can complete a *Gestalt* (stimulus configuration) when parts of it are missing.

Six additional perceptual factors of relevance to this chapter may be noted (Guilford, 1959). They include:

7. auditory discrimination
8. auditory sensitivity
9. length–size estimation
10. musical aptitude
11. visual movement
12. visual sensitivity

Average subpopulation differences attributable to our three biopsychological vectors have been reported for sensory modalities and mechanisms (e.g., visual, auditory, cutaneous, gustatory, and olfactory acuity) and also for the perception of color and visual illusions. A hereditary foundation for most of these traits is well established, although determinations of direct gene → factor linkages are rare (Loehlin *et al.*, 1975; Maccoby & Jacklin, 1974; Malina, 1973; Post, 1962; Spuhler & Lindzey, 1967; Tyler, 1965; Vandenberg, 1971; Waardenburg, 1963). With respect to the importance of genetics for singers and instrumentalists discussed above, twin and family studies provide considerable evidence for the inheritance of special components in musical talent, including tonal memory and the discrimination of duration, loudness, rhythm, and pitch (Guilford, 1959; Scheinfeld, 1965; W. R. Thompson, 1974; Vandenberg, 1971).

A comment on the latter is in order. Despite a widespread, almost reverential belief among musicians in the hereditary phenomenon of *absolute* (perfect) *pitch*, there is no scientific evidence that the ability of some persons to identify and produce musical tones of rather exact frequencies (semitone accuracy) is other than a case of acquiring skilled *relative-pitch* perception. Proper controls for anchor points, tonal memory, and physical cues are rarely imposed on the claimants. Furthermore, numerous experiments indicate that, assuming adequate musical aptitude and efficient learning conditions, one can progressively improve the ability to name or sound a given tone based upon an initial standard or reference tone (Lundin, 1967; Ward, 1963). Most cases of well-developed abilities to identify or produce specific pitches are probably attributable to consistent parental reinforcement of such behavior in children already gifted with fine pitch discrimination. Relative-pitch ability, therefore, is an important age-related skill featuring crucial genetic and learning contingencies. It is a mandatory acquisition for vocal, wind, and string musicians.

Several investigations have found significant differences in perceptual factor loadings among taxonomic groups classified by age, race, and sex, but these have been mostly limited to perceptual speed and spatial orientation (Baughman & Dahlstrom, 1968; Loehlin *et al.*, 1975; Maccoby & Jacklin, 1974). The heritability of spatial orientation appears to be stronger than the other perceptual factors (Vandenberg, 1968, 1971). In the present volume, Osborne (Chapter 6) reports evidence from twin research that both perceptual speed and spatial orientation are substantially inherited, a fact which holds true for males and females as well as for Negroids and Caucasoids.

Naturally, the heritability of a test need not be the same for different sexes, ages, or races. Genetic factors may be more powerful in causing variation within one taxon than within another, or environmental influences may be differentially attenuated in the two groups. Heritability coefficients are not invariants. It would be unreasonable, therefore, to expect that a test of perceptual (or psychomotor or athletic) ability will elicit the same index of genetic determination when different biopsychological comparisons are made.

Athletic Skills

By contrast with perceptual performance, athletic performance is manifestly motor. Popular sports typically call for complex integrations of numerous separate movements, and the emphasis upon flexibility and whole-body coordination is almost as pronounced as that traditionally placed by coaches on physical fitness. Factor-analytic techniques have come up with at least nine *physical fitness factors* (Fleishman, 1964; Singer, 1972) that are independent of the psychomotor and perceptual factors already discussed. They are

1. dynamic flexibility
2. dynamic strength
3. explosive strength
4. extent flexibility
5. gross body coordination
6. gross body equilibrium
7. stamina (cardiovascular endurance)
8. static strength
9. trunk strength

There seems to be no "general physical fitness" factor (Fleishman, 1964).

Varying with the sport or practical skill in question will be the relative emphases (loadings) of these basic aptitudes. For example, one would expect gross body equilibrium to be an important factor in the selection of divers and gymnasts, whereas explosive strength should be relevant to weight lifting and shot putting. It happens that different morphologies are also correlated with particular sports, indeed with outstanding athletic prowess. The somatotypes of 435 male and 722 female Olympic participants in the Tokyo Games of 1964 were generally *ectomorphic* for the male basketball players and female distance runners; *mesomorphic* for the gymnasts, regardless of sex; and *endomorphic* for the male water poloists and female weight throwers (Hirata, 1966). These findings were confirmed and extended in a detailed study of 1117 male and 148 female athletes attending the Mexico City Olympics in 1968 (de Garay *et al.*, 1974).

Other investigations over the past half-century support the hypothesis that great athletes in certain sports exhibit modal anatomical configurations. Among men it is regularly observed that basketball players tend to be ectomorphic, gymnasts mesomorphic, and water poloists endomorphic. Even in a single sport like college football, distinctive somatotypes are typical of the various team positions (Singer, 1972). Race is another potent variable (Codwell, 1949; Jordan, 1969; Singer, 1975; Worthy, 1974) that combines with sex (Broverman, Klaiber, Kobayashi, & Vogel, 1968; Fleishman, 1964; Hirata, 1966; Kay, 1969; Maccoby & Jacklin, 1974; Ounsted & Taylor, 1972; Singer, 1972, 1975) and age (Bayley, 1935, 1965; Birren, 1964; McGraw, 1939; Shirley, 1931; Singer, 1975) to influence physical proficiency. Even IQ, provided its range is not curtailed, is a trait of strong hereditary provenance that is significantly involved in these “fitness” behaviors (Singer, 1975; Sloan, 1951; Thurstone, 1959).

So the data show that the principal organismic variables (race, sex, age, somatotype, intelligence) are consistently associated with athletic performance. Twin research and other genetic investigations also reveal that such physical characteristics have substantial heritabilities (deGaray *et al.*, 1974; Gedda, 1961). Specific ancestries, morphologies, and other endowments are probably not necessary, however, and certainly not sufficient. You do not have to be Caucasoid to excel in rowing, or Mongoloid to excel in gymnastics, or Negroid to excel in basketball; endomorphic masculinity is no guarantee of a *sumo* title, nor is mesomorphic femininity one of a diving medal; a tennis singles champion might be younger than 15 or older than 30; and quarterbacks need not be paragons of intellect. My obvious point is that it would be erroneous to conclude that great athletes are merely born. They are also made.

INTERACTIONS OF ORGANISMIC FACTORS AND PRACTICE VARIABLES

Before consideration of the data on practice × organismic interactions in the domains of age, race, and sex variation, it may be helpful to nonpsychologists for me to explain the meaning of interaction in abstract terms, and then relate it to Hull's (1945, 1952) postulate about individual and group differences by analyzing some actual psychomotor acquisition scores recorded from subjects of different aptitudes.

Interaction, Variation, and Lawfulness

Assume that we are conducting an experiment that deals with complex functional relationships among psychology's three major classes of variables (Noble, 1966b; Woodworth, 1929) such that $R = f(S, O)$, where S is manipulable over a wide range and there are at least two fixed levels or values of the third variable (O_1, O_2). If you like, let R = a response measure of skill, S = the number of stimulations, and O = some organismic factor of anatomical, genetic, or physiological interest. Statistically speaking, an *interaction* between the two jointly operating independent variables (S, O) in determining the mean scores of the dependent variable $(\bar{R})$ would exist if the graphical plot of $\bar{R} = f_1(S, O_1)$ were not parallel to the plot of $\bar{R} = f_2(S, O_2)$. In other words, the lines or curves depicting the geometry corresponding to the two algebraic functions (f_1, f_2) would fail to show equal differences in $\bar{R}$ scores between the two levels of O as we move from low to high S values, or vice versa. Three possible cases of simple linear interactions are shown in Figure 10.3. Such nonadditive outcomes of our hypothetical experiment are represented by the multiplicative formula $R = f(S \times O)$, as contrasted with the additive formula $R = f(S + O)$. In all nonadditive cases, we may conclude that S and O *interact* in their joint effect upon R. The graphical appearance of interaction might be convergence of the O groups with increasing S, or divergence of the O groups with increasing S, or some combination of the two trends as in reversal or crossover data. But are the findings replicable?

The analysis of variance[7] is a statistical method of testing for the significance of main effects and interactions among variables by partitioning the total variance (σ^2) of the obtained scores into components, calculating ratios of systematic σ^2 to error σ^2, and computing their probabilities (p). For example, the *main effects* of S in Figure 10.3 are shown by the fact that $\bar{R}$ for $S_5 > \bar{R}$ for S_1, averaging over both O conditions, in the convergent and divergent cases. The reversal case shows no main effect of S. An *interaction* between S and O for the convergent case is indicated because $\bar{R}$ for $S_5 > \bar{R}$ for S_1 at level O_1, whereas $\bar{R}$ for $S_5 = \bar{R}$ for S_1 at level O_2. Statisticians refer to an additive outcome (parallelism) as the *null hypothesis;* i.e., the magnitudes of the combined effects of S and O upon R are about the same for different categories of O. Alternatively, this null hypothesis asserts that the average differences in $\bar{R}$ scores over the S scale for various pairs of O_1 and O_2 values are approxi-

[7] For the definition of variance (σ^2), see note 1.

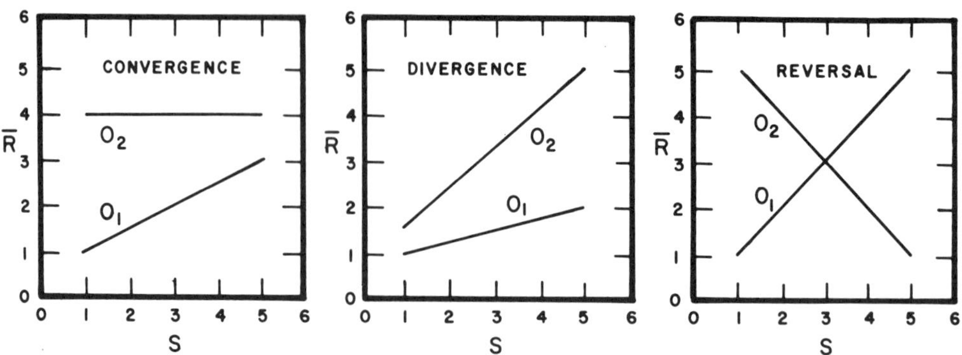

Figure 10.3. Some examples of interaction among linearly-related variables where $R = f(S \times O)$. See text for explanation.

mately constant. I say *about* and *approximately* because random sampling fluctuations usually produce slight deviations from uniformity. Now if this additive hypothesis were to be overthrown by a *significant* $S \times O$ variance ($\sigma_{s \cdot o}{}^2$) relative to the error variance ($\sigma_{error}{}^2$)—e.g., an associated probability less than 5 in 100 ($p < .05$) that the measured departures from additivity are chancy—then we would conclude in favor of an observed $S \times O$ interaction. Interactions may occur between two factors (e.g., practice × age), among three factors (e.g., practice × age × race), among four factors (e.g., practice × age × race × sex), or among more. For brevity, I shall speak of 2-factor, 3-factor, 4-factor, and generally n-factor interactions, but it is difficult for most people to visualize interactions beyond the triple and quadruple cases. Fortunately, higher-order interactions do not need to be grasped perceptually. They do, however, need to be grasped conceptually.

Turning now to the second question, we note that Hull's view of differential and comparative psychology implies that although there are powerful biological constraints on learning and performance, the discovery of general laws of wide applicability is, in principle, still possible. To put it technically, we should be able to combine nomothetic (Wundtian) and idiographic (Galtonian) principles of scientific description. An illustration may clarify the point. Drawing upon unpublished human tracking data collected in this laboratory, I have selected 60 adult, Caucasoid, right-handed males for quantitative analysis. Their mean age was 19.8 years. Each subject received 100 trials of standard practice on the Rotary Pursuitmeter (see Figure 10.2) under the same experimental conditions mentioned in connection with Figure 10.1 (Noble, 1970a). None of the men was familiar with the learning apparatus, and all were treated exactly alike during the demonstration and training phases. Indeed, the subjects were run in subgroups of 3 or 4 at a time with random assignment to the four apparatus copies and to the various experimenters, adopting the "psychomotor line" of Melton's (1947) program, so that possible constant errors associated with our aptitude groups of interest would be minimized. In other words, any extraneous variables (e.g., factors confounded with apparatus, experimenter, time of day, etc.) that tended to have the same influence on all subjects of a particular group but different influences on subjects of other groups were probably neutralized by the research design. Since we are focussing on the numerical constants of the habit process, $H = f(N)$, all nonassociational variables (e.g., D, V, I, age, race, sex) must either be held constant or randomized among the groups presumed to differ mainly in hereditary talent for pursuing the rotor.

Based on their innate tracking aptitudes as reflected by initial proficiency recorded during Block 1 (Trials 1–5) on this task of exceptionally high heritability, I stratified the 60 men into three equal-sized homogeneous levels of High, Medium, and Low aptitudes ($n = 20$ subjects per group). Their time-on-target scores, converted to percentages and averaged ($\overline{R}\%$), were then grouped into 20 blocks of 5 trials each (N). The results of these operations are displayed in Figure 10.4. A 20×3 analysis of variance revealed significant main effects of practice and aptitude, as well as their interaction ($p < .01$). A null hypothesis cannot be sustained for these data. The three groups are widely separated at the outset of training, they converge toward the end of the practice session, and they appear to be approaching different asymptotes (final levels). Their trends are differential; the curves are not parallel. In point of fact, the trend analysis reveals significantly different rates of gain in proficiency with training.

But if the *forms* of the functions in Figure 10.4 are the same, with only different parameters to represent their distinctive origins, rates, and limits, then our theorizing about $H = f(N)$ would receive additional empirical corroboration. From Hull's (1943, 1952) general theory of learning and performance, as elaborated by others (Noble, 1966a, 1966b, 1969a, 1970a, 1970b; Spence, 1956, 1960), it can be shown that the acquisition of pursuit tracking skill with practice may be represented by the following equation, a derivation of which appears on pp. 362–368:

$$R\% = M(1 - e^{-kN}) \pm T, \quad (3)$$

where M = maximum habit gain of $R\%$, $e = 2.718$, k = rate parameter of habit growth, N = number of trials or blocks of reinforced practice, and T = initial transfer value or origin of the function when $N = 0$. Each curve in Figure 10.4 was fitted with Equation (3) by the method of least squares. All three parameters (M, k, T) were allowed to covary until the residual variance was minimized and the fit maximized. The equations for the High, Medium, and Low aptitude groups are

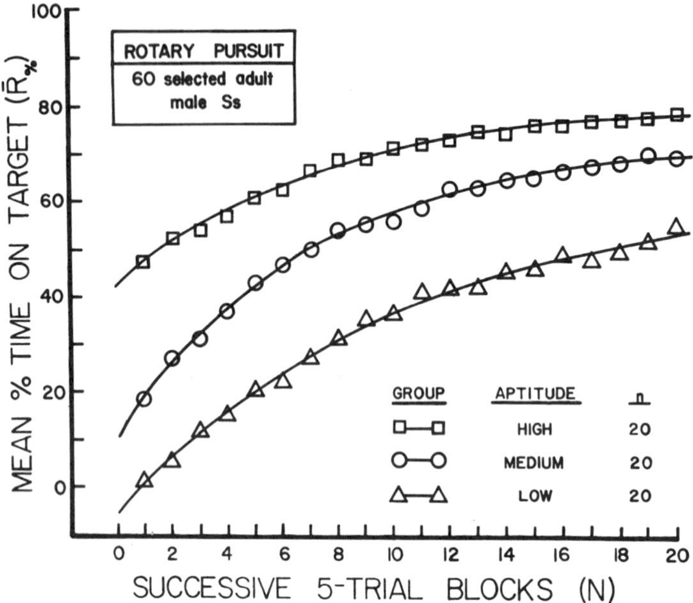

Figure 10.4. Acquisition curves of mean percentage of time on target ($\bar{R}_{\%}$) as functions of successive practice blocks (N), with initial ability on Block 1 as the aptitude parameter. Each group contains 20 selected subjects of homogeneous ability. The curves were fitted by Equation (3) using the method of least squares. (Data from Noble, unpublished.)

$$\text{High: } \bar{R}_{\%} = 37.35(1 - e^{-.1401N}) + 42.30 \tag{4}$$

$$\text{Medium: } \bar{R}_{\%} = 61.77(1 - e^{-.1442N}) + 10.51 \tag{5}$$

$$\text{Low: } \bar{R}_{\%} = 67.43(1 - e^{-.0982N}) - 5.71 \tag{6}$$

Equation (3) accounts for 99.18% of the σ^2 in the case of the High aptitude group, 99.69% of the σ^2 for the Medium group, and 99.50% of the σ^2 for the Low group. The average goodness of fit for Equations (4), (5), and (6) is 99.46%; it follows that the error of prediction from Equation (3) is less than 1%. I conclude that Hull's position on individual and group differences is consistent with these psychomotor data and therefore still a viable proposition, at least for his hypothesis of habit (H) formation. I would infer from his theory that similar extensions could be made to other intervening variables, such as D, V, K, I, O, and L.[8] Later, some evidence will be reviewed that supports Hull's notion for I as well as for H. Now let me proceed to consider age, race, and sex.

[8] As specialists in theoretical psychology know, I have simplified Hull's orthography and omitted the details of certain processes in order to facilitate this presentation. Some of us who are more Spencean than Hullian regard the following general formulation as a closer

Age and Learning Tasks

Age is a biological factor that interacts with practice variables in the determination of human performance on psychomotor, perceptual, and athletic tasks. Following a list of generalizations, I shall discuss the role of age in each group of tasks, present a few salient acquisition phenomena, then close the section with a treatment of certain temporal interactions.

Age Generalizations

Although both Galton and Wundt realized that the psychology of infants, children, adolescents, and adults might be different, it was left for others to investigate systematically the development of human behavior from youth to senescence. An exploratory period extended into the 1920s (Gesell, 1928; Gesell & Thompson, 1929; Thorndike, 1914; Thorndike *et al.*, 1928), when Whipple's *Manual of Mental and Physical Tests* (1921–1924) cited a few reports of age differences on psychomotor tasks. Accompanying Piaget's (1936) genetic emphasis on assimilation (stimulus reception and interpretation) and accommodation (cognitive reorganization) during the sensorimotor stage of infancy, several classic studies prior to World War II illuminated the principles of maturation and ontogeny (Bayley, 1935; Espenschade, 1940; Halverson, 1940; Hicks, 1931;

---

approximation to the data of most laboratory studies of conditioning and learning:

$$R = f\{[H(D + K) - (I + O)] > L\}$$

Research by the Iowa neobehaviorists has modified the Yale position in several respects. They have:

1. subsumed stimulus dynamism, V, under D and K and postulated an additive motivational complex $(D + K)$ rather than a multiplicative one (Logan, 1968; Spence, 1956, 1960);
2. incorporated an associative mechanism for the learning of rewards and punishments within the incentive factor, K (Logan, 1968; Spence, 1956);
3. elaborated the concept of temporary work decrement or reactive inhibition, I_r, in conjunction with the inhibition (or frustration) of nonreinforcement, I_n, in lieu of the now defunct permanent work decrement or conditioned inhibition, $_sI_r$ (Adams, 1956; Adams & Reynolds, 1954; Ammons, 1947, 1970; Amsel, 1967; Spence, 1956, 1960);
4. suggested random threshold variation, L, as a theoretical alternative to the oscillation function, O, and the multiplication of H by D when motivational variables are manipulated (Grice, 1971);
5. radically broadened the theory of a normally distributed, labile threshold, L, to include dependence "on such factors as motivation, incentive, emotional state, set, attention, sequence effects, adaptation, and individual differences (Grice, 1972:3)," thereby minimizing sources of variability arising from parametric differences in $H = f(N)$.

Not all of these revisions are consistent with one another, but they do indicate the extraordinary versatility of the Hull–Spence theory in the hands of capable partisans who are as much at home with mathematics as with experimentation.

McCloy, 1935; McGraw, 1939, 1943; Miles, 1933; Shirley, 1931). Intensive postwar research studies, some longitudinal but most cross-sectional, have provided a continuing flow of information about growth sequences and age changes in psychomotor performance (Ammons, 1958; Ammons & Ammons, 1970; Bayley, 1965; Birren, 1964; Gesell & Amatruda, 1947; Hicks & Birren, 1970; Illingworth, 1966; Kay, 1969; McGeoch & Irion, 1952; Noble, Baker, & Jones, 1964; Singer, 1975; Tyler, 1965; Welford, 1958; Welford & Birren, 1965). On the basis of those studies and surveys, the following generalizations may be offered: *(1)* psychomotor performance is a nonmonotonic function of chronological age, increasing rapidly through infancy, childhood, and adolescence to about ages 18–20, leveling off during the period of early maturity (20–30 years), then decreasing slowly through middle and later maturity into old age (> 75 years); *(2)* the typical declines of the late maturity years are more marked for *speed* tests (having strict time limits and paced throughout) than for *power* tests (having no time limits and self-paced); *(3)* sensory and perceptual abilities show an earlier and steeper impairment than do motor and cognitive abilities; *(4)* complex tasks tend to produce greater differences among age groups than do simple tasks; *(5)* although general trends based on the averages of age groups are clear and replicable, there is considerable overlapping caused by individual differences.

Perceptual skills, ranging from simple discriminations characteristic of the various sensory modalities to complex judgments of relationships among patterned stimuli, are notably influenced by changes in age that progress from infancy through senescence (Birren, 1964; Tyler, 1965; Welford, 1958). Population impairment rates, measured by the relative frequency per 1000 persons of defects in visual and auditory functions, generally rise with age in a positively accelerated fashion. Increased probability of defective sensory abilities is quite precipitous after the age of 70, and the major modalities of seeing and hearing exhibit remarkably congruent functions from 12 years to 80 years. Both absolute and difference thresholds go up with advancing years. Consequently, stimulus energies must be progressively increased, on the average, in order to evoke either an *identifying response* (e.g., "I perceive that a stimulus is present") or a *differential response* (e.g., "I perceive that a change or difference in stimulation has occurred").

Decremental effects of aging in the visual modality are measurable by a number of changes, most of which are irreversible. They include lowered visual acuity, diminishing pupil size, greater failure of accommodation, elevated dark-adaptation thresholds, reduced visibility–illumination functions, and nonmonotonic alterations in critical flicker frequency thresholds. The chief impairment of the auditory modality, also irreversible, is a progressive audiometric hearing loss; this is especially characteristic of persons over 45 years old and in the frequency range above 10,000 Hz. Age-related changes of one's vestibular mechanisms, affecting the senses of bodily position, balance, and motion, appear as reduced postrotational nystagmus (eye-oscillation) time and decreased nausea with vertical stimulation of the semicircular canals. Gustatory sensitivity, defined by taste thresholds for detecting bitter, salty, sour, and sweet substances, undergoes a decline after 50 years of age because of atrophy of the papillae of the tongue. Cutaneous (touch) thresholds on the skin and eye also tend to rise after the midcentury mark; this is equally true of one's sense of vibration, a tactile frequency of 100 Hz being less discriminable in the lower extremities than in the upper. Olfactory (chemical odor) and pain (radiant heat) sensitivities, on the other hand, do not seem to be greatly impaired by the ravages of time.

Many optical illusions are influenced by chronological age: the Delboeuf, Ebbinghaus, Müller–Lyer, and Ponzo illusions are familiar ones. For the first three cases, the typical amount of overestimation (e.g., the segment with outgoing fins in the Müller–Lyer) decreases in childhood and adolescence, levels off around 20, then remains stable until about 45, after which it tends to rise a bit. A similar nonmonotonic trend occurs for the Ponzo, except that its U-shaped function is accomplished earlier in life (Jaeger, 1977).

Turning to visual form perception, we find few age differences in recognizing two-dimensional figures unless the viewing conditions are made difficult by poor illumination, inadequate contrast, or brief exposure durations. Tests of subjects' abilities to perceive hidden or missing figures and to resolve ambiguous drawings reveal consistent differences between those in their 20s and those in their 70s. Although studies of human anatomy confirm the principle of a decreasing number of neurons and receptors in the peripheral and central nervous systems with advancing age, Birren (1964) has suggested that critical biological constraints on perceptual skill probably do not occur before the age of 70 on the average. The besetting problem of senescence is that, unlike the situation in youth—where everything is bright, loud, clear, and tasty—one must in great age try to discriminate and identify sensations of drastically reduced intensity. Comparatively speaking, the perceptual world of our elders is dull, muffled, hazy, and bland.

Athletic skills as pervasive as throwing, balancing, running, catching, reaching, gripping, and jumping exhibit rapid improvements from childhood into adolescence and early maturity, then enter upon the slower decline that characterizes most psychomotor and some perceptual abilities (Bayley, 1935; Espenschade, 1940; Fleishman, 1964; Halverson, 1940; Illingworth, 1966; Kay, 1969; McGraw, 1939, 1943; Shirley, 1931; Singer, 1972, 1975; Welford & Birren, 1965). The majority of athletes attain their prime before 40, but champions in certain sports tend to have different mean ages from those in others (Birren, 1964). Top performers in tennis, baseball, and boxing, for instance, are typically younger (M = 27–30 years) than their counterparts in bowling, marksmanship, and billiards (M = 30–36 years). The causes of the difference lie presumably in the degree of muscular strength, stamina, whole-body exertion, coordination, and fast reactivity required by the former group of sports (i.e., running, jumping, hitting, twisting, sliding, stretching, anticipating). In Olympic contests, distinctive modal ages for champions are well known; e.g., swimmers are usually younger than gymnasts (Hirata, 1966). Old-timers may compete in doubles tennis on an international level, and play thrilling matches at that, but singles tennis is of necessity a young person's game. Golf, on the other hand, is a self-paced, leisurely sport that does not so vigorously stress the cardiovascular and pulmonary systems of the geriatric set.

Sports constitute a universal form of recreation that represents, moreover, one of the finest expressions of humanity's aspirations to excellence. From ancient times, poets have immortalized the victories of great athletes. Although the physiological benefits of a life-span program of physical fitness (e.g., calisthenics) are widely recognized, sports participation by children and adults confers many additional rewards through the development of varied skills, the personal satisfaction of play, and the social reinforcements of group activities (e.g., mixed doubles in tennis). An enormous practical value attaches to psychological research on aging because yesterday's tomorrow is today. Time's inexorable advance waits for no one.

Some Acquisition Phenomena

As mentioned previously, there is evidence in experiments on the learning of psychomotor skills that interactions occur between chronological age and task complexity. The classic study of this phenomenon was conducted by Ruch (1934) using a pursuit tracking apparatus like that pictured in Figure 10.2 (device F). Two training conditions on this modified Rotary Pursuitmeter were administered to 120 subjects in three different age groups whose individuals ranged from 12 to 82 years old. One task was practiced under *direct vision* (DV) that exploited the positive transfer from daily life, a case of S–R compatibility (Fitts, 1964); the other task was practiced under *mirror vision* (MV) that maximized negative transfer, a case of S–R incompatibility. Perceptually, the MV task resembled that of the Mirror Tracer in Figure 10.2 (device D) because the standard target and cursor images were reversed. Subjects of both sexes were included; but Ruch did not give the frequency breakdown and no tests of sex effects were reported. Although race was not mentioned either, the subjects were probably of Caucasoid ancestry and of high socioecconomic status (SES). They all practiced for 25 trials in a self-paced mode, with frequency scores recorded every 30 sec.

Ruch's result are summarized in Table 10.3 in terms of means (M), standard deviations (σ), mean differences (M_d), and significance tests (t). Some important details about the procedure and results are recorded in the footnotes to the table. Ruch found that product–moment correlations (r) of MV with age and intelligence were larger than those of DV; however, intercorrelations of MV and DV increased consistently with age: $r = .24$, .53, and .63 for the younger, middle-aged, and older subjects, respectively. The more complex MV condition apparently called for greater cognitive capabilities as age increased; e.g., IQ scores correlated .34 for the younger, .36 for the middle-aged, and .50 for the older subjects, respectively. A prime conclusion to be drawn from this

TABLE 10.3

Means (M), Standard Deviations (σ), Mean Differences (M_d), and Significance Tests (t) of Pursuit Tracking Scores Made by 120 Subjects in Three Different Age Groups under Direct Vision (DV) and Mirror Vision (MV) Conditions on a Self-paced Rotary Pursuitmeter[a,b]

		Age group (n = 40 each)				M_d and t		
Condition		12–17	34–59	60–82		12–17 versus 34–59	34–59 versus 60–82	12–17 versus 60–82
DV	M =	2857.0	2805.0	2392.0	M_d =	52.0	413.0	465.0
	σ =	244.3	287.2	415.9	t =	0.9	5.4[c]	6.1[c]
MV	M =	771.9	740.0	406.2	M_d =	31.9	333.8	365.7
	σ =	214.2	286.2	166.1	t =	0.6	6.4[c]	8.5[c]
		DV–MV	DV–MV	DV–MV				
M_d and t	M_d =	2085.1	2065.0	1985.8				
	t =	40.1[c]	31.8[c]	27.7[c]				

[a] Adapted from Ruch, 1934.

[b] I call Ruch's DV and MV tasks *self*-paced because the turntable would rotate only if the subject maintained stylus–target contact. This is a radical departure from the standard operating procedure of today, which is *apparatus*-paced (see Figure 10.2). Maximum speed of rotation was 32 rpm. Scores were recorded as the total number of tenth revolutions made in 25 massed trials of 30 sec. each. Two weeks intervened between the 12.5-min. DV and MV sessions. Reliability coefficients, based on odd–even trial correlations and corrected by Spearman's formula, averaged .98 for DV and .96 for MV combining all age groups.

[c] The probability of obtaining t ratios this large or larger by chance is less than 1 in 1000 ($p < .001$).

experiment is that the relative skill margin of younger subjects over the older was greater for the MV than for the DV task. These effects were confounded with level of proficiency, however, because practice was more massed for the younger subjects on DV than on MV. Still, the M_d of $465.0 > $ the M_d of 365.7, and $t_d = 8.5 - 6.1 = 2.4$, which is significant. A graph of the means indicates that Ruch's inference of an age–complexity interaction is tenable. This conclusion is consistent with independent evidence (Thorndike *et al.* 1928) that older people may be more prone than younger ones to performance deficits under conditions of negative transfer or increased complexity (see also McGeoch & Irion, 1952; Miles, 1933; Welford & Birren, 1965).

Another example of acquisition phenomena is provided by an experiment from my laboratory (Noble, Baker, & Jones, 1964) utilizing a three-dimensional practice × age × sex design in which 600 Caucasoid subjects of both sexes, separated into 30 age groups between 8 years and 87 years, were trained on the Discrimination Reaction Timer (Figure 10.2, device B) for 320 trials. This study corroborated the classic non-monotonic age–proficiency function cited earlier, as well as the principle of overlapping individual differences. In addition, we found that the mean acquisition curves (taking reciprocals of reaction time as a measure of speed) followed different trends for the sexes. Our analysis of variance revealed that the practice × sex interaction, computed over 16 blocks of 20 trials each, was significant ($p < .001$), as were the main effects of sex, age, and practice ($p < .001$). Indeed, all of the 2-factor interactions were significant; only the 3-factor practice × age × sex interaction was not (see Figure 10.3 and the associated discussion in the text).

Now what do these data on complex, color–spatial discriminations and multiple-choice reaction times imply about the principle of psychomotor lawfulness mentioned above? Are they, for instance, compatible with Hull's (1943, 1952) general theory of learning and performance, and with the mathematical model stipulated by Equation (3)? The reader will recall that Equations (4), (5), and (6) were applied to the acquisition of skill on the Rotary Pursuitmeter (Figure 10.2, device F), a continuous tracking task, by three groups of adult male Caucasoid subjects who differed primarily in terms of innate aptitude for rotary pursuit training. Equation (3) agreed very well with their group mean curves; an average of 99.46% of the variance in time-on-target scores was accounted for (see Figure 10.4). In the Noble-Baker-Jones (1964) experiment, we discovered that the mean predictability of all speed curves for the $15 \times 2 = 30$ groups of subjects based on Equation (3) was 97.98% , indicating firm support for Hull's hypothesis as similarly applicable to the acquisition of skill on the Discrimination Reaction Timer. In order to focus our attention upon the interaction of practice and age in a manner comparable to that of Figure 10.4, let us consider the mean response speed scores ($\overline{R}_s$) for males between the ages of 19 years and 84 years recorded over 16 blocks of training in the study by Noble, Baker, & Jones (1964). In this age range, of course, one is dealing with the post-peak decline of proficiency. For the present analysis, I have selected six groups of 20 males each for plotting in Figure 10.5. Each curve was fitted with Equation (3) by the same methods I used to secure Equations (4), (5), and (6). The formulas representing the mean acquisition of $\overline{R}_s$ proficiency for the 120 selected men whose median ages were 19, 35, 46.5, 55, 67.5, and 75.5 years were as follows:

$$19.0 \text{ yr.}: \overline{R}_s = 2.21(1 - e^{-.2560N}) + 2.42 \quad (7)$$

$$35.0 \text{ yr.}: \overline{R}_s = 1.84(1 - e^{-.2144N}) + 2.32 \quad (8)$$

$$46.5 \text{ yr.}: \overline{R}_s = 2.10(1 - e^{-.1995N}) + 1.83 \quad (9)$$

$$55.0 \text{ yr.}: \overline{R}_s = 1.79(1 - e^{-.1274N}) + 1.81 \quad (10)$$

$$67.5 \text{ yr.}: \overline{R}_s = 1.45(1 - e^{-.1487N}) + 1.36 \quad (11)$$

$$75.5 \text{ yr.}: \overline{R}_s = 1.21(1 - e^{-.0682N}) + 1.18 \quad (12)$$

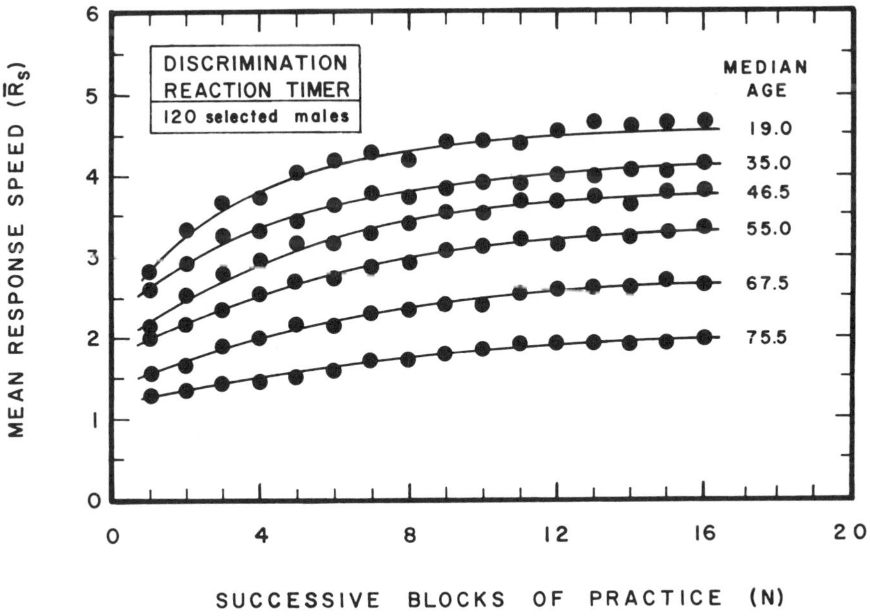

Figure 10.5. Acquisition curves of mean response speed ($\overline{R}_s$) in discrimination reaction as functions of successive practice blocks (N), with median chronological age (19 to 75.5 years of age) as the parameter. Each group contains 20 adult, male Caucasoid subjects of homogeneous age. The curves were fitted by Equation (3) using the method of least squares. (Data from Noble, Baker, & Jones, 1964.)

Equation (3), on the average, accounts for 98.19% of the variance in the six age groups; the mean error of prediction from Equations (7), (8), (9), (10), (11), and (12), therefore, is less than 2%. Once again, I am persuaded that Hull's doctrine of individual and group differences is in accord with the facts. It appears that acquisition curves of groups stratified by age and sex who practice on the Discrimination Reaction Timer for 320 trials can be predicted with considerable accuracy by an exponential formula, Equation (3), whose asymptote, rate, and intercept parameters jointly reflect human taxonomic differences, and whose mathematical form remains invariant over an extended range of chronological ages. I should add that rate of gain (i.e., the k parameter) alone cannot be employed to describe age differences in discrimination–reaction performance; the asymptote ($= M$) and R-intercept ($= T$) parameters must also be considered. The problem of experimentally isolating the variance components attributable to *aptitude* and to *capacity* is of fundamental importance (Adams, 1957; Bechtoldt, 1970; Noble, 1970a, 1970b, 1972, 1976b, 1977a; Noble, Baker, & Jones, 1964).

Temporal Interactions with Sex and Race

Secular trends in age-related phenomena are contingent upon sex and race. Here are some temporal–organismic interactions to consider as we explore further ramifications of the nature–nurture problem wherever it impinges on complex human action.

The members of *Homo sapiens*, especially the females of the species, enjoy the greatest longevity of all primates; their average life span is exceeded by only a few mammals, birds, reptiles, and molluscs. The venerable exceptions are certain species (whales, eagles, vultures, turtles, clams) that tend to live about twice as long as human beings; i.e., a century and a half. Statistically speaking, the heritability of this trait is very high in spite of accidents, climatic extremes, predators, infectious diseases, and other potentially lethal environmental events. According to Birren, "*species differences* in longevity appear to be largely determined by biological controls that are within the animals at birth (1964:57)." As far as within-species differences go, however, environmental factors are probably more powerful than genetic factors, at least in determining human life spans. Thus, an unmarried, city-dwelling, heavy smoker working in a highly stressful job situation who is grossly overweight and unwilling to control his elevated lipoprotein concentrations is running a much greater risk of early death than a married, rural-dwelling, nonsmoker working in a relaxed job situation who maintains a lean weight record and moderate lipoprotein concentrations. On the other hand, the longevities of MZ twins are significantly more alike than those of DZ twins or siblings. Furthermore, the heritage of long-lived parents or grandparents confers a definitely higher life-span probability. So time-bound genotypes as well as situational variables have a bearing on average and maximum longevities (Scheinfeld, 1965).

Sex is important too. In all civilized societies, females have lower mortality rates than males at most ages of life, and these have generally improved over the past few hundred years. For example, the expectation of life at birth (i.e., mean length of life) in England and Wales during the middle of the nineteenth century was about 39.9 years for males and 41.9 years for females. By World War II, the British statistics had risen to 60.2 years and 64.4 years, respectively. For Caucasoids in the United States, the life expectancies in 1901 were roughly 48.2 years for males, 51.1 years for females. By 1944, these figures had been elevated to 63.6 years and 69 years respectively, and by 1975, to 69.1 years and 77 years. Between 1977 and 2050, the Census Bureau projects means of 71.8 years and 81 years for males and females. A cautious extrapolation of the American data by sex thus indicates that the average expectation of life from the beginning of the twentieth century will have increased at a faster rate for women than for men by the middle of the twenty-first century. The sexes appear to be following a *diverging* trend over time. Technically, there is a sex × time interaction for life expectancy (see Figure 10.3). Though both will continue to improve, women are predicted to outstrip men in this respect.

Race is also a moderating longevity variable. In the United States, for instance, the mean life expectancy of infants in 1966 (averaging the sexes) was about 71.5 years for Caucasoids and 65.5 years for non-Caucasoids (Malina, 1973). At the turn of the century, however, the two subpopulations were averaging roughly 50 years versus 35 years respectively (Birren, 1964), a mean racial difference that decreased from 15 years to 6 years by 1966. From 1900 to 1977, the gap in average longevity for the two largest American ethnic groups has been reduced by approximately two-thirds. Such ameliorative changes are usually attributed to early diagnosis and improved treatment of major diseases (infectious, cardiovascular, etc.). Unlike the life expectancy statistics for the sexes, therefore, those for the Caucasoid and Negroid races in this country appear to be following a sharply *converging* trend over time (Birren, 1964; Malina, 1973). Technically, there is a race × time interaction for life expectancy (see Figure 10.3). Afro-Americans are rapidly approaching their Caucasoid countrymen in terms of average length of life. Barring wars, epidemics, or other major catastrophes, both will continue to improve, but Negroids' rate of gain will probably be greater.

Parenthetically, it is interesting to contrast these twentieth-century racial longevity data with earlier demographic, health, birth, and mortality statistics for Afro-Americans, slave and free. According to Weyl

(1970), the Negroid population in the United States increased during the decade prior to secession by 12.3% for freedmen and 28.8% for bondsmen, the difference probably reflecting better nutrition and medical care on the plantations. Although birth and death rates varied greatly by *era* (e.g., eighteenth versus nineteenth century) and by *assignment* (e.g., cotton or tobacco workers versus rice or sugar workers versus house servants or craftsmen), it appears that the average health of the slaves in southern states was considerably better than that of emancipated Afro-Americans in northern states. In Connecticut, Massachusetts, Pennsylvania, and Rhode Island, mortalities consistently exceeded natalities within the manumitted subpopulation. By contrast, death statistics for persons in bondage below the Mason and Dixon Line were reportedly no different from those of Caucasoids living in the same regions. Moreover, the infant slave mortality rate on plantations was about 153 per 1000, as compared with an average death rate (up to 1915) of 180 per 1000 for Negroid infants in Massachusetts, New York, and Pennsylvania.

There can be no justification for a system of involuntary servitude, but contemporary historical analysis suggests that the true state of antebellum demographic affairs doubtless lay somewhere between the extremes of propaganda issued by the proslavery and antislavery factions during the impending crisis of 1848–1861 (see Potter, 1976). Modern infant mortality rates in the United States are significantly lower, of course, for both Caucasoids and non-Caucasoids. During the 4-year period from 1963 through 1966, yearling deaths in the former group averaged 21.5 per 1000 whereas in the latter group (racially about 91.7% Negroid) the mean was 40.4 per 1000. The disadvantageous Negroid mortality rate is offset by advantageous natality and fertility rates relative to those of the American Caucasoid population. Between 1959 and 1967 the mean *birth* rates of Caucasoids and non-Caucasoids were 20.3 per 1000 persons and 62.5 per 1000 persons, respectively. The comparable mean *fertility* rates were 101.3 per 1000 women and 142 per 1000 women, respectively (computed from tables in Malina, 1973). Judging from the excess of births over deaths at the present writing, Caucasoids have nearly stabilized their relative population strength whereas Negroids are increasing theirs, especially among the youngest women of childbearing age. For females aged 15–19 years in 1967, for example, the number of live births per 1000 Negroid women exceeded that for Caucasoid women by a factor of 2.4 to 1. For those aged 10–14 years in the same birth cohort, the comparative fertility ratio was 13.7 to 1 (Malina, 1973).

To resume our discussion of the role of genetic factors in sickness and in health, it is pertinent to observe that strong correlations have been demonstrated between subspecific taxonomic memberships and inherited susceptibilities to various diseases. For example, Mediterranean

Caucasoids and Sephardic Jews are prone to β-thalassemia and G6PD-deficiency; Ashkenazic Jews to Niemann–Pick and Tay–Sachs diseases; African Negroids to hypertension and various hemoglobinopathies (e.g., sickle-cell anemia); Japanese and Korean Mongoloids to acatalasia and Oguchi's disease; American and North European Caucasoids to phenylketonuria and pernicious anemia; Amerinds to rheumatoid arthritis and congenital hip dislocation; Hawaiian Polynesians to diabetes and coronary disease (Coon, 1965, 1974; Damon, 1971; Garn, 1971, 1974; Goldsby, 1971; Malina, 1973; Osborne, 1971; Scheinfeld, 1965; Stern, 1973). One's racial origins may thus crucially affect one's disease susceptibility, and hence one's longevity, although the actual contraction and manifestation of a particular disease will depend upon local environmental variables.

In short, as Lord Russell once remarked in his anecdotage, it pays to choose your ancestors with care. At the same time, one can maximize the chances of survival by selecting optimal environments and modes of living. There is obvious value to society for its intellectuals to have the capability of healthy and extended life spans. The splendid cultural achievements of gifted persons during their so-called "postretirement" years are legendary. How fortunate for mankind that Carver, Casals, Confucius, Darwin, Galton, Goethe, Grandma Moses, Michelangelo, Russell, Shaw, Titian, Tolstoy, Toscanini, Verdi, and Voltaire lived and worked far beyond the arbitrary three score and five of which modern bureaucrats are so enamored.

Race and Learning Tasks

Race, like age, is a biological factor that interacts with practice variables in the determination of human performance on psychomotor, perceptual, and athletic tasks. After an excursus on taxonomics, and a recital of basic principles, I shall discuss the role of race in each group of tasks then present a few salient acquisition phenomena. Included will be a treatment of certain ethnic interactions.

Race generalizations

Scientific conceptions about the manifold similarities and differences among the major races of humankind with respect to psychomotor, perceptual, and athletic skills (as well as intellectual abilities) have varied greatly during the past 220 years, so a little background is called for. In 1758, the Swedish naturalist Carolus Linnaeus (1707–1778) correctly classified us, in his seminal book *Systema Naturae,* as members of the order Primates, the genus *Homo*, and the species *sapiens*. This taxonomy survives today, although he did not use the modern term *subspecies*.[9] Mainly on the basis of ancestral homelands, facial characteristics, pigmentation, and behavior, Linnaeus identified four geographical races of *H. sapiens* thus: *afer*, *americanus*, *asiaticus*, and *europaeus*. In modern terminology (Baker, 1974; Comas, 1960; Coon, 1965, 1974; Dobzhansky, 1965; Garn, 1971, 1974; Goldsby, 1971; Hooton, 1956; Howells, 1960; Laughlin, 1964, 1966; Loehlin *et al.*, 1975; Mayr, 1970; Osborne, 1971; Stewart, 1952), these grand divisions of humanity correspond approximately to the living ethnic taxa of Negroids (*afer*), Amerinds (*americanus*), Mongoloids (*asiaticus*), and Caucasoids (*europaeus*), respectively. (Linnaeus mentioned no equivalents of contemporary Australoids or Capoids.) In French, these would be called *grand'races*; in German, *Hauptrassen*. How insightful the Linnaean taxonomy was is attested by the fact that "all classifications have consistently drawn lines at the major geographical boundaries (Osborne, 1971:168)." This is a natural consequence of the role of geography in producing variation among subspecies. The other evolutionary processes governing race formation are selection, genetic drift, mutation, and selective migration. Not only are intercontinental genetic differences maximal for mankind, but also clinal populations (groups that are phenotypically and geographically intermediate) tend toward genetic hybridity. Well-known hybrids today include Afro-Americans, Cape Coloreds, Eurasians, Ladinos, and Neo-Hawaiians. I need not point out that the phenomenon of intermediates does not make of race a myth. On the contrary, clines and hybrids are characteristics of races.

Shortly after Linnaeus' work, the German anatomist and physiologist, Johann F. Blumenbach (1752–1840), combined cranial measurements with the other criteria to identify five races, adding Malayans. Blumenbach emphasized the unity of mankind, advancing what today's physical

[9] Baker (1974, p. 5) has coined a new taxonomic term for all zoological categories that are minor to the species *H. sapiens*. He recommends grouping subspecific human taxa that denote individuals who are genetically related under the plural label *stirpes*, taken from the Latin (the singular being *stirps*). The advantages of talking about "stirpal" differences, instead of "racial" or "ethnic" differences, reside in comprehensiveness and univocality of terminology. First, all human taxa below the species (e.g., geographical races, local races, microraces) are generically comprised by *stirpes*. Second, the fundamental meaning of this term is biological (i.e., genetic and anthropological); it is not basically cultural, economic, linguistic, national, political, religious, or social. Behavior is important, too. For a treatment of the evolutionary role of mammalian behavior in general, and of behavorial shifts as selection pressures among hominids in particular, see Mayr (1970, Chapters 19 & 20). Washburn (1960) has made a strong case for anatomy and behavior forming an interacting set of evolutionary processes. Coon (1965) has distinguished between environmental selection and behavioral selection in a discussion of racial differences in adaptive tratis (Chapter 8). Baker (1974, Chapter 7) has commented on the behavioral differences documented among races of mice, birds, bees, apes, and men. Few specialists, however, are willing to include behavior as a criterion of subspecies.

anthropologists call the single-species doctrine, and he deplored the categorization of certain races as "inferior" or "subhuman." (These pejorative terms are eschewed in this book. In point of fact, primatologists do not even countenance such language in reference to anthropoid apes, our closest living kinfolk in the hominoid superfamily.) Not that all races are necessarily equal, said Blumenbach, only that they are necessarily human. With the advent of the theories of evolution (Darwin) and genetics (Mendel), scientists' reliance upon geographic, morphological, and ethnographic methods of classifying human subspecies declined. They are still useful, however, as we shall see. Developments in paleoanthropology, geochronology, biochemistry, and population genetics (Fisher) during the twentieth century tended to harmonize the polarized viewpoints of the "anatomists" and "genticists" as anthropometry and serology, for instance, began to complement each other in physical anthropology. The upshot is that the contemporary definition of human races regards them as mutually interfertile subspecies; i.e., breeding populations that differ in the relative frequencies of one or more genes (Coon, 1965; Garn, 1971; Mayr, 1970; Stern, 1973).

Among the living geographical races of mankind (see the "Note on Taxonomy," p. xvii), there are undeniable anatomical, physiological, and psychological differences of genotypic origin. Many of these traits are visible for all who have eyes to see. Invisible differences also exist, differences that require biochemical tests and electron microscopes to detect. One of the major developments in racial genetics has been the gene-frequency method of classifying subspecies by the analysis of serological data (blood samples). Modern blood-group classifications (e.g., the ABO, MNSs, P, and Rh systems) are based upon certain properties of human red cells as determined by the reactions of antigens on their surfaces to the serum of other people. When these antigenic substances, or immunogens, produce antibodies, the result is agglutination of the red cells. Consequently, this clumping effect is a standard test in serology for determining a person's blood group. Researchers have found distinctively different frequency patterns of these blood group antigens among the major races. Taking the three largest subspecies, for example, we see from Table 10.4 that just six blood-group systems divide the three largest racial populations of the world into broad geographical regions.[10]

[10] Australoids comprise about .4% of the world's population, Capoids about .004%. Including all hybrids, their total is less than .5% (Coon, 1965). Classic Australoids (the aborigines of Australia) differ from the major subspecies of Table 10.4 in absence of blood groups B and MS, and in maximum frequencies of group N and the $F_y{}^a$ gene. The most homogeneous Capoids (Bushmen of South Africa) are low in blood group B, lack the A_2 group and the r gene, are high in group O, and lead the world in frequency of the R° gene. Other Capoids (the hybrid Hottentots) are distinctive in exhibiting cytogenetic deviations from the modal human chromosome count of 46; nearly half of the Hottentot samples tested ranged above or below that number (Coon, 1965, 1974; Garn, 1971, 1974).

TABLE 10.4

Comparative Gene Frequencies among the Major Human Races for Different Blood-group Systems[a]

System	Gene	Caucasoids	Mongoloids	Negroids
ABO	A_2	Moderate	Absent	Moderate
	B	Low	High	Medium
MNSs	MS	High	Low	Moderate
Rh	r	High	Absent	Medium
Duffy	F_y	Absent	Absent	High
Diego	Di^a	Absent	High	Absent
Sutter	Js^a	Absent	Absent	High

[a] Adapted from Garn, 1971; Osborne, 1971.

Naturally, it is possible to group more coarsely or to divide more finely, with the number of subspecific taxa ranging from 5 (Coon, 1965) to 34 (Dobzhansky, 1965). Debates between anthropological "lumpers" and "splitters" will undoubtedly continue for a long time, but whether the taxonomists eventually settle upon 5 or 50 races is biologically unimportant provided one does not lose sight of the genetic continuity of the human subspecies.

It will come as a surprise to some that fingerprints reveal modal pattern differences among the races of mankind, and these dermatoglyphics (see Chapter 3 by Rife) are not influenced, so far as we know, by the environment (Cummins & Midlow, 1961). In general, loop patterns predominate over whorls and arches in Capoids, Caucasoids, and Negroids; whorls and loops are about equally common in Mongoloids; whorls reach their highest frequency in Australoids; arches are rare among Mongoloids and Australoids but common among Capoids and Negroids. Like the data from ethnic studies of morphology and serology, the facts of dermatoglyphics tend to separate the world's native populations geographically (Coon, 1974).

It is important to understand that the inventory of a person's genetic and morphological traits (serological profile, disease susceptibility, somatotype, pigmentation, facial bones, eye shape or size, dentition, bone density, nasal form, fingerprint patterns, color perception, endocrine hormones, earwax, hair characteristics, taste sensitivity, altitude and temperature adaptations, etc.) is not the *basis* of membership in a given race. It is the *result* (Ginsburg & Laughlin, 1966; Laughlin, 1966). No individual human being can be precisely "typical" of a given race. Typological views about human races, such as those held by racial supremacists (e.g., the German Nazis), are erroneous. Rather than being "types," races are dynamic human populations manifesting a considerable degree of intergradation through space and time. Nevertheless, the statistically defined races of modern physical anthropologists and human geneticists exist among the basic laws of biology (Laughlin, 1968; Mayr, 1970).

Although the variety of possible anatomical, physiological, and behavioral traits is theoretically limitless, panmixia (random mating) among human beings does not occur. Because of strong inbreeding tendencies, it is statistically improbable that any two human races have the same means (M) and variances (σ^2) for all psychological traits. We should expect, therefore, that significant differences in psychomotor, perceptual, and athletic behavior will be found among ethnic taxa throughout the world.

Looming large in the field of differential psychology, although not always appreciated throughout their subtler methodological ramifications, are the twin problems of defining and measuring differences among individuals and groups. By *individual differences* one stipulates, for a given set of single scores on Trait A recorded from N persons, that $\sigma_a^2 > 0$; similarly for Trait B, that $\sigma_b^2 > 0$. By *group differences* one refers to the fact that, for at least two sets of scores on the same Trait A recorded from persons classified in different taxa X and Y, either $\overline{X}_a \neq \overline{Y}_a$, or $\sigma^2_{x_a} \neq \sigma^2_{y_a}$, or both may be true statements. These problems have close kinship because empirical psychological differences within and between groups of subjects are believed to manifest similar distribution forms, which can be analyzed by the same statistical techniques.

Measurements of any definitive class of criterion responses (e.g., rotary pursuit skill) having $\sigma^2 > 0$ constitute an *individual-difference variable* as well as an ability variable. Furthermore, if two or more group means or variances differ with respect to this same trait, say when subjects are classified by age, then the criterion response class is also a *group-difference variable.* Of course, both variables are dependent rather than independent, and the group differences may not arise from the same sources as the individual differences do. At any rate, skills exhibiting these organismic characteristics often have a transsituational property that supports the hypothesis of quantitative variations among distinctive human taxa: e.g., groups differentiated genetically, anthropometrically, or psychometrically by age, sex, or race. Whatever relationships are discovered, replicated, and determined to be statistically significant are matters of empirical truth. They cannot be denied on a priori or ideological grounds. Nevertheless, some well-meaning people are reluctant to admit the possibility of genetic race differences in behavior. They fear the vague and unpredictable social consequences of conceding that group-difference variables with significant heritabilities may exist for human subspecies. Such persons are less commonly troubled by statistical evidence of sex or age differences whose origins may be equally genotypic. This is a question of value judgments and of personal attitudes toward an uncertain future. Although I do not know of any social or biological maladies for which scientific ignorance is a remedy, I do know of numerous instances where science and technology have benefitted mankind enormously. Our wisest course, therefore, is to pursue the development of scientific facts and hypotheses in this field with all the determination and objectivity we can muster—because knowledge is cognitively superior to belief (see Preface, p. xiv). Scientific laws and theories are preferable to dogmatic propositions, and especially to misbeliefs, for the simple reason that there is a strict logical dependence of scientific concepts upon empirical data. Induction, deduction, and abduction are interwoven in the process (Noble, 1976a). Knowledge not only confers rational understanding of (human) nature, it also leads through practical applications to a more effective adjustment to the world in which we live.

Race and Behavior

In keeping with my historical sketch above, it is fitting that the first psychomotor experiment on racial groups was conducted by psychologists influenced jointly by the Wundtian and Galtonian traditions. Cattell was the link. In 1895, at the University of Pennsylvania, R. Meade Bache and Lightner Witmer measured simple reaction times to auditory, galvanic, and visual stimuli in small samples of Mongoloid Amerinds (*M*), Caucasoids (*C*), and Negroids (*N*). The results, expressed in terms of mean speed scores generalized over the three tasks, were $M > N > C$; i.e., American Indians of Mongoloid ancestry tended to respond faster than Afro-Americans of Negroid ancestry, who appeared to be speedier in turn than Americans of Caucasoid ancestry (Garth, 1931; Spuhler & Lindzey, 1967). Unfortunately, these pioneering data were not statistically tested for significance. Results are often different when simple reaction times are compared with complex reaction times; i.e., tasks involving the more cognitive processes of discrimination and choice. A recent study of college students belonging to these same racial groups who practiced on the Discrimination Reaction Timer (Fig. 10.2) for 240 trials produced the following relative mean errors: $N > M > C$. Speed scores, on the average, were the reverse of error scores : $C > M > N$. In this experiment, the psychomotor differences among the races were significant (Noble & Vithakamontri, 1975). We also observed that Asiatic Mongoloids and Indo-Dravidian Caucasoids performed at about the same average level as American Caucasoids (see also p. 350).

Rivers' (1901, 1905) ethnopsychological research in Southeast Asia and India unearthed evidence of higher visual acuity and less color blindness among the Australoid Papuans of Murray Island and New Guinea than among Caucasoid Europeans but lower performance on auditory discrimination tasks. Noting a deficiency in the perception of blue tints by dark-skinned people, Rivers hypothesized that pigmentation instead of culture might account for the differences. During the year 1902–1903 he worked among the Todas of Southern India; they are a hybrid people of mixed Australoid–Caucasoid ancestry (Coon, 1965). Rivers discovered that they, too, were keener than Caucasoids at visual and cutaneous discrimination, and gave opposite results for audition, pain, and olfaction. The fact of significant racial differences in refractive error has been amply confirmed by the investigations of Post (1962), who has reported that primitive groups of Mongoloids (Amerinds), Caucasoid–Mongoloid hybrids (Ainus), Negroids (both full-sized and dwarfed Africans), and Australoids (Negritos) manifest higher mean visual acuities than Caucasoids from Europe and America. Variations in refraction have a strong hereditary component (Loehlin *et al.*, 1975; Post, 1962; Waardenburg, 1963).

It was shown in Rivers' (1905) work with the Todas that these nonliterate people were less susceptible to certain optical illusions than either Melanesians (a clinal population of Australoid–Mongoloid hybrids) or Caucasoid Europeans. Judging from research reviewed by Spuhler and Lindzey (1967), the incidence of the Müller–Lyer illusion among different races is: Caucasoids > Negroids > Australoids > Capoids. The range is great. Caucasoids from North America and South Africa are about four times more likely than Capoids (Bushmen) from South Africa to judge the "feathered" segment of this familiar old optical puzzle as longer than the "arrowed" segment. Although social-learning explanations couched in terms of the "carpentered world" of Western civilization have been vigorously advanced, Pollack's revival and extension of Rivers' hypothesis is more plausible. In a series of careful investigations of the role of fundus pigmentation (i.e., density of melanin concentrations in the *fundus oculi* or macular region of the retina), Pollack and his associates (Mitchell & Pollack, 1974; Mitchell, Pollack, & McGrew, 1977; Pollack & Silvar, 1967; Silvar & Pollack, 1967) have employed ophthalmoscopic, tachistoscopic, and psychometric techniques to establish the following propositions.

1. There is a high correlation between fundus pigmentation and skin color, hence race, in children of Caucasoid and Negroid ancestry.
2. Negroids and darkly pigmented Caucasoids are significantly less sensitive to the Müller–Lyer illusion than lightly pigmented Caucasoids are.
3. Although IQ block-design performance on standard red–white discriminations does not differ for the two races, tests on blue–yellow discriminations cause impairment for children with dark pigmentation;
4. Reduction of available light intensity produces a comparable performance decrement in lightly pigmented Caucasoid adults.

5. The density of retinal pigmentation in Negroids under normal levels of illumination reduces their sensitivity to short (blue) wavelengths, thereby interfering with the perception of blue–yellow contrast and lowering their spatial abilities relative to Caucasoids.

Pollack's nativistic (genetic) approach to racial differences in illusion-judgment and color-form perception appears to be more promising than the ecological (environmentalistic) approach preferred by social scientists. Although he has not yet extended his illusion-detection methods to a variety of human taxa, the retinal pigmentation hypothesis has received substantial corroboration and consequently merits international testing.[11]

At the St. Louis World's Fair of 1904 (Woodworth, 1910), it was noted that subjects from civilized societies tended to surpass subjects from primitive societies on auditory tasks but not on visual tasks. As Rivers and others since have reported, Woodworth found wide discrepancies in refractive error among different races. Groups of Mongoloid Amerinds and Australoid–Mongoloid Filipinos outperformed Caucasoids in visual acuity whereas the reverse was true of a simple psychomotor skill that has come to be a standard component of intelligence testing. Form Boards are performance versions of Ebbinghaus' completion test, and Woodworth included them in his St. Louis project. From those data, as presented in detail by Klineberg (1928) but rearranged with the aid of current anthropological theory, I have classified his 16 ethnic groups of 734 adult males into four of the subspecies cited above (*A*, *C*, *M*, *N*), along with two hybrid categories (*CM*, *MA*) to represent samples from clinal populations. Each subject had the task of fitting nine geometric forms into their correct holes, on the basis of perceived shape, with maximum speed and minimum error. Table 10.5 presents weighted mean time and error scores after three trials on the apparatus for each of these six taxa. The order of average ability ranked by either measure is $C > CM > M > MA > A > N$. Although many of Woodworth's samples are small and overlap each other at the category boundaries, the locations of the hybrid groups' central tendencies are consistent with a primarily genetic interpretation of psychomotor aptitudes. Why otherwise would Ainus and Singhalese fall between the ranks of Caucasoids and Mongoloids while Bagobos and Moros are intermediate to Mongoloids and Australoids?

Readers are cautioned that these samples were not necessarily repre-

[11] There are, of course, other variables affecting light transmission and central nervous system functions. Pollack's hypothesis is most relevant to the perceptual skills of Caucasoids and Negroids operating under special viewing conditions. It does not satisfactorily account for all interracial differences in illusory perception (see also Spuhler & Lindzey, 1967).

TABLE 10.5

Average Psychomotor Performances by Young Males of Different Races on Form-board Tests[a]

Taxon	Number of subjects	Mean time	Mean errors	Rank
Caucasoids	74	27.80	1.60	1
Caucasoid–Mongoloid hybrids	515	31.17	1.86	2
Mongoloids	51	34.24	1.94	3
Mongoloid–Australoid hybrids	75	44.77	2.10	4
Australoids	12	63.30	4.17	5
Negroids	7	82.20	5.33	6
	N = 734			

[a] The data were collected in 1904 at the St. Louis World's Fair and first reported by Woodworth (1910). Later they were analyzed in unsystematic ethnic groupings by Klineberg (1928), then reanalyzed by Noble (1974a) in terms of modern subspecies classifications of mankind. Each central tendency is a weighted arithmetic mean for the taxa shown. No measures of variability were reported in the original publications, so tests of statistical significance cannot be computed.

sentative of their populations. Woodworth did not attempt to compensate for the differences in psychomotor experiences and cultural values that characterize different societies. Nevertheless, I am inclined to regard these Form-Board tests as culture-fair, if not culture-free, evaluations of most of the individuals who took the tests. Lynn presents a discussion of culture-fair tests in Chapter 9 (see also Jensen, 1972, 1973). On the matter of inadequate cross-cultural research by such pioneers as Rivers and Woodworth, a remark by Spuhler and Lindzey (1967) is in order at this point: "While one might argue that the shortcomings in design and method vitiate the findings, it is difficult to see how these studies could be used as the basis for claiming an *absence* of race differences. And yet this is actually what has occurred (p. 378)."

Further evidence of considerable race differences in psychomotor skills may be found in research using the Porteus Maze. As Lynn (Chapter 9) has pointed out, there are modal discrepancies between primitive and civilized groups in this test of maze-tracing ability. I have prepared Table 10.6 to illustrate the probable range of average talent on the Maze among 2454 adults drawn from the five major subspecies around the world. The sample of Capoids is too small for reliable generalizations, like those of the dwarfed Negroids and Australoids in Table 10.5, but the other samples are of adequate size. Again, the Caucasoids and Mongoloids tend to excel, whereas the natives of Africa, Australia, India, and the Philippines appear to be less proficient. I doubt that Table 10.6

TABLE 10.6

Average Psychomotor Performances by Adults of Different Races on Porteus Maze Tests[a]

Taxon	Number of subjects	Mean score (test age)	Test quotient[b]	Rank
Caucasoids	1275	13.99	100	1
Mongoloids	430	13.59	97	2
Negroids	201	10.52	75	3
Australoids	523	9.91	70	4
Capoids	25	7.56	55	5
	N = 2454			

[a] These data were collated from a variety of studies and reviews by Porteus and his associates (Porteus, 1950, 1967; Porteus & Gregor, 1963), checked against the reports of the original investigators, then classified in terms of modern subspecies taxonomy. Hybrid groups were excluded. Each central tendency is a weighted arithmetic mean of the test ages in years for the taxa shown. These scores correspond roughly to mental ages on intelligence tests.

[b] A Test Quotient (TQ) on the Porteus Maze has approximately the same meaning as a performance IQ. These values were obtained for a fixed chronological age of 15 years by interpolation in the older normative tables published by Porteus (1959, Appendix D). His later norms (Porteus, 1959, Appendix E) show elevated TQ scores but do not alter the rank order shown above.

reflects mainly the ecological distinction of agricultural–technological versus hunter–gatherer societies, however, because the Mongoloid Carolinians and Chamorros of Saipan Island exceeded Caucasoid subjects from the United States. So did Amerinds from North America. It may be noted, also, that full-sized Negroids from Africa performed relatively on the Maze than full-sized Australoids from India and Australia.

An even larger set of 10,603 Porteus Maze cases is available in the literature. Nearly all of these subjects were schoolchildren (Porteus, 1950, 1967), so an independent and more highly selected group of samples is possible. I have calculated their average TQs as follows: Caucasoids = 100.34 (n = 1437), Mongoloids = 96.55 (n = 6469), Mongoloid–Australoid hybrids = 93.11 (n = 1697), Australoids = 87.03 (n = 1000). It is interesting that the rank order of Caucasoids, Mongoloids, and Australoids is the same as those shown in Tables 10.5 and 10.6, with the children of mixed ancestry falling appropriately between the Mongoloid and Australoid groups. Once more, certain subsamples from culturally deprived environments surpassed the Caucasoid norms: *(1)* 720 Amerindian youngsters earned a mean TQ of 108.1; *(2)* 200 Saipanese children obtained an average TQ of 102.1. Reviewing some of their research on cross-cultural testing, Porteus and Gregor

(1963) concluded that "this variance in ethnic group performances in the Maze, just as it does in individuals, arises mainly from differences in native ability, the inference reached by the Indian anthropologists . . . cultural or nurtural advantages have considerable influence, but as determining factors come far short of outweighing the natural and probably hereditary differences in mentality (p. 722)." More recently, Porteus (1967) reaffirmed the triple virtues of his Maze. It is, he said, a test that exhibits: *(1)* culture-fair properties; *(2)* intrinsic motivation; and *(3)* validity for predicting a rather specialized aspect of nonverbal human adaptability. Porteus (1959) has consistently regarded it as providing an index of "planfulness, foresight, vigilance, mental alertness, anticipation, and prerehearsal (pp. 152–153)." Despite its significant positive correlation with standard IQ tests ($r \approx .6$), the Maze "is not an adequate measure of scholastic brightness (Porteus, 1967, p. 416)."

Before proceeding to a discussion of race-related athletic skills, I wish to emphasize the fact that no subspecific taxon has been found to be uniformly or consistently "superior" in all perceptual and psychomotor abilities of relevance to human action (Noble, 1974a). Bluntly, there is no "master race." Sometimes in college lectures I confound students, most of whom have Caucasoid ancestry, by presenting a slide of two dozen biopsychological comparisons in which the majoritarian taxon brings up the rear about as often as it leads the other taxa in the excellence or desirability of those attributes. The information has a sobering influence upon the unreflective ethnocentrism of many young scholars. Here are some examples.

1. Negroids usually outperform Caucasoids at rhythmic discrimination, ability to taste phenylthiocarbamide (PTC), speed of skeletal and postural maturation, auditory and visual acuity, freedom from phenylketonuria (PKU), rate of dizygotic twinning, proficiency in numerous athletic tests and sports (e.g., dashes, hurdles, jumps, boxing, grip strength), resistance to certain optical illusions, high incidence of trichromatic (normal) color perception, immunity to malaria, and tolerance of humid heat.

2. Mongoloids characteristically exceed the average ability of Caucasoids on tests of spatial visualization, PTC taste sensitivity, long-distance running, virtual immunity to Rh-negative disease, normal color perception, high-altitude survival, resistance to PKU, and tolerance of extreme cold.

3. Australoids typically surpass Caucasoids in visual acuity, resistance to certain optical illusions, virtual immunity to Rh-negative disease, and freedom from dichromatic (red, green) color blindness.

4. Capoids generally eclipse the performance of Caucasoids, as well as the other races, in desert survival, resistance to certain optical illusions, and primitive hunting skills (vying for the latter honor with the

aborigines of central Australia). Who bests whom in the biopsychological contests of life depends as much on the tasks, tests, or traits in dispute as it does on the races, sexes, or ages of the individuals.

Turning now to the athletic arena, we recall that body size and form are important factors in sports achievement. From Allen's Rule in zoology it follows that racial and subracial characteristics will be implicated to some extent (Coon, 1965, p. 252). Stature and ectomorphy provide an illustration. One would not expect a superb basketball team composed of diminutive Bushmen, Japanese, Negritos, or Pygmies to win any laurels playing against a merely excellent team of lanky Nilotes, Patagonians, Scandinavians, or Turkomans. When the taller team wins, furthermore, it would be inappropriate to call their opponents "inferior," although they may in fact be *shorter* by 24 inches, a formidable handicap in basketball. Similarly, there are natural advantages in various sports that derive from racial attributes of mesomorphy or endomorphy, as mentioned earlier. Since height, weight, and morphology are interrelated, however, being short and light may have compensations. Take, for instance, the Olympic records of Mongoloids in free-style wrestling. Between 1896 and 1968, the Japanese never produced a champion in the middle-to-heavy weight categories, yet Japanese flyweights, bantamweights, and featherweights won three bronze, three silver, and six gold medals during that period (Baker, 1974, p. 554). Mesomorphic cyclists, gymnasts, and swimmers of the Mongoloid race have also starred in recent Olympic Games (de Garay *et al.*, 1974). Mongoloids have other natural gifts: Amerinds of the Tarahumara tribe of northern Mexico are world famous for their long-distance running ability. Eskimos excel at maintaining high basal metabolic levels for optimum psychomotor activity under low-temperature stress. Andeans and Tibetans are notable for genotypes determining their cardiovascular and pulmonary systems that permit them to thrive and reproduce at much higher elevations than other races (or subraces) can tolerate (Baker & Weiner, 1966; Coon, 1965; Laughlin, 1968).

Studies of the athletic abilities of Negroids have turned up several interesting findings, especially among Afro-American subjects. Codwell (1949) recruited a group of 505 male high school students in Texas having different degrees of Negroid and Caucasoid ancestry to whom he administered standard tests of physical fitness from McCloy's battery. By dividing his subjects into three groups on the basis of several anthropometric criteria, Codwell found that the predominantly Negroid boys' proficiency was greater than that of the hybrids in 5 out of 6 tests. There were significant increments in the subjects' ability to execute the Sargent Jump and the Burpee Test as their degree of Caucasoid admixture declined, and no statistical evidence of hybrid vigor appeared.

Investigations of the skeletal and postural maturation rates of younger subjects of these two races have generally concurred in showing precocious physical and motor development in Negroid, as compared with Caucasoid, infants and children (Bayley, 1965; Jensen, 1972, 1973; Jordan, 1969; Malina, 1969, 1973; Noble, 1969b). Mongoloids, incidentally, tend to lag behind the Caucasoids, a fact that led Jensen (1973) to hypothesize that the three taxa may be ordered in terms of their evolutionary ages, with such characteristics as an extended developmental period, a greater relative frequency of single births, and a higher average IQ being determined in part by the subspecies' earlier transition from the grade of *H. erectus* to that of *H. sapiens*. It is true that infant motor precocity scores are inversely correlated with adult intelligence test scores, and that the incidence of multiple births is negatively related to IQ (thus placing Caucasoids intermediate to Mongoloids and Negroids), but whether the three taxa crossed the threshold from a brutal to a sapient state in the order $M > C > N$ has yet to be established. Some intriguing computations of genetic distances between the three pairs of races, developed from analyses of gene frequencies in several blood-group systems in different countries, agree in locating Caucasoids between Mongoloids and Negroids with respect to the lengths of time their ancestral populations have been separated (see Loehlin *et al.*, 1975, pp. 39–40).

Whichever anthropological theory may turn out to be correct, it appears that certain morphological and behavioral differences between Negroids and Caucasoids in the United States may confer some advantages on the former race in several sports calling for power, agility, and explosive bursts of speed: e.g., basketball, football, baseball, boxing, and short foot races. Among the beneficial traits exhibited, on the average, by Afro-Americans are greater muscular strength, heavier bones, less body fat, shorter trunks, larger necks, shallower chests, longer forearms, longer hands, narrower hips, longer lower legs, longer and wider feet, greater sprinting speed, higher jumping abilities, and more efficient heat-dissipation mechanisms (Jordan, 1969; Malina, 1969, 1973). To this catalog of differences between the two races in the mean proportions, composition, and functions of the body may be added the related phenomenon of outstanding athletic records. According to a recent report (*Time*, May 9, 1977), 71% of the American medals in track and field events and 100% of our gold medals in boxing at the Montreal Olympics were won by Negroids. In professional sports, Afro-American athletes exceed their population percentage in at least three categories: They comprise 19% of all major-league baseball players, 42% of the National Football League, and 65% of the National Basketball Association. Although such remarkable evidence of proficiency and overrepresentation in these athletic skills is undoubtedly conditioned by sociological and economic variables, I doubt that the morphological–behavioral correlations cited earlier are merely fortuitous. A reasonable hypothesis is that genetic race differences play an important role.

One psychologist has proposed that it is not race per se but rather inherited iris pigmentation (eye color) that is the more fundamental correlate of proficiency in different psychomotor, athletic, and perceptual skill situations. The range of human eye colors is caused by varying melanin deposits, and Worthy (1974) postulates that differential filtering of the wavelengths of light entering the eye somehow modulates (perhaps via pineal gland and hypothalamus) the striped-muscle behavior of individuals with low, medium, or high concentrations of the pigment. Dark-eyed people should perform better on tasks requiring "sensitivity, speed, and reactive responses," he maintains, whereas light-eyed people should perform better on tasks requiring "hesitation, inhibition, and self-paced responses (p. 11)." Applying his hypothesis to sports psychology, Worthy assembled evidence from American professional baseball, football, bowling, and basketball statistics (including both Caucasoids and Negroids) that seemed to be consistent with the notion that eye color *(1)* is a significant factor in athletic proficiency; *(2)* accounts for some of the well-known racial differences; and *(3)* interacts with the nature of the sport. He also cited data indicating that dark-eyed people have better visual acuity, less color blindness, and more accurate perception of optical illusions (Worthy, 1974), but the latter propositions are not at issue in this section (see Malina, 1973; Mitchell & Pollack, 1974; Mitchell *et al.*, 1977; Pollack & Silvar, 1967; Post, 1962; Rivers, 1901, 1905; Silvar & Pollack, 1967; Spuhler & Lindzey, 1967).

Worthy and his associates have extended his hypothesis to other perceptual skills in a study of 80 Caucasoid college students using printed tests of speed and accuracy purportedly calling for "reactive" (paced) rather than "self-paced" (nonreactive) abilities. When averaged and plotted, the findings revealed a slight tendency for brown-eyed women ($n = 20$) to be more proficient than blue-eyed women ($n = 20$), and for brown-eyed men ($n = 20$) to be more proficient than blue-eyed men ($n = 20$). From graphical evidence (Worthy, 1974, p. 154), the mean sex difference (females > males) appeared to be greater than the mean eye-color difference (browns > blues), but detailed statistical evaluations of these comparisons were not reported. That women often do better than men on tests of perceptual speed and finger dexterity is a familiar fact (Broverman *et al.*, 1968; Maccoby & Jacklin, 1974; Tyler, 1965); the unanswered question concerns the significance of iris pigmentation.

Recent work by Landers, Obermeier, and Wolf (1977) does not provide much corroboration of Worthy's hypothesis. Neither their review of the literature, their examination of a college football team's statistics, nor their laboratory experiments on reaction-time and pursuit-tracking tasks produced any consistent trend or practical evidence in its favor. There were a few statistically significant differences, but most of the crucial tests failed to support the notion of eye color as a broad skill determinant. Nor was SES an important factor; the variance ratios were less than unity.

Some Acquisition Phenomena

Now that we have covered tests of perceptual and athletic skills, it is time to consider psychomotor learning experiments. I shall begin by pitting eye color against race.

Additional data embarrassing to Worthy's hypothesis are available in the archives of human learning research conducted in this laboratory, some still unpublished (Noble, 1968, 1969b, 1971; Noble, Buie, & Wilkerson, 1977). In one large-scale investigation of the acquisition and transfer of psychomotor skills, my colleagues and I recruited a representative sample of 500 right-handed children from rural elementary schools in several northeastern counties of Georgia. Both sexes and the two major races of the state (Caucasoid and Negroid) were included, the subjects' ages ranging from 9 years to 12 years. All data were collected under standardized conditions in our Mobile Psychomotor Skills Research Laboratory by a biracial team of assistants. Subspecies classification was initially based upon morphology then confirmed by more exact anthropometric and genetic indexes. We used an inventory that recorded each subject's iris color, eyeball size and form, interpupillary distance, skin pigmentation, nasal width, lip thickness, ear dimensions, type of earwax, hair color and form, dental characteristics, prognathism of jaws, cheekbone type, handedness, fingerprint patterns, fingernail form, and ability to taste phenylthiocarbamide (PTC).[12] These biometric measurements were shown to have excellent reliability and validity (Noble, 1971); the taxonomic outcomes agreed closely with those of research specialists in physical anthropology and human genetics (Baker, 1974; Comas, 1960; Coon, 1965, 1974; Cummins & Midlow, 1961; Day, 1932;

[12] The Bioinventory for Human Subjects (BHS) was developed in this laboratory for use with a wide range of subspecific taxa, so not all the phenotypes listed were recorded for all groups. Certain BHS indexes were more appropriate for Asiatic Mongoloids than for African Negroids (e.g., incidence of epicanthic eyefolds, high cheekbones, dry earwax, shoveled incisors, curved fingernails). Whenever two racial groups were compared experimentally, however, the BHS was applied equally to all subjects. Generally, we also employed sexually and racially integrated teams of research assistants to collect the data. For instance, in a recent investigation of possible interactions among experimenters and adult subjects, who were performing on the Selective Mathometer (Figure 10.2), all combinations of experimenters and subjects were programmed in order to test for psychomotor effects of sex (male versus female) and race (Caucasoid versus Mongoloid). We found no evidence of overall proficiency differences attributable to the sex or race of either subjects or experimenters in these samples (Noble & Yeh, 1977).

Dobzhansky, 1965; Garn, 1971, 1974; Gates, 1949; Goldsby, 1971; Gottesman, 1968; Harrison, Owen, DaRocha, & Salzano, 1967; Herskovits, 1930; Hooton, 1956; Howells, 1960; Jordan, 1969; Lasker & Lee, 1957; Laughlin, 1964; Malina, 1973; Mayr, 1970; Mazess, 1967; Osborne, 1971; Pollitzer, Boyle, Cornoni, & Namboodiri, 1970; Reed, 1969; Sanghvi, 1953; Spuhler & Lindzey, 1967; Stewart, 1952).

Our research program featured three of the learning devices pictured in Figure 10.2: the Discrimination Reaction Timer (device B), the Rotary Pursuitmeter (device F), and the Selective Mathometer (device H). Limiting my attention to the Rotary Pursuitmeter, I examined the protocols of a cohort of about 300 children matched in age, sex, and race who practiced this targeting skill for 50 successive trials with a 20:10 work:rest ratio per trial (i.e., a 20-sec. work, 10-sec. rest cycle). Half of the subjects used the right hand, half the left, and in each of those groups a subgroup changed their stylus hands (left→right or right→left) halfway through the practice session, then continued in that mode to the end of training. In order to obtain numerically comparable samples of the different eye colors, I drew a group of 80 Negroids and 120 Caucasoids so that 40 subjects appeared in each iris category: two for the Negroid children (browns versus blacks) and three for the Caucasoid children (grayish blues versus bluish greens versus light and dark browns). This design gave 20 subjects for each practice condition (left versus right hand) and insured equal representations by sex (10 males, 10 females) and age ($M = 10.25$ years per cell). Considering only Trials 1–24 before the change of hands, and grouping the time-on-target scores by 2-trial blocks along the practice dimension, our $12 \times 2 \times 2$ analysis of variance for the 80 selected Negroid children revealed significant main effects attributable to amount of practice ($p < .001$) and conditions of practice ($p < .025$), but neither eye color nor any other source of variance was significant. Grouping the 120 selected Caucasoid children in the same fashion, except that three iris categories were available, a $12 \times 3 \times 2$ analysis of variance revealed significant main effects attributable to amount of practice and conditions of practice ($p < .001$), as well as significant interactions of practice with hand conditions ($p < .001$) and with iris pigmentation ($p < .05$). There was no main effect of eye color, however; and no other interactions were significant. As a matter of fact, not even the rank orders of final proficiency in the two racial samples for a given condition were consistent with Worthy's expectation. Among Caucasoids, blue eyes tended (nonsignificantly) to outperform brown eyes; among Negroids, brown eyes tended (nonsignificantly) to outperform black eyes. Darker-eyed children, in short, did not exhibit any overall tracking advantage relative to lighter-eyed children in this experiment. While it is true that the Caucasoid subjects showed a significant trend difference in their practice × iris interaction, it was in the opposite direction to Worthy's forecast. I suspect this phenomenon may have more to do with *subraces* (e.g., Nordic, Alpine, or Mediterranean local races) than with *subspecies*. One should remember that Caucasoids are not a homogeneous race; local races and microraces abound (Garn, 1971).

I am tempted now to consider testing a racial hypothesis instead of the eye-color hypothesis. Evaluating data from the same cohort by race (Caucasoid versus Negroid) and by conditions (right versus left), and disregarding eye color, sex, and age for the moment, a $4 \times 2 \times 2$ analysis of variance on a companion group of 152 subjects for 24 trials (blocked by sixes) indicated that Caucasoids not only tracked the target with a generally higher level of proficiency than Negroids but also were acquiring this skill at a faster rate. The two racial samples were so different behaviorally that the average Negroid *right*-hand ability was consistently below the average Caucasoid *left*-hand ability. It is as if the race difference were being potentiated by some other biopsychological variable, say brain dominance. From what is known about the differential heritability of preferred (right) hand versus nonpreferred (left) hand performance on this task and its unusually high heritability (Noble, 1969b), it is not surprising to find a race $\times$ hand interaction under these conditions. Indeed, all main effects and their 2-factor interactions were significant ($p < .025$). Practice, handedness, and race are the major variables (Noble, 1969b). It is clear, therefore, that Worthy's interesting views are disverified by our research. When race was held constant there were no main effects attributable to iris pigmentation in samples matched for age, sex, and practice conditions. Furthermore, when an extended range of eye colors was available, as among Caucasoids, the rank order of proficiencies was not what Worthy expected for this paced, reactive psychomotor task. Our results are consistent with those of others who have employed the Rotary Pursuitmeter (Landers *et al.*, 1977).

An obvious deduction from our biopsychological theory of psychomotor performance is that a sample of hybrid children from this experiment, who could reliably be classified as having mixed Caucasoid and Negroid ancestry, should exhibit a level of tracking proficiency on the Rotary Pursuitmeter intermediate to the acquisition curves of the more homogeneous racial samples. Proceeding to apply a trichotomy to our BHS data (see footnote 12) on 81 children ($n = 27$ cases each, matched for age, sex, and practice conditions), I found (Noble, 1971) that the majority of our measurements agreed with the genealogical, anthropometric, and genetic data of other research workers (e.g., Comas, 1960; Coon, 1965; Day, 1932; Harrison *et al.*, 1967; Herskovits, 1930; Hooton, 1956; Mazess, 1967; Osborne, 1971; Pollitzer *et al.*, 1970; Reed, 1969; Stewart, 1952). Our pigmentation ratings, which were defined operationally by skin-color judgments on Gates' (1949) Scale, turned out to be one of the most reliable predictors in the Bioinventory. No two observers' measurements deviated by more than one scale numeral. Each rating was a median of two independent judgments recorded from the volar surface of a subject's forearm under constant illumination by trained assistants in the Mobile Laboratory. A histogram of the pigmentation data for 920 children appears in Figure 10.6. Subjects who were identified, on the basis of all BHS phenotypes, as predominantly Negroids (n = 136) had ratings that ranged from 1 to 2.5, Hybrids (n = 204) from 3 to 7.5, and Caucasoids (n = 580) from 8 to 9. A 4 × 3 analysis of variance carried out on the tracking scores of these three groups of subjects over 24 trials (blocked by sixes) detected significant main effects of practice, race, and their interaction ($p < .025$). Consistent with my pre-experimental hypothesis, the rank order of average proficiency was Caucasoids > Hybrids > Negroids. Their mean time-on-target scores, in percentage terms, were 4.6%, 2.6%, and 2.1%, respectively (Noble, 1968), and the groups were spaced appropriately.

An alternative hypothesis might be formulated on the basis of postulated differences in SES, or unequal extralaboratory familiarization expe-

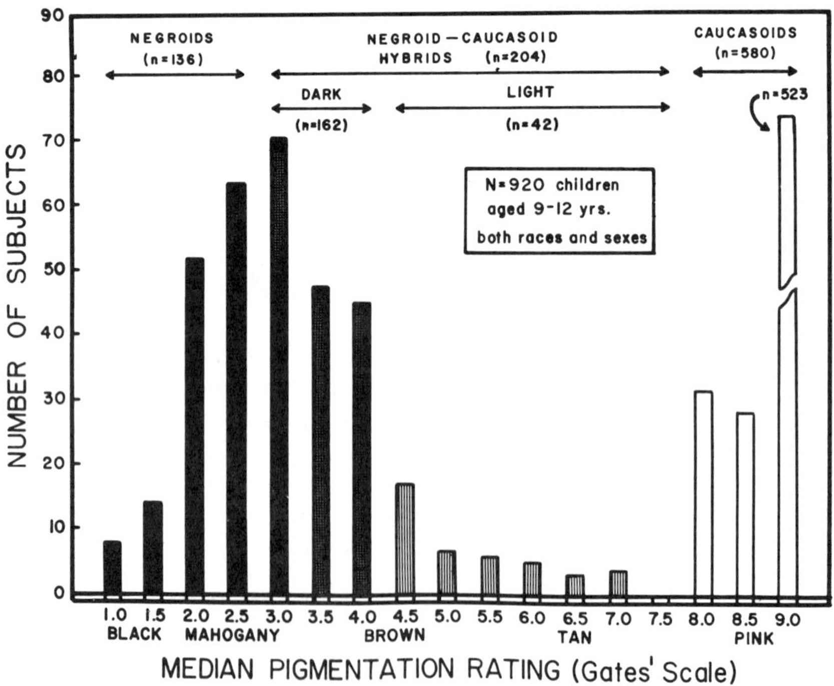

Figure 10.6. Histogram of median pigmentation ratings of 920 grade school children of Caucasoid and Negroid ancestry as determined by an empirical application of Gates' Scale. Classification of the subjects as predominantly Caucasoid (n = 580), Hybrid (n = 204), and predominantly Negroid (n = 136) was based on the Bioinventory for Human Subjects. See text for explanation. (Data from Noble, 1971.)

riences, among the three taxa. However, this viewpoint would first have to explain how the large initial differences in proficiency came to be produced between the left and right hands of our original two groups of youngsters described above. One might entertain the *outré* notion that rural Caucasoid children receive more cultural benefits for their *non-preferred* hands than rural Negroid children receive for their *preferred* hands, and that there is a differential right–left transfer of training caused by environmental deprivation (Noble, 1977a, 1977b). This possibility, which I have considered elsewhere (Noble, 1969b) along with malnutrition, would be more attractive if the Rotary Pursuitmeter were not such an esoteric task, and if it were more highly correlated with extratask variables. As McNemar (1933) has shown, early performance on the rotor is primarily determined by hereditary factors (see also Jones, 1966, 1969, 1972). Incidentally, an extension of this experiment with 186 children in three comparable groups ($n = 62$ cases each) practicing for 50 trials (blocked by tens) corroborated the results. A 5×3 analysis of variance showed that both main effects and the practice × race interaction were significant ($p < .001$). As before, there were different average rates of gain among the three taxa, with the Caucasoid–Hybrid mean difference exceeding the Hybrid–Negroid difference. This unequal spacing is also reflected in the average pigmentation values, thereby buttressing the evidence for a genetic hypothesis.

Drawing further from the unpublished data of this laboratory, I have attempted to overthrow the biopsychological theory by effecting an even finer titration among the Afro-American children of mixed ancestry. Simply put, this was a test of the hypothesis that genotypic factors are more powerful in determining the variance in psychomotor performance than socioeconomic factors are within the hybrid subpopulation of Afro-Americans. Statistically, SES variables are inadequate predictors of behavior on culture-fair tests as compared with biometric variables. Referring to Figure 10.6, I subdivided the Hybrids into "dark" and "light" categories based on their pigmentation ratings and correlated BHS traits. Dark Hybrids ranged from 3 to 4 on Gates' Scale, Light Hybrids from 4.5 to 7.5. From each of these new subdivisions, I selected 26 subjects who were matched for age, sex, and conditions of practice. In SES terms, the two cohorts were indistinguishable. They were then compared in tracking proficiency (mean time, $\overline{R}_{tt}$, and percentage time, $\overline{R}_{\%}$) over 24 trials (blocked by eights) with equal-sized samples from the predominantly Negroid (1–2.5) and predominantly Caucasoid (8–9) groups ($n = 26$ each, matched as above). The outcome of this analysis is plotted in Figure 10.7. Statistical tests of the time-on-target scores of these 104 children arrayed in a 3×4 design revealed significant variances attributable to practice, race, and their interaction ($p < .01$). Remarkably, the four trends are quite distinctive; none of the groups' mean scores in Figure 10.7 are

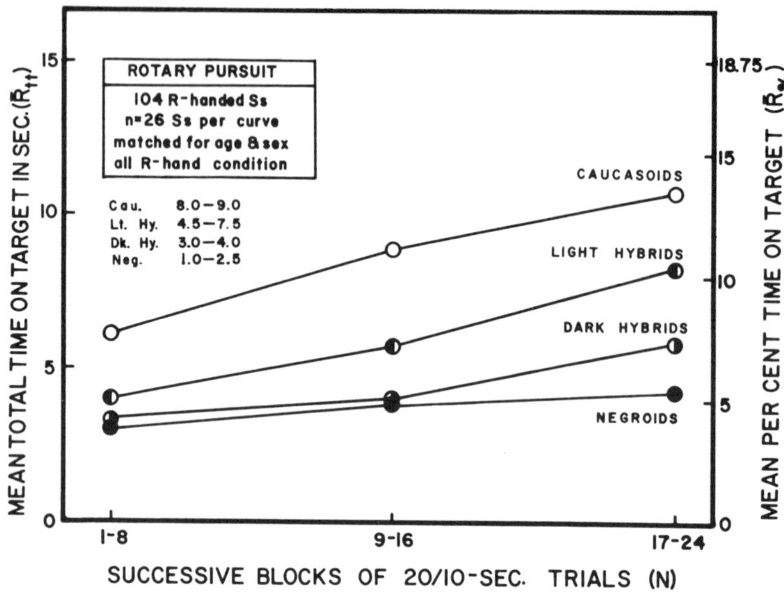

Figure 10.7. Acquisition curves of mean time on target ($\overline{R}_{tt}$) and mean percentage time on target ($\overline{R}_{\%}$) in rotary pursuit as functions of successive practice blocks (N), with race and hybridity as the parameter. Each group contains 26 grade school children matched for age, sex, and practice conditions. (Data from Noble, unpublished.)

intertwined. Once again, the spacing is roughly proportional to the anthropometric indexes.[13] There is considerable overlapping by individual subjects, of course, but this is typical of youngsters' performance early in training on psychomotor devices; it is more characteristic of the two lower curves in the graph than of the two upper ones. All in all, the preceding analyses summate to a considerable degree of support for a biopsychological theory of skill acquisition. To argue that such delicate quadruple distinctions in aptitude for a laboratory-controlled learning task as those portrayed in Figure 10.7 are programmed mainly by ambient social and economic conditions in a rural southeastern environment would appear to be a strained interpretation.

[13] Although it is often treated with doubt or derision, the variable of skin color among Caucasoids, Negroids, and Afro-American hybrids of mixed ancestry is both reliable and valid (Day, 1932; Gates, 1949; Harrison *et al.*, 1967; Herskovits, 1930; Mazess, 1967; Noble, 1971; Pollitzer *et al.*, 1970; Stern, 1973). Natural selection is the main reason for pigmentation differences among human races and subraces. The hue and reflectance of the epidermis are adaptive traits of polygenic inheritance (Stern, 1973). Darker skins characteristic of hot, humid, tropical zones (e.g., African Congo, Southern India) are protected from sunburn by day as well as from chills by night; lighter skins typical of extreme latitudes (e.g., Northern Europe) facilitate the intake of vitamin D_2, and the consequent avoidance of rickets, in spite of deficient sunlight (Coon, 1965). Selective mating is another variable affecting the skin

Moving along briskly to other acquisition phenomena, I should like now to consider the learning and performance of elementary and high school students on the Selective Mathometer (see Figure 10.2, device H). This experiment was conducted within the same geographical region as previously described. It also utilized the Mobile Laboratory and the BHS, but the subjects were recruited from the wider age range of 9 to 19 years. We (Noble & Artley, 1977) sampled a group of 88 Caucasoids and 88 Negroids, who were divided equally by sex ($n = 44$ males, 44 females per taxon). Mean age for the Caucasoids was 12.67 years; that for the Negroids was 12.98 years. All subjects were required to practice four-choice serial learning problems by the self-paced noncorrection method for 40 trials (see Noble, 1966a, 1969a). After grouping the response probability (R_p) scores of the 176 subjects by cohorts ($n = 44$) into 5-trial blocks, Artley and I observed the results exhibited in Figure 10.8. Caucasoids acquired skill on the Mathometer more rapidly than Negroids did, and they attained a higher level of accuracy after 40 practice trials. There was a tendency for males to outperform females in each taxon, this being more marked among Negroids than among Caucasoids.

---

color of progeny (see Jensen, Chapter 4); lighter-skinned brides are often preferred by urban Afro-American, Indo-Dravidian, Japanese, and Syrian grooms. With respect to Afro-Americans, several investigators have reported a tendency for females to be lighter than males, especially in cities and suburbs (Day, 1932; Harrison *et al*., 1967; Mazess, 1967; Pollitzer *et al*., 1970). For children, however, the sex difference in pigmentation is usually less pronounced than for adults. From the archives of this laboratory I drew a sample of 345 Negroid subjects (191 boys, 154 girls) in the age range from 9 to 12 years old. On the Gates Scale their ratings varied from 1.0 through 7.0. Applying the chi-square (χ^2) statistical formula, I tested for the independence of sex and pigmentation. To avoid small theoretical frequencies at the extremes of the two distributions, I combined the tails to produce 8 categories instead of 13. In line with adult data, the girls' mean rating was slightly higher (3.05) than the boys' mean (2.90), but $\chi^2 = 11.24$, which for 7 degrees of freedom (df) is not significant ($p > .10$). The null hypothesis of independence, therefore, cannot be rejected for these children. A close association between genealogy and skin color for Afro-Americans is shown by Day's (1932) research. She recorded the hybridities of 139 men and 233 women by proportions of Caucasoid admixture (in eighths), along with pigmentation ratings of each subject (on von Luschan's Scale). Unfortunately, her rating device does not exhibit transitivity throughout its 36 numerical values. By transforming Day's skin-color ratings into their ordinal equivalents on the 9-point Gates Scale, however, I was able to construct a 7×7 color $\times$ hybridity matrix for each sex group separately ($df = 36$). Testing for the independence of genealogy and pigmentation, $\chi^2 = 135.09$ for the males ($p < .001$); $\chi^2 = 229.07$ for the females ($p < .001$). There is, in other words, a significant association between ancestry and skin color. Finally, since Day's subjects were all adults, a χ^2 test of the independence of sex and pigmentation was expected to overthrow the null hypothesis. From a 7×2 matrix of the transformed skin-color ratings of the same 372 subjects separated by sex, I computed $\chi^2 = 23.34$, which for $df = 6$ is significant ($p < .001$). Due apparently to greater Caucasoid admixture, the women in Day's study were palpably lighter than the men. This sex difference in the pigmentation distributions of adult Negroids and hybrids is of sufficient statistical magnitude to reject the notion of independence at the 0.1% point. In short, skin-color variations are meaningful biopsychological phenomena.

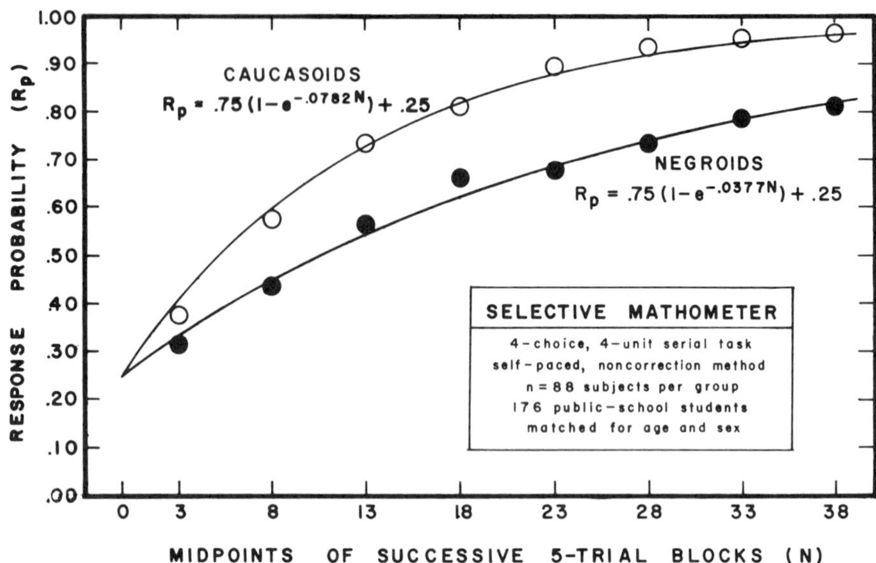

Figure 10.8. Acquisition curves of response probability (R_p) in selective mathometry as functions of successive practice blocks (N), with race as the aptitude parameter. Each group contains 88 public-school students matched for age and sex. The curves were fitted by Equation (3) with $M = .75$, $T = .25$, and only k free to vary. The origin (.25) and limit (1.00) of the R_p scores are rational values and identical for both taxa. (Data from Noble & Artley, 1977).

An $8 \times 2 \times 2$ analysis of variance of R_p scores indicated significant main effects of practice and race ($p < .001$), but neither sex nor any of the interactions were significant. We also computed measures of response speed (R_s). Other statistical tests revealed a significant age difference in proficiency, with the high school students ($M = 14.99$ years) surpassing the elementary school students ($M = 10.71$ years) in traits of both accuracy and speed ($p < .001$). Within age groups, there was significant evidence of practice $\times$ race interactions ($p < .05$); i.e., differential trends for Caucasoids and Negroids of comparable ages. We concluded that average public school students of the former race have greater aptitude for nonverbal multiple-choice learning tasks than those of the latter, especially if they are older and male. This generalization is more applicable, however, to accuracy scores than to speed scores. The theoretical curves for the pooled data of each race, disregarding age and sex, have been fitted by Equation (3) in Figure 10.8. These formulas differ only in the rate parameters (k) of the two taxa, that for Caucasoids being larger than the one for Negroids. Their R_p scores have a mean predictability of 98.96%, adding further corroboration of the theory to be outlined in the final section (see p. 362).

Continuing with the Selective Mathometer (Figure 10.2), we have conducted a similar experiment (Noble & Vithakamontri, 1975) that differs in two respects from the foregoing: *(1)* it employed college students from several Oriental countries in Asia as well as students from India, Pakistan, and the United States; *(2)* the task, a 10-choice serial learning problem, was more complex than the 4-choice task shown in Figure 10.8, thus more suitable for the collegians' greater cognitive abilities. Examination of the literature on international comparisons of psychomotor skills indicates that investigators have always encountered difficulties of interpretation caused by inadequate matching, language barriers, or uncontrolled disparities among the cultures being surveyed. One never knows how great is the confounding of biology, linguistics, and sociology in these interpopulation studies. English-speaking students attending American colleges from abroad provide a potential source of subjects who are probably less heterogeneous than those sampled by the pioneering ethnopsychologists who were often forced to contrast literate urban dwellers with nonliterate hunter–gatherers. Self-selection by the voyaging scholars is another source of variability that was not controlled, for we were unable reliably to determine how representative they were of their parent populations. Nevertheless, some information is better than none, and we do have a unique laboratory task for them to attack. Perhaps we can regard this experiment as a small step in the direction of reducing scientific uncertainty about the degree of human variation in comparative investigations of psychomotor learning and performance around the world.

We (Noble & Vithakamontri, 1975) recruited samples of 20 college men and 20 college women from each of the following four subpopulations at the University: *(1)* American Caucasoids; *(2)* American Negroids; *(3)* Asiatic Mongoloids from Hong Kong, Japan, South Korea, Taiwan, and Thailand; *(4)* Indo-Dravidian Caucasoids from India and Pakistan. After being administered the BHS (see footnote 12), all subjects received 40 trials of a 10-choice paced learning problem on the Mathometer. The R_p scores of the 160 subjects were grouped in eight 5-trial blocks, subdivided by the four taxa and the two sexes. Then we performed an $8 \times 4 \times 2$ analysis of variance, which showed significant main effects of practice and taxa as well as their interaction ($p < .001$). Sex was not a significant factor, nor were any of the other interactions.

Combining the sexes, Figure 10.9 presents the average R_p acquisition curves of the four taxa ($n = 40$ cases each). Vithakamontri and I concluded that, under these conditions, American Caucasoids and Asiatic Mongoloids probably have greater aptitude and capacity for nonverbal multiple-choice learning tasks than either American Negroids or Indo-Dravidian Caucasoids ($p < .001$). The upper two taxa in Figure 10.9 do not differ significantly from each other ($p > .05$), nor do the lower two

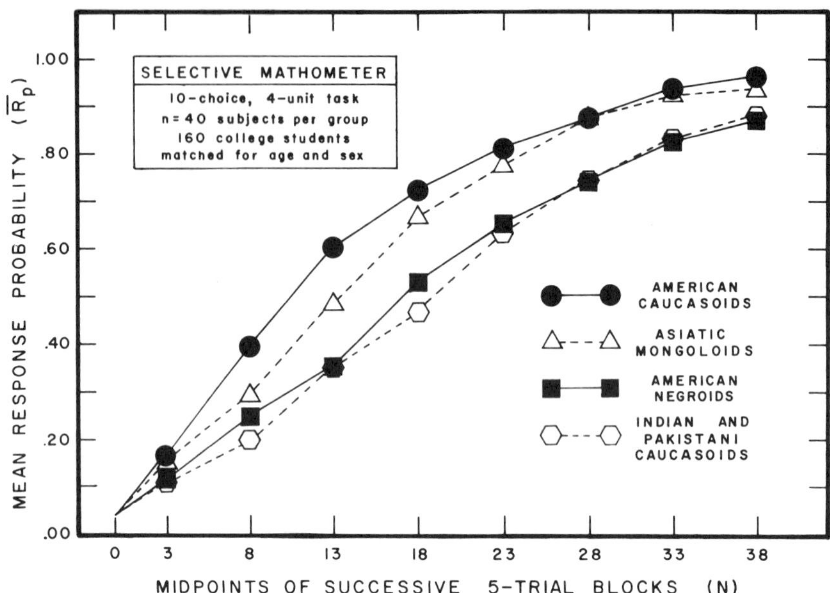

Figure 10.9. Acquisition curves of mean response probability ($\overline{R}_p$) in selective mathometry as functions of successive practice blocks (N), with a combined ethnic-racial classification as the parameter. Each group contains 40 college students matched for age and sex. The origin of all curves is rational: $R_p = (.40)\cdot(.10) = .04$. (Data from Noble & Vithakamontri, 1975.)

taxa in that graph ($p > .20$). As a caveat, it should be understood that our findings cannot be generalized very widely. The data pertain to R_p scores on this particular apparatus achieved by highly selected college students, the foreigners among whom may not be typical of students in their native lands. It is just as important to mention qualifications about the learning device as it is about the representativeness of the nonindigenous subjects because we found, in the course of the same project, that average performance scores (R_s, $R-$) on the Discrimination Reaction Timer (Figure 10.2) ranked the taxa differently. On that apparatus, the mean proficiencies of 275 college students classified in five taxa were ranked thus: (Asiatic Mongoloids) = (American Caucasoids) = (Indo-Dravidian Caucasoids) > (Amerinds) > (American Negroids). Because of unequal numbers of cases and incomplete sex matching, we plan to continue this project until more comparable data are collected, especially from the Amerindian population.

Commentary

To close this section, I should like to offer some additional remarks about the proper interpretation of racial data and the salience of racial research. Human beings, especially the members of civilized societies, have an inveterate tendency to perceive simple hierarchies where none exist, to assume single dimensions where the variables are multiple and the interactions complex. The races of mankind, although everywhere interrelated by some of the same genes, cannot be ordered along a common scale of values. No particular taxon contains, at least not in manifest or usable form, the entire range of genotypic variability necessary for the optimal evolution of our species in all possible habitats. Latent hereditary potential in sufficient quantity *may* exist in every one of the geographical and local races, but such a ubiquitous reservoir of genotypic talent is presently only a theoretical probability (Dobzhansky, 1965; Ginsburg & Laughlin, 1966). Empirically, our subspecies do differ in many fundamental traits, as I have shown above.

Nor is it by chance that the various human taxa have developed distinctive aptitudes, capacities, and susceptibilities in their respective ecological niches. Remove an Aleut mariner from his bidarka, the Ona mother from her portable lean-to, or stalking aborigines from the outback, then transport them to Metropolis, and a serious maladjustment will be the probable result. Still, there will eventually evolve an adaptation, for all are human beings. Population gene pools are complexly organized, and they are continually adjusting to the flux of local conditions. The record indicates that morphology and behavior have evolved together (see Chapter 11 by Darlington), so I am persuaded that there is no true dichotomy between biology and psychology, only a biopsychological continuum.

It is in this larger sense that life scientists should investigate human nature in all its diverse complexity. To quote Laughlin (1968):

> Valid generalizations about "man" should be based upon studies of many groups representing the entire species, and not wholly or even primarily upon only one subdivision of it. All human groups, small or large, are informative for scientific studies, and all are entitled to the benefits of such studies [p. 12]
>
> Disappearing peoples offer a clear case for prompt attention. Each population unit, whether a deme, tribe, local race, or a more inclusive population system, is a major evolutionary experiment in human adaptability. It cannot be reduplicated and it will be forever lost unless recorded and analyzed [p. 16].

When we realize that the Chono and Kalapuya Amerinds, the Norsemen of West Greenland, the Sadlermiut Eskimos, and the Tasmanians are no more, then we begin to understand the urgency as well as the humanity of a broad-scale approach to the study of *H. sapiens*.

Sex and Learning Tasks

Sex, like age and race, is a biological factor that interacts with practice variables in the determination of human performance on psychomotor, perceptual, and athletic tasks. Following a list of a dozen generalizations, I shall discuss the role of sex in psychomotor tasks, present a few salient acquisition phenomena, then close the section with a treatment of certain gender interactions.

Sex Generalizations

Biopsychological research concerning differences between the sexes relevant to our focus on human learning and performance has not been as prolific as the studies of age and race. Among the major sources of information about sex differences are the writings of Baughman and Dahlstrom (1968), Broverman *et al.* (1968), Fleishman (1964), Maccoby and Jacklin (1974), McGeoch and Irion (1952), Ounsted and Taylor (1972), Scheinfeld (1965), Singer (1975), Thorndike (1914), Tyler (1965), and Whipple (1921–1924). Readers may also wish to consult related matters in the essays of this volume by Jensen (Chapter 4), Lehrke (Chapter 7), and Shuey (Chapter 8). On the basis of those studies and surveys, as well as my previous discussions of age and race, the following generalizations may be offered.

1. Girls and women are, on the average, more skillful than boys and men on such perceptual and psychomotor tests as color perception, aiming and dotting, finger dexterity, inverted alphabet printing, and card sorting.
2. On speeded tasks, girls tend to reach their maximum proficiency earlier in life than boys do; males continue to gain over a longer period, and they typically surpass the ability of females for approximately 50 years.
3. Following puberty boys generally excel in the performance of athletic skills requiring strength and stamina (e.g., throwing, jumping, running).
4. Olympic records of women for swimming and track-and-field events are lower than those of men, and are won by females who are younger, on the average, than male champions in the same events.
5. On standardized printed achievement tests calling for perceptual and cognitive skills (e.g., mathematics, science), girls tend to lag behind boys between the prepubertal and late teen ages; on verbal aptitude tests, girls begin to excel during adolescence, whereas boys lead in numerical and spatial abilities.
6. Significant sex differences have been found on most of the psychomotor tasks investigated to date; that includes the Seguin Form Board, the Porteus Maze, and nearly all of the learning devices illustrated in Figure 10.2.
7. In some psychomotor tasks, sex interacts with the variables of prac-

tice, race, and age; e.g., Discrimination Reaction Timer, Mirror Tracer, Rotary Pursuitmeter, and Selective Mathometer.

8. Sex effects are revealed by differences in initial proficiency, rates of gain, and asymptotic levels of performance, but not always within the same task.

9. Much overlapping of the response distributions has been observed, such that within-sex variances are frequently greater than between-sex variances.

10. There is no consistent ascendancy of either sex over the other on all psychomotor tasks; i.e., no warrant exists for using terms like "superiority" or "inferiority" in an honorific sense.

11. The roles of heredity, environment, and their covariance are largely unknown at the present time; e.g., the heritability (h^2) of sex differences in psychomotor skills is, with the exception of spatial visualization, mostly terra incognita.

12. Not all perceptual, psychomotor, or athletic differences that are correlated with sex are intrinsically biological; undoubtedly the learning and execution of skills by males and females are modulated by distinctive social learning, sex-role playing, unequal economic opportunities, and other culturally conditioned influences (e.g., pejorative labels of "sissy" or "tomboy").

Earlier in the chapter, I mentioned the paucity of female composers and conductors. To that curious deficit, which happily is now being mitigated, may be added the fact that women rarely excel in the creation of serious drama or epic poetry, although they are outstanding in writing prose fiction (e.g., Lady Murasaki, Margaret Mitchell) and lyric poetry (e.g., Sappho, Emily Dickinson). Comparative psychologists, no less than comparative *littérateurs*, find such phenomena to be puzzling because they cannot easily be ascribed to incquities, discrimination, societal expectations, or mere accident. For further discussions of this problem see Maccoby and Jacklin (1974), Ounsted and Taylor (1972), and Tyler (1965).

The role of sex in psychomotor skills is of practical as well as theoretical importance. Consider the practical issue. According to the Department of Labor, about 99% of the secretarial jobs in the United States are held by women. Females make up 96% of the typists and 71% of the office-machine operators. However, only 9% of all physicians and dentists, and fewer than 1% of the engineers and construction craftsmen are women. A person does not have to know much about the psychology of human performance to realize that all of these positions entail the acquisition and utilization of behavioral skills that are loaded on the perceptual and motor dimensions. Why the dramatic male–female difference in the labor-pool proportions for these jobs? That information is not available to me, but there is considerable anecdotal evidence that American women have been the victims of "typecasting" for positions in the secretarial, clerical, child-care, nursing, and cosmetology areas. They hold over 90% of such positions.

Three years ago, the Women's Bureau of the Labor Department announced that it had launched a campaign to encourage women to break away from stereotyped "female" positions and train for some of the higher-salaried careers now dominated by men. Across the land, job counselors urged girls and women to prepare for such fields as engineering and the sciences. Positions in systems analysis, computer programming, drafting, and other technical specialties are still growing fast in 1977. If women have the requisite psychomotor aptitudes and capacities, they are in line for a financial bonanza during the near future. Do they possess these traits, and if so will they exploit them? Only the future will tell.

Some Acquisition Phenomena

Returning now to theoretical issues, let us examine briefly the most conspicuous laboratory evidence of the role of sex in psychomotor learning and performance. Beginning with rotary tracking skills in children, there are numerous studies documenting the greater average proficiency of boys over girls (Ammons, 1958; Ammons & Ammons, 1970; Noble, 1968, 1969b, 1974b; Noble *et al.*, 1977). Drawing again from the archives of this laboratory, I have constructed Figure 10.10 from the unpublished records of 370 right-handed children aged 9–12 years representing the Caucasoid and Negroid races residing in rural northeast Georgia, who practiced on the Rotary Pursuitmeter (Figure 10.2) in our Mobile Laboratory for 50 trials. Matching the boys and girls separately for race, age, and conditions of practice, as described earlier, I grouped the time-on-target scores of these two cohorts ($n = 185$ cases per sex) into five 10-trial blocks of training. Figure 10.10 presents their acquisition data in terms of both mean time ($\overline{R}_{tt}$) and mean percent ($\overline{R}_{\%}$) targeting scores as a function of the number of successive practice blocks (N). Clearly, the average curve for the males exceeds that of the females, and the two sexes appear to be diverging with continued training. A 5×2 analysis of variance on the R_{tt} scores of the two groups confirmed our preexperimental hypothesis with significant main effects of practice ($p < .001$) and sex ($p < .01$) as well as a significant practice $\times$ sex interaction ($p < .001$). I conclude that the boys, on the average, are initially more skillful than the girls, are acquiring proficiency at a faster rate, and are probably approaching a higher final level of performance under these conditions. This statement may be generalized to right-handed children of both races within the 9–12 year age range, and it applies to practice with either the preferred or the nonpreferred hand (Noble, 1974b; Noble

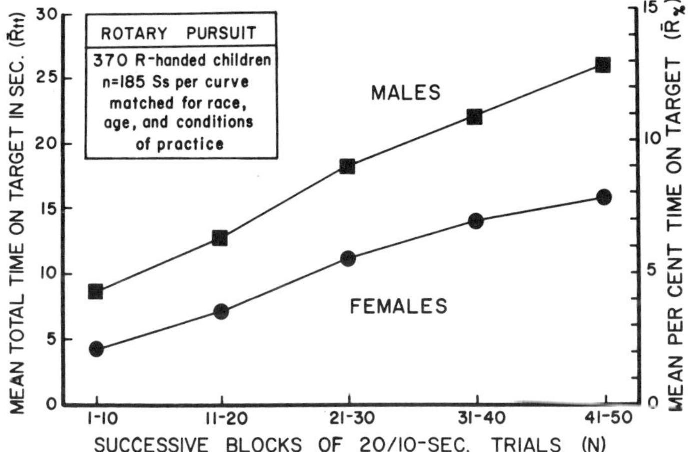

Figure 10.10. Acquisition curves of mean time on target ($\overline{R}_{tt}$) and mean percentage time on target ($\overline{R}\%$) in rotary pursuit as functions of successive practice blocks (N), with sex as the parameter. Each group contains 185 grade school children matched for race, age, and practice conditions. (Data from Noble, unpublished.)

et al., 1977). Before leaving the children I might note that Ammons and Ammons (1970), in their extensive investigations of tracking skill, have not observed large sex differences on the rotor below the third grade. After then, however, the gap widens and gender becomes complicated with aging. "Sex-related variables are of sufficient magnitude that we must control for sex of subject in most rotary pursuit studies (1970: 213)."

At the adult level, there is equally solid evidence that males surpass females in pursuit tracking ability (Ammons & Ammons, 1970; Bilodeau & Bilodeau, 1961; Huang & Payne, 1975; Noble, 1968, 1970a, 1974b; Noble *et al.*, 1977; Wilkerson, Noble, & Skelley, 1975). Indeed, this masculine ascendancy with respect to the simple, rhythmic, eye–hand skill of keeping a lightweight, hinged, metal cursor in steady contact with a rotating, dime-sized, silver disc extends well into middle adulthood. For a group of 500 Caucasoid college students in the age range from 17 to 41 years, I found (Noble, 1970a) that the 256 men in the project were not only more accurate, on the average, but also less variable than the 244 women during a session of 100 trials, defined in our standard units of work: rest cycles (20 sec. working, 10 sec. resting). The acquisition curves of both sexes were highly predictable from Equation (3) (average fit = 99.76%), and every source of variability except age was significant ($p < .001$). In short, I found that practice, sex, and their interaction are all potent factors in this common old laboratory task on which human skill is so strongly determined by inheritance. A discussion of the implications of this experiment for the theory of psychomotor skills is provided elsewhere (Noble, 1970a). In addition, readers may be willing to record yet another "plus" for Hull's (1943, 1952) general theory of learning and performance, and specifically for his viewpoint about the parametric role of group differences in a mathematical treatment of behavior (see also Figure 10.4 and associated text). In point of fact, the subjects' initial ability, rate of acquisition, and asymptotic proficiency were all positively related in these data.

Next, I should like to examine studies of gender differences in performance on the Discrimination Reaction Timer (Figure 10.2). The first of these experiments (Noble, Baker, & Jones, 1964) has already been summarized. Suffice it to reinforce the point that Equation (3) was an excellent predictive model for 30 different acquisition functions ($n = 20$ cases each) representing the average speed with which Caucasoid youngsters, adults, and oldsters from 8 to 87 years old made complex, color–spatial discriminations and correct multiple-choice reactions. The overall mean predictability of Equation (3) was 97.98%. For 300 males in the project, the average fit was 97.91%; for 300 females, it was 98.05%. Here, as in Figure 10.5, we encounter further substantiation of Hull's theory. Nor does the evidence end there. Four other investigations of the role of sex in the acquisition of skill on the Discrimination Reaction Timer, all with different experimental designs, have produced confirmatory results in college students varying in sex, race, or personality (Noble & Hays, 1966; Noble & Kalivoda, 1977; Noble & Skelley, 1976; Noble & Vithakamontri, 1975). Two of these projects, undertaken in collaboration with Mrs. Diane M. Kalivoda and Mrs. Cherie S. Skelley, extended our conventional 320-trial practice session under nonspecific instructions to at least 960 trials within four consecutive days of training.[14] In Figure 10.11 I have combined the response speed (R_s) scores of 45 men and 45 women from one experiment (Noble & Kalivoda, 1977) with the R_s scores of 40

[14] Proficiency on the Discrimination Reaction Timer as measured by speed of response (R_s) is importantly related to the nature of the instructions. For a given age, race, and sex group, nonspecific instructions make the task more difficult than standard AAF instructions (Melton, 1947) because they do not reveal the relation between the red and green lights. When subjects have to discover the solution for themselves (e.g., Noble, 1969b; Noble & Hays, 1966; Noble & Kalivoda, 1977; Noble & Skelley, 1976; Noble & Vithakamontri, 1975), their $\bar{R}_s$ scores are significantly lower than those of comparable subjects who have been given explicit color-spatial directions (e.g., Noble, Baker, & Jones, 1964). I have now cited considerable evidence of the importance of age and sex in discrimination reaction. Race is a crucial factor too, especially under nonspecific instructions. Aptitude (k) as well as capacity ($M + T$) differences were found in an experiment in which 106 Caucasoid and 106 Negroid children were compared over 160 trials (Noble, 1969b). Although both groups were well matched at the outset of training, the Caucasoids gained skill more rapidly than the Negroids and reached a higher level of proficiency ($p < .01$). Both acquisition curves were exponential, with an average of 96.52% of the variance in speed scores accounted for by Equation (3). All these results are consistent with the theory presented on page 362.

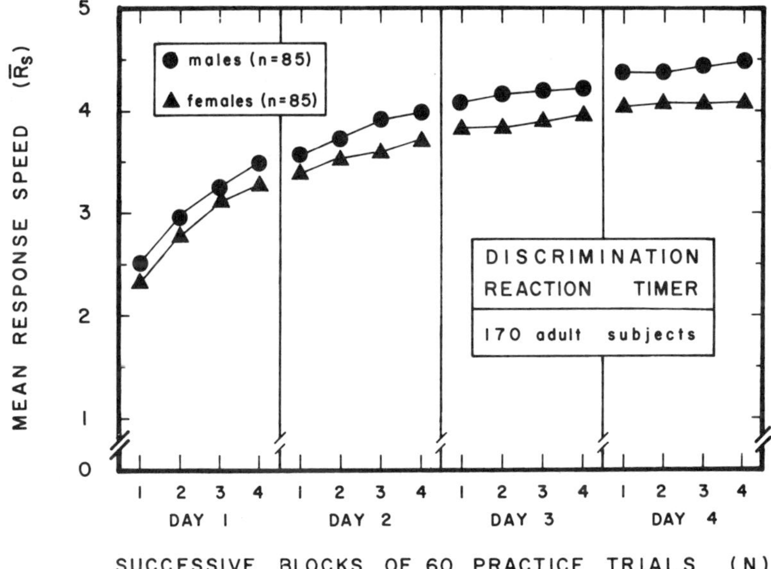

Figure 10.11. Acquisition curves of mean response speed ($\overline{R}_s$) in discrimination reaction as functions of successive practice blocks (N), with sex as the parameter. Each group contains 85 Caucasoid college students of comparable ages. (Data from Noble & Kalivoda, 1977; Noble & Skelley, 1976.)

men and 40 women from the other experiment (Noble & Skelley, 1976) to produce a pair of mean acquisition curves ($\overline{R}_s$) representing the comparative proficiency of 170 adult Caucasoid subjects ($n = 85$ per sex group) practicing for 16 blocks of 60 trials each *(N)*. The modal age in both studies was 19 years. From 4×2 analyses of variance applied to the data arranged in days × sex or blocks × sex matrices, one may conclude briefly that both main effects and the practice × sex interaction are generally significant ($p < .05$). From Figure 10.11 it is apparent that the males began Day 1 with speedier reactions than the females, and they continued to increase their proficiency at a faster rate. By the end of Day 4 the women started to level off whereas the men were still gaining. Analyses of error scores ($R-$) failed to detect any main effect of sex ($p > .05$), but both groups eliminated their incorrect responses at significantly rapid rates ($p < .001$). The comparability of $\overline{R}-$ scores between males and females suggests that they were well matched in average cognitive ability. As Kalivoda, Skelley, and I interpret our results, the sexes differed essentially in factors of *spatial-visualization aptitude* and *motor-speed capacity*, not general intelligence (IQ). In fact, the mean IQs of the men and women, based on the majority's Scholastic Aptitude Test scores, were not significantly different ($p > .10$).

Gender Interactions with Age, Race, and Practice

Genetic and hormonal variables seem to be implicated in the foregoing experiments on sex differences in complex spatial perception and choice reaction. Research by a number of specialists (Bock & Kolakowski, 1973; Broverman *et al.*, 1968; Hartlage, 1970; Loehlin *et al.*, 1975; Maccoby & Jacklin, 1974; Ounsted & Taylor, 1972) indicates that the ability of people to visualize spatial relations and to "manipulate" visual images symbolically is influenced by an X-linked major gene. Bock and Kolakowski (1973) gave a test of spatial relations to parents and children in a sample of 167 Midwestern families (predominantly Caucasoid). Correlation of age-corrected data among the family members identified by sex revealed the expected cross–sex, parent–child pattern for a recessive sex-linked gene. The results showed a consistent sex difference favoring boys and men, as is generally found for tests of spatial ability (Vandenberg, 1968, 1971; see also Osborne, Chapter 6). Statistical analyses of 296 families tested on similar items enabled Bock and Kolakowski to reject an autosomal polygenic model for the inheritance of spatial ability; i.e., a theory of multiple-gene determination.

That kind of non-X chromosome theory has been useful in accounting for verbal ability and for general intelligence. But these newer findings point to a single gene locus on the X chromosome. Since the gene-enhancing spatial ability is recessive, and because females inherit XX, this perceptual skill will appear in women only if both chromosomes bear the *alleles.*[15] If they do not, then the phenotypic trait will be that of the dominant allele. Males inherit XY, and their Y chromosomes carry no dominating alleles. Hence, one recessive allele in a male is enough to provide spatial competence. In other words, females have to inherit these recessive genes from both parents in order to manifest good spatial ability. Males inherit them only from their mothers, and that accounts for their greater average proficiency on such tests. As shown in Table 10.7, the pattern of parent–child correlations (r) is quite different for sex-limited and sex-linked traits. To form a quick background, take the case of general intelligence. For IQ, based on nearly 1000 cases (Tyler, 1965), the correlations of fathers with children and mothers with children are practically identical: $r = .49$. Table 10.7 indicates the same thing for physical stature; the values of $r_{ms} \simeq r_{md} = r_{fd} = r_{fs} \simeq .50$. Height is a sex-limited trait, and so is IQ. Parent–child correlation coefficients around .50 are consistent with an autosomal theory of inheritance (Jensen, 1973). For serum immunoglobulin concentration, however, Table 10.7 exhibits obvious inequalities. The r values in the middle row are such that $r_{ms} > r_{md}$ and $r_{fd} > r_{fs}$. A similar pattern is found for spatial relations ability in the

[15] In genetics, alternative forms of a gene occurring at a single gene-locus on a chromosome are known as *alleles* or *allelomorphs* (see Stern, 1973).

TABLE 10.7

Parent–Child Correlations for Sex-Limited and Sex-Linked Traits[a]

Trait measured	Number of families	Correlations[b]			
		r_{fs}	r_{fd}	r_{ms}	r_{md}
	Sex-limited				
Physical stature	>1000	.51	.51	.49	.51
	Sex-linked				
Serum immunoglobulin concentration	64	.10	.31	.36	.19
Spatial relations ability	296	.11	.28	.25	.14

[a] Adapted from Bock & Kolakowski, 1973.
[b] f = father, s = son, m = mother, d = daughter.

bottom row. The latter two traits are both sex-linked. These data are inconsistent with the polygenic model. Instead, they point to recessive X-linked inheritance.

In discussing some research on cross-cultural differences in spatial perception between primitive Mongoloids and Negroids, wherein it was found that the former greatly exceeded the abilities of the latter, by behaving like European Caucasoids, Loehlin *et al.* (1975) entertain (but not seriously) the notion that the spatial-visualizing gene of Bock and Kolakowski might have a lower incidence in African than in European or Asiatic populations. They conclude that "ethnic group and socioeconomic status are separate but correlated variables, that both crosscut the heredity–environment distinction, and that ability differences among existing groups can rarely be unambiguously attributed to genetic or environmental differences between them (Loehlin *et al.*, 1975: 195)." I wonder, though, whether our own Discrimination Reaction Timer experiments that involved Caucasoid and Negroid children (Noble, 1969b) and Caucasoid, Negroid, and Mongoloid adults (Noble & Vithakamontri, 1975) do not revive the biopsychological stratagem, albeit on another front. For instance, the largest mean differences *between the sexes* in proficiency ($\overline{R}-$) that Miss Vithakamontri and I recorded on that apparatus were for American Negroids and Indo-Dravidian Caucasoids. The mean sex differences were smallest for American Caucasoids and Asiatic Mongoloids. The first two taxa had the darkest pigmented epidermis of all our subjects; the last two had the lightest skins. Now, the Rivers–Pollack fundus-pigmentation hypothesis may apply here because a high correlation exists between those two melanin systems of the body. Alternatively, it may be that the comparatively low proficiency (high $\overline{R}-$ scores) and high sex-differential of the Afro-Americans and Indo-

Dravidians is deducible from genetic theory on the assumption of a sex-linked recessive trait being involved in discrimination-reaction performance. Other hints of possible X-linkage in spatial ability can be found in the work of Baughman and Dahlstrom (1968), Porteus (see Table 10.6), and Tenopyr (1967). In reference to Figure 10.9, perhaps it is worth repeating that the rank order of proficiency ($\overline{R}_p$) scores of the four taxa on the Selective Mathometer was not the same as the rankings by speed ($\overline{R}_s$) or error ($\overline{R}-$) scores on the Discrimination Reaction Timer. It is the latter that calls for color–spatial perceptions and quadruple-choice disjunctive reactions, and that is why the data are of such singular appeal. We hope to return to this problem in the laboratory.

Fleishman (1960) found two task-specific factors involved in learned performance on the Rotary Pursuitmeter, one of which may be genetic (Noble, 1970a). Based on studies of MZ and DZ twins, as we have already learned, the heritability of pursuit tracking is close to 90%, about the same as the genetic determination of physical stature. It also happens that the heritability of the preferred (right) hand is greater than that of the nonpreferred (left) hand (Vandenberg, 1966).[16] However, these data come mainly from Caucasoid males and small samples. Little is known about tracking heritabilities for other races, or for Caucasoid females. The heritability of choice-reaction tasks is quite high, but lower than that of tracking tasks. There are also significant h^2 values for hand steadiness, speed drill, card sorting, and spool packing, as mentioned above. The average of all these psychomotor heritabilities is 62%. For spatial ability the h^2 is greater among males (75%) than among females (32%), a fact that Bock and Vandenberg gleaned from 187 twin pairs using the Differential Aptitude Tests (Vandenberg, 1968). Psychology urgently needs a systematist to put this whole domain into proper order. More functionalistic studies of large scope will undoubtedly be required, but there are numerous phenomena already established that are potentially susceptible to theoretical integration. The challenge is great.

Several gender interactions with age, race, and practice have already surfaced in the preceding sections, so I should like to terminate this final discourse on sex and learning with a treatment of the eximious research being conducted by Payne and his colleagues (Huang & Payne, 1975; McCaffrey & Payne, 1977; Payne & Huang, 1977; Zegoib & Payne, 1977) on the subtle interactions among age, sex, task factors, practice distribu-

[16] According to a recent survey (Hicks & Kinsbourne, 1976), about 90% of humans are right-handed. There is convincing direct as well as indirect evidence for the inheritance of handedness whereas the case for learning is weak. Some studies reveal correlations between either left-handedness or cross-dominance (mixed-handedness) and such pathologies as mental retardation, epilepsy, cerebral palsy, aphasia, and apraxia. However, these reports may suffer from sampling bias. Large, unselected numbers of subjects ($n > 7900$) in Britain and the USA do not exhibit symptoms of greater impairment among left-handers.

tion, and reminiscence. The latter term may be unfamiliar to some readers. *Reminiscence* is merely a gain in proficiency without practice. When subjects perform trial after trial on some psychomotor learning task without any scheduled rest periods, and are then given a short break, say midway through the training session, their mean scores on the very next trial will usually exhibit a marked improvement. The name for the condition of unrelieved work preceding the break is *massed* practice; if rests are interpolated between the work periods it is called *distributed* practice. There is a sense, therefore, in which reminiscence and trial spacing are related. In order for reminiscence to be measured in an *experimental* group of subjects, there is ordinarily provided a massed practice *control* group against whose performance we compare the experimental group's postrest trial scores. Reminiscence effects (the measured gains) are most prominent in tasks demanding continuous attending and responding (e.g., Rotary Pursuitmeter on a 30:0 cycle); the effects are least often observed with discrete responding apparatus (e.g., Complex Coordinator, Discrimination Reaction Timer, Selective Mathometer).[17]

The theoretical importance of this concept derives from its role in testing Hull's (1943) hypothesis of reactive inhibition, symbolized I_r. Simply put, Hull speculated that a decremental process cumulates in the organism as a positive function of working and a negative function of resting. The phenomenon of reminiscence also manifests bilateral transfer of skill (e.g., from the right to the left hand), suggesting that the locus of the work decrement lies in the central nervous system rather than in the peripheral nervous system or in specific effector mechanisms.

In connection with Hull's (1945, 1952) doctrine of individual and group differences, I mentioned that extensions of his theory in that domain could be made to intervening variables other than habit (H). As I interpret Payne's work, he is achieving that for Hull's I_r. One basic question he has asked is whether males and females reminisce alike. Another is concerned with the role of age; are reminiscence gains different in boys and men, girls and women? Still another question is raised about the nature and influence of task factors; e.g., whether they interact with sex or age, or both.

Briefly, what Payne's team has discovered about sex is that the proclivities to reminiscence "shift from one sex to the other across pubertal years (Payne & Huang, 1977, p. 31)." Experiments on preadolescent children employing the Mirror Tracer and the Rotary Pursuitmeter (Figure 10.2) show that *boys reminisce more than girls* (Zegoib & Payne, 1977), whereas experiments on young adults using the same two ap-

[17] In several experiments on the Selective Mathometer (Noble, 1968, 1969a) we have observed significant effects due to practice distribution, but no reminiscence has appeared yet. Unlike rotary pursuit, selective learning is a discontinuous, discrete-responding task.

paratus indicate that *men reminisce less than women* (Huang & Payne, 1975; Payne & Huang, 1977). Thus, sex dominance in those postrest gains of skill we call reminiscence undergoes a reversal with biological maturation. Payne theorizes that this reversal comes about because the two sexes release I_r at different rates from one developmental stage to another. Hence, the dominant sex should enjoy the greater advantage from having practice distributed. This prediction was given marginal support with prepubescent subjects (Zegoib & Payne, 1977) as well as strong support with postpubescent subjects (McCaffrey & Payne, 1977). Adults of the two sexes performed mirror tracing about the same under massed practice, but females surpassed males when practice was distributed, an unusual result indeed.[18] The complicating problem of task differences is indicated by the finding that sex effects do not appear with inverted alphabet printing but do so consistently, depending on the subjects' ontogenetic phase, with mirror tracing and rotary pursuit. I venture to predict that sex differences in reminiscence will also fail to occur on the Complex Coordinator, the Discrimination Reaction Timer, the Manual Lever, the Selective Mathometer, the Star Discrimeter, and the Rudder Control (Figure 10.2). To suggest these dimensional limitations in no way detracts from the theoretical interest of Payne's research. Quite the contrary, he is doing for I_r what Logan did for K or Grice for L (see footnote 8). Besides, my forecast could be wrong.

THEORY OF SKILL ACQUISITION

In the modest conceptual scheme to follow I shall employ the terms *habit*, *learning*, and *association* as equivalent *theoretical* notions. The term *acquisition*, by contrast, is an *empirical* notion. *Skill* and its synonyms (*proficiency*, *ability*, *performance*, etc.) have been delineated earlier in the chapter in conjunction with psychomotor, perceptual, and athletic behavior. The theory to be outlined, therefore, is primarily oriented to the attainment of proficiency by healthy human organisms performing on nonverbal psychomotor tasks. Although it need not be limited to *H. sapiens* or to the domain of psychomotor behavior, that is my present focus. This miniature theory, moreover, is an effort to explain in symbolic language how "habit formation," or the "growth of learning," or

[18] The outcome was unexpected because numerous investigations of massed versus distributed practice using adults on the Rotary Pursuitmeter have indicated that men are significantly more proficient than women over a wide range of work:rest cycles (Noble, 1970a, 1974b; Wilkerson, Noble, & Skelley, 1975). Perhaps the Mirror Tracer's *linear* pursuit tracking properties make it distinctive as compared with *rotary* pursuit tracking. Task factors are emerging more often today as significant sources of interaction variance in studies of organismic variables.

the "development of associations" may be used to account for the observable acquisition of skill by normal people of either sex and all subspecies throughout the life span.

As a conceptual scheme of deliberately limited scope, the learning system adumbrated here should not be mistaken for a general behavior theory in the Hullian or Spencean mode (Hull, 1943, 1952; Spence, 1956, 1960). The theory is *petite* rather than *grande*. Nor should my expository interest in this elementary deductive system be regarded as mirroring a belief in its uniqueness. Alternative postulate sets are well known in logic and science. However radically such models in psychology may appear to differ, if they yield the same logical structures (and therefore the same empirical behavior theorems) then they are formally equivalent. Idiosyncratic notations are, in principle, eliminable. What follows now, with attempted greater concision, is an embodiment of methodology, experimentation, and theory presented elsewhere (Noble, 1966a, 1966b, 1969a, 1970b, 1972, 1975, 1976a, 1976b, 1977a).

Speculation

I should like to begin with a set of seven definitions. These are presented in Table 10.8. Let us assume that a hypothetical S–R relation, or habit (H) connection, of the instrumental type exists in an idealized human subject at the beginning of the learning experiment. Assume further that this nonzero associative tendency is weak because of such

TABLE 10.8

Definitions of Concepts in the Theory[a]

D1	N =	ordinal number of reinforced practice trials (or uniform time periods); regarded as the major experimentally manipulated independent variable.
D2	R =	empirical response scores measured in units of time, amplitude, or frequency; regarded as the major class of dependent variables.
D3	H_n =	hypothetical variable quantity denoting the habit strength of an ideal subject after N trials (or time periods); H_n ranges from 0 to 1.
D4	M =	constant multiplying factor that converts H to the empirically measured scale of R; regarded as the maximum proficiency attainable by the subject under prevailing experimental conditions.
D5	T =	empirical additive (or subtractive) constant of transfer of training, measured in units of R; regarded as jointly determined by past experience and hereditary factors.
D6	ρ =	hypothetical amount of resistance to the formation of an S–R association, ranging from 0 to 1; ρ is a reaction threshold.
D7	k =	hypothetical rate parameter governing the decay of ρ; related to organismic factors such as aptitude, sex, race, and age of the subject.

[a] Adapted from Noble, 1970b.

contingencies as inadequate training or transfer, interference from incompatible habits, low genetic aptitude, or other characteristics of the organism that produce an inertial resistance in the learning process.[19] The net effect of these initial and boundary conditions may be conceptualized as combining in some unspecified manner to produce marked

[19] As I have suggested elsewhere (Noble, 1970b, 1972, 1976b, 1977a), habits and memories are congruent processes of adaptation that are intimately related to general intelligence (g). To *learn* is to acquire one or more permanent reaction tendencies through reinforced practice; to *remember* is to activate one or more of those memories at a later time by means of recall or recognition tests. All of this is cognitive behavior, whether it be psychomotor skills, verbal associations, or problem solutions. Not surprisingly, proficiency in such tasks is positively correlated with IQ (Allison, 1960; Carver & DuBois, 1967; Distefano, Ellis, & Sloan, 1958; Duncanson, 1964; Ellis, Barnett, & Pryer, 1957; Ellis, Pryer, & Barnett, 1960; Ellis & Sloan, 1957; Francis & Rarick, 1959; Glaser, 1967; Grice, 1955; Jensen, 1963, 1972, 1973; Noble, 1972; Noble & Artley, 1977; Noble, Buie & Wilkerson, 1977; Noble & Kalivoda, 1977; Sloan, 1951; Stake, 1961; Wright & Hearn, 1964). A more viable alternative to the popular doctrine (McGeoch & Irion, 1952; Tyler, 1965; Woodrow, 1946) that learning and intelligence are practically orthogonal is the hypothesis of commonality among diverse cognitive processes. My own version of this hypothesis (Noble, 1972) is that a theoretical monotonic relationship exists between Hull's H and Spearman's g. More precisely, g is some positive function of the aptitude parameter k in Equation 3 and Table 10.8. If the other variables of the exponential formula (i.e., M, T) are held constant, then for unselected heterogeneous subjects training on complex tasks, significant correlations (r) should appear between their mean acquisition rates and their mean intelligence quotients. The magnitudes of r would vary, of course, with task complexity (Jensen, 1972, 1973). Empirically, this implies that uniform practice conditions would produce differential rates and levels of psychomotor skill in groups of subjects stratified by IQ but matched in initial proficiency. Because of their probabilistic (multiple-choice) nature, selective learning devices are the most appropriate test situations. When such apparatus have been used (Jensen, Collins, & Vreeland, 1962; Noble, 1966a, 1969a) the outcomes were as predicted (Jensen, 1963; Noble & Artley, 1977). Memories, as habits, are stable intellectual dispositions that persist over time and provide the capability of responding appropriately to subsequent stimulation; i.e., with accuracy, promptness, and vigor. The hypothesis of habit-memory kinship is consistent with a variety of observations in the basic and applied behavioral sciences, but the relation between *original* cognition and *re*-cognition is not that of identity. For example, there may be minimal values that our associations must exceed in order to be memorable (Noble, 1976b, 1977a). For another, the defining operations for acquisition and retention are distinctive, albeit intermingled in most experiments (McGeoch & Irion, 1952). Finally, there are the effects of age to consider. Examination of the relevant evidence from Caucasoid populations (Pakkenberg & Voigt, 1964; Vierordt, 1890) indicates nonmonotonic alterations in brain weight from birth to nearly 90 years of age that closely parallel the perceptual, athletic, and psychomotor changes already reviewed (e.g., Birren, 1964; Miles, 1933; Noble, Baker, & Jones, 1964; Welford, 1958; Welford & Birren, 1965). The main biological bases of the rise and decline of brain weight with time are *(a)* growth of neurons in size, increase of glia cells, and myelination of fibers up to about 20 years of age; and *(b)* atrophy afterward due to loss or disruption of protein molecules and the accumulation of abnormal products in cells and tissues. Between 20 and 90 years of age approximately 50,000 neurons deteriorate daily, producing a loss of several grams per year. Probably the primary damage is to DNA, thus corrupting the genetic code, but another locus of aging is the immunological system. The progressive impairment of middle and great age is more easily detected by complex psychomotor tasks than by cellular assays. Our results in

opposition of the criterion response class (R) to evocation by the relevant stimulus pattern (S), as required by a given psychomotor task. Having defined this hypothetical associative inertia in Table 10.8 as varying along a scale from 0 to 1, assume that $H = 1 - \rho$. Now, depending on the quantitative manner in which ρ decreases as N increases without bound, it follows that H will grow in the converse manner with N. Suppose further that H_n approaches a maximum value M as a limit. The corollary of this proposition is that $\Delta H_n = f\{M - H_n\}$. That is, the finite difference after N trials would be computed theoretically by subtracting the habit strength (H) on that particular trial (H_n) from the subject's theoretical maximum value (M); ΔH_n varies as a function of that difference, being large for small N and decreasing later.

If we select by analogy the continuous, negatively accelerated decay function common to many physical and biological processes (from cooling curves and the damping of springs to radioactive emission and condenser discharges), then we would postulate that $\rho = e^{-kN}$. Here the exponential factor ($e = 2.718$) is the base of the natural system of logarithms. The quantity e^{-kN} is a transient function for all phenomena in which the velocity of change (rate of decrease) is proportional to the magnitude of whatever remains (to decrease). For instance, atmospheric pressure is an exponential function of altitude relative to sea level. Meteorologically, the rate constant k depends on such factors as surface area, energy, and electrical resistance. Suppose that a physicist has experimental reasons to expect that the rate of heat transfer between a small, warm object cooling in a large ambient medium is proportional to the difference in temperature between the object and the medium. This hypothesis would be represented mathematically by specifying a constant k such that if $f(t)$ denotes the temperature difference at time t, then $d\,f(t)/d\,t = k\,f(t)$. The differential equation when integrated implies that $f(t) = Ce^{kt}$; hence C is the initial difference (i.e., $C = f\{0\}$). Since the temperature difference is decreasing, k must be negative. By analogy, $\Delta H \propto (M - H)$; i.e., proportional habit increments are assumed.

Returning to the domain of psychomotor skills, I shall argue now that by adding (or subtracting) the transfer constant T in Table 10.8, we can stipulate that empirical response scores R will be constituted as follows:

$$R = M(H) + T \tag{13}$$

which by substitution is:

$$R = M(1 - \rho) + T \tag{14}$$

---

the Noble-Baker-Jones (1964) experiment provide a clue that aging chiefly affects the M and T parameters of Equation 3, thereby raising or lowering the asymptotes ($M + T$) of the acquisition functions (see Figure 10.6 and Equations 7–12 for a regular decline of capacity indices). These are some of the considerations that entered into definitions D3 through D7 of Table 10.8.

and by postulation is:

$$R = M(1 - e^{-kN}) + T \tag{3}$$

Equation (14) is offered as the fundamental equation for the acquisition of skill in psychomotor learning. The resemblance of Equation (3) to functions proposed by Hull (1943) and Spence (1956), inter alia, is obvious but its rationale is different. We must now derive it mathematically. I begin by formalizing the above preliminary assumptions as explicit postulates.

Postulates

Our first step is to idealize the variables H_n and ΔH_n so that they are functions of a continuous variable x, where $0 \leqslant x < \infty$, rather than of the discrete variable N, where $N = 0, 1, 2, \ldots,$ indefinitely. The function must be capable of being differentiated, such that $f(x)$ is defined for all nonnegative real numbers x, and such that for $N = 0, 1, 2, \ldots,$ the value of $f(N) = H_n$. If we were to let $g(x) = m - f(x)$, where m is the asymptotic value of habit, then $g(x)$ is defined and differentiable for $x > 0$. Thus for each nonnegative integer N, $g(N) = \Delta H_n$. Thus if $f(x)$ and $g(x)$ can be found, the behavior of H_n and ΔH_n will be determined. Since we have assumed that the rate of increase of H_n is proportional to ΔH_n, it is reasonable to impose it also on $f(x)$ and $g(x)$. Only then can mathematical operations be carried out with relevance to the empirical situation. The derivation follows in a series of 16 steps.

Assuming that

$$R = M(H) + T \tag{13}$$

postulate the derivative

$$f'(x) = kg(x) \tag{15}$$

The difference equation is

$$g(x) = m - f(x) \tag{16}$$

hence its derivative is

$$g'(x) = -f'(x) \tag{17}$$

$$\therefore g'(x) = -kg(x) \tag{18}$$

and

$$\frac{g'(x)}{g(x)} = -k \tag{19}$$

Integrating Equation 19, we have

$$\int - kdx = \int \frac{g'(x)}{g(x)}\, dx = -kx + c_1 \tag{20}$$

whose indefinite integral is

$$\log_e g(x) = -kx + c_1 \tag{21}$$

Taking antilogarithms

$$g(x) = e^{-kx+c_1} = e^{c_1} \cdot e^{-kx} \tag{22}$$

$$\therefore g(x) = c \cdot e^{-kx}, \text{ where } c = e^{c_1}. \tag{23}$$

Since $f(x) - m \quad g(x)$ by Equation (16), then

$$f(x) = m - c \cdot e^{-kx}. \tag{24}$$

If R_0 denotes the initial proficiency of the subject sample (due to transfer conditions or organismic characteristics), then $R_0 = R(0) = f(0) = m - ce^0 = m - c$, hence $c = m - R_0$. Thus,

$$f(x) = m - (m - R_0)e^{-kx}. \tag{25}$$

Changing symbols, let $x = N$, $m - R_0 = M$, and $R_0 = T$.

Now since $f(N) = R_n$ for $N = 0, 1, 2, \ldots$, then we have

$$R_n = m - (m - R_0)e^{-kN}. \tag{26}$$

Substituting:

$$R = (M + T) - (M)e^{-kN}. \tag{27}$$

Rearranging:

$$R - M \quad Me^{-kN} + T \tag{28}$$

and therefore:

$$R = M(1 - e^{-kN}) + T. \tag{3}$$

Q.E.D.

This must be the form of the equation for H_n provided our initial speculation is correct; i.e., provided that the rate of increase of $H_n \propto \Delta H_n$.

Evidence

Although space limitations do not allow me to present further evidence of data supporting the theory, it has been shown in this chapter and elsewhere (Noble, 1970b) that Equation (3) is consistent with pooled acquisition curves from such diverse psychomotor tasks as the following

10: typewriting, complex reaction time, rotary pursuit, discrimination reaction, star discrimetry, two-hand coordination, mirror tracing, inverted alphabet printing, complex coordination, and compensatory tracking. The indices of predictability (goodness of fit), measured in terms of the vertical deviations of empirical points from the theoretical curves generated by Equation (3), were all in excess of 95%. On the average, more than 98% of the variance in the observed mean scores could be accounted for by Equation (3).

I conclude that the present conceptual system is appropriate, beyond reasonable doubt, for at least 10 distinctively different human psychomotor learning situations, with an average error variance less than 2%. The theory can be extrapolated beyond the class of tracking tasks and half a dozen other situations to certain studies of age, sex, and race, but exactly how far it can be generalized is yet to be determined.[20] We must discover the initial and boundary conditions for different psychomotor skill problems so that the empirical parameters M, k, and T can be evaluated. Then it will be possible to test specific implicates of the postulate set. The exponential law represented by Equation (3) is explained by the theory in the logical sense that the law is a deductive consequence of the theory. However, one should remember that the recording of empirical events cannot *prove* a theoretical generalization with certainty. But an explanation, however imperfect, has pragmatic value if it aids scientists in formulating hypotheses, taking novel observations, performing crucial experiments, and arranging their facts into a systematic framework.

SUMMARY

The biopsychological factors of age, race, and sex are powerful determinants of psychomotor learning and performance. That is the central message of this chapter. In review, I have considered the following four major topics.

The Domain and History of Skills

Three categories of human skills, ranging from sensory abilities at one extreme to motor abilities at the other, were identified: perceptual,

[20] In several publications (Noble, 1966a, 1968, 1969a, 1974a; Noble, Noble, & Alcock, 1958) we have employed a rational, double-exponential equation to account for R_p acquisition curves on the Selective Mathometer. Predictability indices calculated from more than two dozen experiments in this laboratory average about 98%, roughly the same as Equation 3. The double-exponential model is not refuted by the data or speculations of the present essay. To the contrary, it is still quite useful for explaining behavior on multiple-choice learning tasks. Moreover, the availability of rival quantitative formulations is a continuing heuristic invitation to theoretical integration. In the future we intend to pit these two models against each other more frequently.

psychomotor, athletic. Primary emphasis was placed upon research in the mainstream of *psychomotor skills.* I defined these behaviors as organized patterns of neuromuscular activities that are usually sequential and continuous in form, and that operate under the guidance of changing feedback signals from the learner and the environment. The history of research in this domain was traced from the origins of opposing nomothetic and idiographic traditions (Wundt, Galton) through a stage of independent development (Cattell, Pearson, Rivers, Spearman, Ebbinghaus) to the beginnings of eclecticism (Woodworth, Thorndike), thence to the evolution of the modern synthesis (Woodrow, Hull, Melton). On the synthetic view, as outlined by Hull, individual and group differences affect the values of parameters used in quantitative descriptions of general trends in learned behavior, but not the mathematical forms of the equations. If true, the integration of differential–correlational psychology with general–experimental psychology would be facilitated.

Analysis of Skills

Rather than being simple reflections of a single, underlying aptitude or capacity, human skills appear to be complex and multifaceted. The technique of factor analysis, shorn of certain procedural and interpretive errors, has revealed at least ten psychomotor factors. None can properly be called "general." Some typical laboratory learning devices (or apparatus tests) were described, all of which have high reliability and validity. Twelve perceptual and nine athletic skill factors were also listed, yet no "common" factor is identifiable in either category. The meaning of *heritability* was critically examined and its limitiations noted. My overall conclusion about the acquisition of skills was that innate aptitudes and capacities as well as efficient practice conditions are jointly necessary for competent performance.

Interactions of Organismic Factors and Practice Variables

Following a nontechnical introduction to the statistical analysis of variance in experimental studies of psychomotor skills, a test of Hull's theory was applied to the learning of pursuit tracking by homogeneous groups of subjects varying in initial ability. The test supported the theory, revealing similar acquisition equations and significant effects of practice, aptitude, and their interaction.

Attention was then directed specifically to the variables of age, race, and sex. In each case, a number of empirical generalizations, acquisition phenomena, and practice interactions were presented. Age trends usually turned out to be nonmonotonic, with a fast ascent of proficiency up to the late teens followed by a slow descent into the middle and advanced ages. Racial differences were pervasive, but there was no consistent hierarchy of the human subspecies in psychomotor performance. A special effort was made to convey the objective, biological meaning of race. As for sex, males tended to excel on some tasks and females on other tasks.

All of these organismic variables interacted mutually in certain situations, and generally with practice variables as well. Despite significant average differences in performance attributable to age, race, or sex, there was much overlapping of the various taxa in the experiments reported. The role of genetic factors in most of these investigations appeared to be substantial. For instance, in several well-controlled and sizable studies of subspecies' performance, there was unequivocal evidence that the average psychomotor behavior of hybrid taxa occupied positions intermediate to those of more homogeneous taxa at the extremes. In some investigations, however, alternative hypotheses may be entertained. A few of the possibilities were cited in conjunction with the phenomena of age, race, and sex.

A Theory of Skill Acquisition

Finally, a deductive theory of limited scope was proposed to account for the learning process in healthy people of either sex and all subspecies throughout the life span. The core assumptions are that the strengthening of habits (H), which are congruent to memories, must overcome an inertial associative resistance (ρ) in the course of numerous reinforced practice trials (N) in order to produce responses (R) to task stimuli (S) with accuracy, promptness, and vigor. The resulting exponential formula has the capability of explaining a variety of skill acquisition phenomena with an average predictability of 98%. The form of the law derived from the theory is, to a first approximation, invariant with respect to organismic factors. These parameters appear as numerical constants in the rational equation thus generated. What is needed now, I suggested, is rigorous experimental testing of the system's implicates.

In closing, I should like to express the hope that this essay may have contributed to the intellectual confederation of the psychology of individual and group differences with the psychology of learning and performance.

REFERENCES

Adams, J. A. Some implications of Hull's theory for human motor performance. *Journal of General Psychology*, 1956, 55, 189–198.

Adams, J. A. The relationship between certain measures of ability and the acquisition of a psychomotor criterion response. *Journal of General Psychology*, 1957, 56, 121–134.

Adams, J. A. Motor skills. *Annual Review of Psychology,* 1964, *15,* 181–202.

Adams, J. A., & Reynolds, B. Effect of shift in distribution of practice conditions following interpolated rest. *Journal of Experimental Psychology*, 1954, *47*, 32–36.

Allison, R. B. *Learning parameters and human abilities.* Princeton, N.J.: Educational Testing Service, 1960.

Ammons, R. B. Acquisition of motor skill: I. Quantitative analysis and theoretical formulation. *Psychological Review*, 1947, *54*, 263–281.

Ammons, R. B. "Le mouvement." In G. H. Seward & J. P. Seward (Eds.), *Current psychological issues.* New York: Holt, 1958.

Ammons, R. B., & Ammons, C. H. Decremental and related processes in skilled performance. In L. E. Smith (Ed.), *The psychology of motor learning.* Chicago: The Athletic Institute, 1970.

Amsel, A. Partial reinforcement effects on vigor and persistence. In K. W. Spence & J. T. Spence (Eds.), *The psychology of learning and motivation.* (Vol. 1). New York: Academic, 1967.

Baker, J. R. *Race.* London: Oxford Univ. Press, 1974.

Baker, P. T., & Weiner, J. S. (Eds.), *The biology of human adaptability.* London: Oxford Univ. Press, 1966.

Bartlett, F. C. *Thinking.* New York: Basic Books, 1958.

Baughman, E. E., & Dahlstrom, W. G. *Negro and white children.* New York: Academic, 1968.

Bayley, N. Development of motor abilities during the first three years. *Monographs of the Society for Research in Child Development*, 1935, *1* (1, Serial No. 1).

Bayley, N. Comparisons of mental and motor test scores for ages 1–15 months by sex, birth order, race, geographical location, and education of parents. *Child Development*, 1965, *36*, 379–411.

Bechtoldt, H. P. Factor analysis and the investigation of hypotheses. *Perceptual and Motor Skills*, 1962, *14*, 319–342.

Bechtoldt, H. P. Motor abilities in studies of motor learning. In L. E. Smith (Ed.), *The psychology of motor learning.* Chicago: The Athletic Institute, 1970.

Bilodeau, E. A. (Ed.). *Principles of skill acquisition.* New York: Academic, 1969.

Bilodeau, E. A., & Bilodeau, I. McD. Motor-skills learning. *Annual Review of Psychology*, 1961, *12*, 243–280.

Birren, J. E. *The psychology of aging.* Englewood Cliffs, N.J.: Prentice-Hall, 1964.

Bock, R. D., & Kolakowski, D. Further evidence of sex-linked major gene influence on human spatial visualizing ability. *American Journal of Human Genetics*, 1973, *25*, 1–14.

Boring, E. G. *A history of experimental psychology* (2nd ed.). New York: Appleton-Century-Crofts, 1950.

Broverman, D. M., Klaiber, E. L., Kobayashi, Y., & Vogel, W. Roles of activation and inhibition in sex differences in cognitive abilities. *Psychological Review*, 1968, *75*, 23–50.

Carver, R. P., & DuBois, P. H. The relationship between learning and intelligence. *Journal of Educational Measurement*, 1967, *4*, 133–136.

Codwell, J. E. Motor function and the hybridity of the American Negro. *Journal of Negro Education*, 1949, *18*, 452–464.

Comas, J. *Manual of physical anthropology*. Springfield, Ill.: Thomas, 1960.

Coon, C. S. *The living races of man.* New York: Knopf, 1965.

Coon, C. S. Human populations. *Encyclopaedia Britannica*, 1974, *14*, 839–848.

Cummins, H., & Midlow, C. *Fingerprints, palms, and soles.* New York: Dover, 1961.

Damon, A. Race, ethnic group, and disease. In R. H. Osborne (Ed.), *The biological and social meaning of race*. San Francisco: Freeman, 1971.

Day, C. B. A study of some Negro–white families in the United States. In E. A. Hooton &

N. I. Bates (Eds.), *Harvard African Studies* (Vol. 10). Cambridge, Mass.: Harvard Univ. Press, 1932.
de Garay, A. L., Levine, L., & Carter, J. E. L. (Eds.). *Genetic and anthropological studies of Olympic athletes*. New York: Academic, 1974.
Distefano, M. K., Jr., Ellis, N. R., & Sloan, W. Motor proficiency in mental defectives. *Perceptual and Motor Skills*, 1958, *8*, 231–234.
Dobzhansky, T. *Mankind evolving*. New Haven: Yale Univ. Press, 1965.
DuBois, P. H. *A history of psychological testing*. Boston: Allyn & Bacon, 1970.
Duncan, C. P. Transfer in motor learning as a function of degree of first-task learning and inter-task similarity. *Journal of Experimental Psychology*, 1953, *45*, 1–11.
Duncanson, J. P. *Intelligence and the ability to learn*. Princeton, N.J.: Educational Testing Service, 1964.
Ellis, N. R., Barnett, C. D., & Pryer, M. W. Performance of mental defectives on the mirror drawing task. *Perceptual and Motor Skills*, 1957, *7*, 271–274.
Ellis, N. R., Pryer, M. W., & Barnett, C. D. Motor learning and retention in normals and defectives. *Perceptual and Motor Skills*, 1960, *10*, 83–91.
Ellis, N. R., & Sloan, W. Rotary pursuit performance as a function of mental age. *Perceptual and Motor Skills*, 1957, *7*, 267–270.
Epstein, W. Perceptual learning. *Encyclopaedia Britannica*, 1974, *10*, 746–748.
Espenschade, A. Motor performance in adolescence, including the study of relationships with measures of physical growth and maturity. *Monographs of the Society for Research in Child Development*, 1940, *5* (1, Serial No. 24).
Fitts, P. M. Perceptual-motor skill learning. In A. W. Melton (Ed.), *Categories of human learning*. New York: Academic, 1964.
Fleishman, E. A. Testing for psychomotor abilities by means of apparatus tests. *Psychological Bulletin*, 1953, *50*, 241–262.
Fleishman, E. A. Dimensional analysis of psychomotor abilities. *Journal of Experimental Psychology*, 1954, *48*, 437–454.
Fleishman, E. A. Abilities at different stages of practice in rotary pursuit performance. *Journal of Experimental Psychology*, 1960, *60*, 162–171.
Fleishman, E. A. *The structure and measurement of physical fitness*. Englewood Cliffs, N.J.: Prentice-Hall, 1964.
Fleishman, E. A. Human abilities and the acquisition of skill: Comments on Professor Jones' paper. In E. A. Bilodeau (Ed.), *Acquisition of skill*. New York: Academic, 1966.
Fleishman, E. A. Structure and measurement of psychomotor abilities. In R. N. Singer (Ed.), *The psychomotor domain*. Philadelphia: Lea & Febiger, 1972.
Fleishman, E. A., & Hempel, W. E., Jr. The relation between abilities and improvement with practice in a visual discrimination reaction task. *Journal of Experimental Psychology*, 1955, *49*, 301–312.
Francis, R. J., & Rarick, G. L. Motor characteristics of the mentally retarded. *American Journal of Mental Deficiency*, 1959, *63*, 792–811.
Garn, S. M. *Human races* (3rd. ed.). Springfield, Ill.: Thomas, 1971.
Garn, S. M. Races of mankind. *Encyclopaedia Britannica*, 1974, *15*, 348–356.
Garth, T. R. *Race psychology*. New York: McGraw-Hill, 1931.
Gates, R. R. *Pedigrees of Negro families*. Philadelphia: Blakiston, 1949.
Gedda, L. Sports and genetics: A study on twins (351 pairs). In L. Larson (Ed.), *Health and fitness in the modern world*. Chicago: The Athletic Institute, 1961.
Geldard, F. A. *Fundamentals of psychology*. New York: Wiley, 1962.
Gesell, A. *Infancy and human growth*. New York: Macmillan, 1928.
Gesell, A., & Amatruda, C. S. *Developmental diagnosis: Normal and abnormal child development* (2nd ed.). New York: Hoeber, 1947.
Gesell, A., & Thompson, H. Twins T and C from infancy to adolescence: A biological study

of individual differences by the method of co-twin control. *Genetic Psychology Monographs*, 1929, *6*, 1–124.

Gibson, E. J. *Principles of perceptual learning and development.* New York: Appleton-Century-Crofts, 1969.

Ginsburg, B. E., & Laughlin, W. S. The multiple bases of human adaptability and achievement: A species point of view. *Eugenics Quarterly,* 1966, *13*, 240–257.

Glaser, R. Some implications of previous work on learning and individual differences. In R. M. Gagné (Ed.), *Learning and individual differences.* Columbus, Ohio: Merrill, 1967.

Goldsby, R. A. *Race and races.* New York: Macmillan, 1971.

Gottesman, I. I. Biogenetics of race and class. In M. Deutsch, I. Katz, & A. R. Jensen (Eds.), *Social class, race, and psychological development.* New York: Holt, Rinehart, Winston, 1968.

Grice, G. R. Discrimination reaction time as a function of anxiety and intelligence. *Journal of Abnormal and Social Psychology*, 1955, *50*, 71–74.

Grice, G. R. A threshold model for drive. In H. H. Kendler & J. T. Spence (Eds.), *Essays in neobehaviorism.* New York: Appleton-Century-Crofts, 1971.

Grice, G. R. Conditioning and a decision theory of response evocation. In G. H. Bower (Ed.), *The psychology of learning and motivation* (Vol. 5). New York: Academic, 1972.

Guilford, J. P. *Personality.* New York: McGraw-Hill, 1959.

Hakstian, A. R., & Cattell, R. B. The checking of primary ability structure on a broader basis of performances. *British Journal of Educational Psychology*, 1974, *44*, 140–154.

Halverson, H. M. Motor development. In A. Gesell (Ed.), *The first five years of life.* New York: Harper, 1940.

Harrison, G. A., Owen, J. J. T., DaRocha, F. J., & Salzano, F. M. Skin colour in southern Brazilian populations. *Human Biology*, 1967, *39*, 21–31.

Hartlage, L. C. Sex-linked inheritance of spatial ability. *Perceptual and Motor Skills*, 1970, *31*, 610.

Hearnshaw, L. S. *A short history of British psychology: 1840–1940.* London: Methuen, 1964.

Henry, F. M. Specificity vs. generality in learning motor skill. In R. C. Brown & G. S. Kenyon (Eds.), *Classical studies on physical activity.* Englewood Cliffs, N.J.: Prentice-Hall, 1968.

Herskovits, M. J. *The anthropometry of the American Negro.* New York: Columbia Univ. Press, 1930.

Hicks, J. A. The acquisition of motor skill in young children. *Univ. of Iowa Studies in Child Welfare*, 1931, *4*, No. 5.

Hicks, L. H., & Birren, J. E. Aging, brain damage, and psychomotor slowing. *Psychological Bulletin*, 1970, *74*, 377–396.

Hicks, R. E., & Kinsbourne, M. On the genesis of human handedness: A review. *Journal of Motor Behavior*, 1976, *8*, 257–266.

Hinrichs, J. R. Ability correlates in learning a psychomotor task. *Journal of Applied Psychology*, 1970, *54*, 56–64.

Hirata, K. I. Physique and age of Tokyo Olympic champions. *Journal of Sports Medicine and Physical Fitness*, 1966, *6*, 207–222.

Hooton, E. A. *Up from the ape* (Rev. ed.). New York: Macmillan, 1956.

Horn, J. L. Human abilities: A review of research and theories in the early 1970s. *Annual Review of Psychology*, 1976, *27*, 437–485.

Howells, W. W. *Mankind in the making.* London: Secker & Warburg, 1960.

Huang, K. L., & Payne, R. B. Individual and sex differences in reminiscence. *Memory and Cognition*, 1975, *3*, 252–256.

Hull, C. L. *Principles of behavior.* New York: Appleton-Century-Crofts, 1943.

Hull, C. L. The place of innate individual and species differences in a natural-science theory of behavior. *Psychological Review*, 1945, *52*, 55–59.
Hull, C. L. *A behavior system*. New Haven: Yale Univ. Press, 1952.
Illingworth, R. S. *The development of the infant and young child.* London: Livingstone, 1966.
Irion, A. L. A brief history of research on the acquisition of skill. In E. A. Bilodeau (Ed.), *Acquisition of skill.* New York: Academic, 1966.
Jaeger, T. B. *An extent interaction theory of illusion formation.* Unpublished doctoral dissertation, University of Georgia, 1977.
Jensen, A. R., Collins, C. C., & Vreeland, R. W. A multiple S-R apparatus for human learning. *American Journal of Psychology*, 1962, *75*, 470–476.
Jensen, A. R. Learning ability in retarded, average, and gifted children. *Merrill-Palmer Quarterly*, 1963, *9*, 123–140.
Jensen, A. R. *Genetics and education.* New York: Harper & Row, 1972.
Jensen, A. R. *Educability and group differences.* New York: Harper & Row, 1973.
Jones, M. B. Individual differences. In E. A. Bilodeau (Ed.), *Acquisition of skill.* New York: Academic, 1966.
Jones, M. B. Differential processes in acquisition. In E. A. Bilodeau (Ed.), *Principles of skill acquisition.* New York: Academic, 1969.
Jones, M. B. Individual differences. In R. N. Singer (Ed.), *The psychomotor domain.* Philadelphia: Lea & Febiger, 1972.
Jordan, J. H. Physiological and anthropometrical comparisons of Negroes and whites. *Journal of Health, Physical Education, and Recreation*, 1969, *40*, 93–99.
Kay, H. The development of motor skills from birth to adolescence. In E. A. Bilodeau (Ed.), *Principles of skill acquisition.* New York: Academic, 1969.
Klineberg, O. An experimental study of speed and other factors in racial differences. *Archives of Psychology*, 1928, *15*, (No. 93), 1–109.
Landers, D. M., Obermeier, G. E., & Wolf, M. D. The influence of external stimuli and eye color on reactive motor behavior. In R. W. Christina & D. M. Landers (Eds.), *Psychology of motor behavior and sport.* Champaign, Ill.: Human Kinetics Publications, 1977.
Lasker, G. W., & Lee, M. M. C. Racial traits in the human teeth. *Journal of Forensic Sciences*, 1957, *2*, 401–419.
Laughlin, W. S. Physical anthropology. *Encyclopaedia Britannica*, 1964, *2*, 45–54.
Laughlin, W. S. Race: A population concept. *Eugenics Quarterly*, 1966, *13*, 326–340.
Laughlin, W. S. Adaptability and human genetics. *Proceedings of the National Academy of Sciences*, 1968, *60*, 12–21.
Loehlin, J. C., Lindzey, G., & Spuhler, J. N. *Race differences in intelligence.* San Francisco: Freeman, 1975.
Logan, F. A. Incentive theory and changes in reward. In K. W. Spence & J. T. Spence (Eds.), *The psychology of learning and motivation* (Vol. 2). New York: Academic, 1968.
Lundin, R. W. *An objective psychology of music* (2nd ed.). New York: Ronald Press, 1967.
Maccoby, E. E., & Jacklin, C. N. *The psychology of sex differences.* Palo Alto: Stanford Univ. Press, 1974.
Malina, R. M. Growth and physical performance of American Negro and white children. *Clinical Pediatrics*, 1969, *8*, 476–483.
Malina, R. M. Biological substrata. In K. S. Miller & R. M. Dreger (Eds.), *Comparative studies of blacks and whites in the United States.* New York: Seminar Press, 1973.
Mayr, E. *Populations, species, and evolution.* Cambridge, Mass.: Harvard Univ. Press, 1970.
Mazess, R. B. Skin color in Bahamian Negroes. *Human Biology*, 1967, *39*, 145–154.
McCaffrey, R. J., & Payne, R. B. Interaction of sex and practice distribution effects. *Bulletin of the Psychonomic Society*, 1977, *10*, 382–384.

McCloy, C. H. The influence of chronological age on motor performance. *Research Quarterly*, 1935, *6*, 61–64.
McGeoch, J. A., & Irion, A. L. *The psychology of human learning* (2nd ed.). New York: Longmans, Green, 1952.
McGraw, M. B. Later development of children specially trained during infancy: Johnny and Jimmy at school age. *Child Development*, 1939, *10*, 1–19.
McGraw, M. B. *The neuromuscular maturation of the human infant.* New York: Columbia Univ. Press, 1943.
McNemar, Q. Twin resemblances in motor skills, and the effect of practice thereon. *Journal of Genetic Psychology*, 1933, *42*, 70–99.
Melton, A. W. (Ed.). *Apparatus Tests.* Washington: U.S. Government Printing Office, 1947. (AAF Aviation Psychology Program, Research Report No. 4).
Melton, A. W. Individual differences and theoretical process variables: General comments on the conference. In R. M. Gagné (Ed.), *Learning and individual differences.* Columbus, Ohio: Merrill, 1967.
Michael, W. B. Factor analyses of tests and criteria: A comparative study of two AAF pilot populations. *Psychological Monographs*, 1949, *63*, 1–55 (Whole No. 298).
Miles, W. Age and human ability. *Psychological Review*, 1933, *40*, 99–123.
Mitchell, N. B., & Pollack, R. H. Block-design performance as a function of hue and race. *Journal of Experimental Child Psychology*, 1974, *17*, 377–382.
Mitchell, N. B., Pollack, R. H., & McGrew, J. F. The relation of form perception to hue and fundus pigmentation. *Bulletin of the Psychonomic Society*, 1977, *9*, 97–99.
Noble, C. E. Verbal learning and individual differences. In C. N. Cofer (Ed.), *Verbal learning and verbal behavior.* New York: McGraw-Hill, 1961.
Noble, C. E. Selective learning. In E. A. Bilodeau (Ed.), *Acquisition of skill.* New York: Academic, 1966. (a)
Noble, C. E. S-O-R and the psychology of human learning. *Psychological Reports*, 1966, *18*, 923–943. (b)
Noble, C. E. The learning of psychomotor skills. *Annual Review of Psychology*, 1968, *19*, 203–250.
Noble, C. E. Outline of human selective learning. In E. A. Bilodeau (Ed.), *Principles of skill acquisition.* New York: Academic, 1969. (a)
Noble, C. E. Race, reality, and experimental psychology. *Perspectives in Biology and Medicine*, 1969, *13*, 10–30. (b)
Noble, C. E. Acquisition of pursuit tracking skill under extended training as a joint function of sex and initial ability. *Journal of Experimental Psychology*, 1970, *86*, 360–373. (a)
Noble, C. E. A theory of psychomotor skill: Derivation and data. *Psychonomic Science*, 1970, *21*, 344. (Abstract) (b)
Noble, C. E. *Reliability and validity of anthropometric measurements for psychomotor skill research.* Unpublished manuscript, 1971.
Noble, C. E. Psychomotor learning and general intelligence. *Psychonomic Science*, 1972, *29*, (4B), 268. (Abstract)
Noble, C. E. Psychomotor learning. *Encyclopaedia Britannica*, 1974, *10*, 748–754. (a)
Noble, C. E. *Sex and psychomotor skills.* Paper delivered at the meeting of the Southern Society for Philosophy and Psychology, Tampa, April 1974. (b)
Noble, C. E. On Johnson's paradox: Hypothesis verfication. *American Journal of Psychology*, 1975, *88*, 537–547.
Noble, C. E. Induction, deduction, abduction. In M. H. Marx & F. E. Goodson (Eds.), *Theories in contemporary psychology* (2nd ed.). New York: Macmillan, 1976. (a)
Noble, C. E. *On learning and memory.* Presidential address delivered at the meeting of the Southern Society for Philosophy and Psychology, Atlanta, April 1976. (b)
Noble, C. E. Practice. In B. B. Wolman (Ed.), *International Encyclopedia of Psychiatry,*

Psychology, Psychoanalysis, and Neurology (Vol. 9). New York: Van Nostrand-Reinhold-Aesculapius, 1977. (a)

Noble, C. E. Pre-practice conditions in learning. In B. B. Wolman (Ed.), *International Encyclopedia of Psychiatry, Psychology, Psychoanalysis, and Neurology* (Vol. 9). New York: Van Nostrand-Reinhold-Aesculapius, 1977. (b)

Noble, C. E., & Artley, C. W. *Sex, race, and age differences in performance on the Selective Mathometer.* Manuscript submitted for publication, 1977.

Noble, C. E., Baker, B. L., & Jones, T. A. Age and sex parameters in psychomotor learning. *Perceptual and Motor Skills,* 1964, *19,* 935–945.

Noble, C. E., Buie, C. W., & Wilkerson, H. R. *Bilateral transfer in tracking by children and adults of different races.* Unpublished manuscript, 1977.

Noble, C. E., & Hays, J. R. Discrimination reaction performance as a function of anxiety and sex parameters. *Perceptual and Motor Skills,* 1966, *23,* 1267–1278.

Noble, C. E., & Kalivoda, D. M. *Time and error scores of men and women of varied ability during four days' training on the Discrimination Reaction Timer.* Manuscript submitted for publication, 1977.

Noble, C. E., Noble, J. L., & Alcock, W. T. Prediction of individual differences in human trial-and-error learning. *Perceptual and Motor Skills,* 1958, *8,* 151–172.

Noble, C. E., & Skelley, C. S. Performance of men and women during extended practice in discrimination reaction. *Bulletin of the Psychonomic Society,* 1976, *8,* 241. (Abstract)

Noble, C. E., & Vithakamontri, K. Some international comparisons of psychomotor performance. *Bulletin of the Psychonomic Society,* 1975, *6* (4B), 434–435. (Abstract)

Noble, C. E., & Yeh, M. C. *Subject × experimenter × sex × race interactions in complex learning by Caucasians and Orientals.* Manuscript submitted for publication, 1977.

Osborne, R. H. The history and nature of race classification. In R. H. Osborne (Ed.), *The biological and social meaning of race.* San Francisco: Freeman, 1971.

Ounsted, E., & Taylor, D. C. (Eds.). *Gender differences: Their ontogeny and significance.* London: Churchill, 1972.

Pakkenberg, H., & Voigt, J. Brain weight of the Danes. *Acta Anatomica,* 1964, *56,* 297–307.

Payne, R. B., & Hauty, G. T. Effect of psychological feedback upon work decrement. *Journal of Experimental Psychology,* 1955, *50,* 343–351.

Payne, R. B., & Huang, K. L. Interaction of sex and task differences in reminiscence. *Journal of Motor Behavior,* 1977, *9,* 29–32.

Pearson, K. On the laws of inheritance in man. II. On the inheritance of the mental and moral characters in man, and its comparison with the inheritance of the physical characters. *Biometrika,* 1904, *3,* 131–190.

Piaget, J. *La naissance de l'intelligence chez l'enfant.* Neuchâtel: Delachaux et Niestlé, 1936.

Pollack, R. H., & Silvar, S. D. Magnitude of the Müller–Lyer illusion in children as a function of pigmentation of the fundus oculi. *Psychonomic Science,* 1967, *8,* 83–84.

Pollitzer, W. S., Boyle, E., Jr., Cornoni, J., & Namboodiri, K. K. Physical anthropology of the Negroes of Charleston, S.C. *Human Biology,* 1970, *42,* 265–279.

Porteus, S. D. *The Porteus Maze Test and intelligence.* Palo Alto, Calif.: Pacific Books, 1950.

Porteus, S. D. *The Maze Test and clinical psychology.* Palo Alto, Calif.: Pacific Books, 1959.

Porteus, S. D. Ethnic groups and the Maze Test. In R. E. Kuttner (Ed.), *Race and modern science.* New York: Social Science Press, 1967.

Porteus, S. D. & Gregor, A. J. Studies in intercultural testing. *Perceptual and Motor Skills,* 1963, *16,* 705–724.

Post, R. H. Population differences in vision acuity: A review, with speculative notes on selection relaxation. *Eugenics Quarterly,* 1962, *9,* 189–212.

Potter, D. M. *The impending crisis: 1848–1861.* New York: Harper & Row, 1976.

Reed, T. E. Caucasian genes in American Negroes. *Science,* 1969, *165,* 762–768.

Reynolds, B. The effect of learning on the predictability of psychomotor performance. *Journal of Experimental Psychology*, 1952, *44*, 189–198.

Rivers, W. H. R. Physiology and psychology: Vision. In A. C. Haddon (Ed.), *Reports of the Cambridge Anthropological Expedition to Torres Straits* (Vol. 2, Part 1). Cambridge: Univ. Press, 1901.

Rivers, W. H. R. Observations on the senses of the Todas. *British Journal of Psychology*, 1905, *1*, 321–396.

Roff, M., & Payne, R. B. The relation between learning ability scores obtained in different situations. *Journal of Psychology*, 1956, *41*, 47–54.

Roff, M., Payne, R. B., & Moore, E. W. *A statistical analysis of the parameters of motor learning*. Randolph Air Force Base, Tex.: USAF School of Aviation Medicine, February 1954.

Ruch, F. L. The differentiative effects of age upon human learning. *Journal of General Psychology*, 1934, *11*, 261–285.

Sanghvi, L. D. Comparison of genetical and morphological methods for a study of biological differences. *American Journal of Physical Anthropology*, 1953, *11*, 385–404.

Scheinfeld, A. *Your heredity and environment*. Philadelphia: Lippincott, 1965.

Shirley, M. M. *The first two years: A study of twenty-five babies*. Vol. 1. *Postural and locomotor development*. Minneapolis: Univ. of Minnesota Press, 1931.

Silvar, S. D., & Pollack, R. H. Racial differences in pigmentation of the fundus oculi. *Psychonomic Science*, 1967, 7, 159–160.

Singer, R. N. (Ed.). *The psychomotor domain*. Philadelphia: Lea & Febiger, 1972.

Singer, R. N. *Motor learning and human performance* (2nd ed.). New York: Macmillan, 1975.

Sloan, W. Motor proficiency and intelligence. *American Journal of Mental Deficiency*, 1951, *55*, 394–406.

Snoddy, G. S. *Evidence for two opposed processes in mental growth*. Lancaster, Penn.: Science Press Printing Co., 1935.

Spearman, C. *The abilities of man*. London: Macmillan, 1927.

Spence, K. W. The methods and postulates of "behaviorism." *Psychological Review*, 1948, *55*, 67–78.

Spence, K. W. *Behavior theory and conditioning*. New Haven: Yale Univ. Press, 1956.

Spence, K. W. *Behavior theory and learning*. Englewood Cliffs, N.J.: Prentice-Hall, 1960.

Spence, K. W. Foreword to *Principles of behavior* (7th print.) by C. L. Hull. New York: Appleton-Century-Crofts, 1966.

Spuhler, J. N., & Lindzey, G. Racial differences in behavior. In J. Hirsch (Ed.), *Behavior-genetic analysis*. New York: McGraw-Hill, 1967.

Stake, R. E. Learning parameters, aptitudes, and achievements. *Psychometric Monographs*, 1961, No. 9.

Stern, C. *Principles of human genetics* (3rd ed.). San Francisco: Freeman, 1973.

Stewart, T. D. (Ed.) *Hrdlička's practical anthropometry* (4th ed.). Philadelphia: Wistar Institute of Anatomy and Biology, 1952.

Sward, K. Jewish musicality in America. *Journal of Applied Psychology*, 1933, *17*, 675–712.

Tenopyr, M. L. *Race and socioeconomic status as moderators in predicting machine-shop training success*. Paper delivered at the meeting of the American Psychological Association, Washington, September 1967.

Thompson, R. F. *Foundations of physiological psychology*. New York: Harper & Row, 1967.

Thompson, W. R. Innate factors in human behavior. *Encyclopaedia Britannica*, 1974, *8*, 1146–1151.

Thorndike, E. L. *Educational psychology: Mental work and fatigue and individual differences and their causes* (Vol. 3). New York: Teachers College, Columbia Univ. Press, 1914.

Thorndike, E. L. *Human learning.* New York: Century, 1931. (MIT reprint, 1966)
Thorndike, E. L. *et al. Adult learning.* New York: Macmillan, 1928.
Thorndike, E. L., & Woodworth, R. S. The influence of improvement in one mental function upon efficiency of other functions. *Psychological Review,* 1901, *8,* 247–261; 384–395; 553–564.
Thurstone, L. L. *Multiple-factor analysis.* Chicago: Univ. of Chicago Press, 1947.
Thurstone, L. L., & Thurstone, T. G. *Factorial studies of intelligence.* Psychometric Monograph No. 2. Chicago: Univ. of Chicago Press, 1941.
Thurstone, T. G. *An evaluation of educating mentally handicapped children in special classes and in regular classes.* Chapel Hill, N.C.: Univ. of North Carolina, School of Education, 1959.
Tyler, L. E. *The psychology of human differences* (3rd ed.). New York: Appleton-Century-Crofts, 1965.
Vandenberg, S. G. Contributions of twin research to psychology. *Psychological Bulletin,* 1966, *66,* 327–352.
Vandenberg, S. G. (Ed.). *Progress in human behavior genetics.* Baltimore: Johns Hopkins Press, 1968.
Vandenberg, S. G. Genetics of intelligence. In L. C. Deighton (Ed.), *The Encyclopedia of Education* (Vol. 5). New York: Macmillan, 1971.
Vierordt, H. Das Massenwachsthum der Körperorgane des Menschen. *Archiv für Anatomie und Entwickelungsgeschichte (Anatomische Abteilung des Archives für Anatomie und Physiologie,* 1890, Supplement, 62–94.
Waardenburg, P. J. *Genetics and ophthalmology* (Vol. 2). Springfield, Ill.: Thomas, 1963.
Ward, W. D. Absolute pitch. *Sound,* 1963, *2,* 14–21; 33–41.
Washburn, S. L. Tools and human evolution. *Scientific American,* 1960, *203* (3), 62–75.
Welford, A. T. *Ageing and human skill.* London: Oxford Univ. Press, 1958.
Welford, A. T., & Birren, J. E. (Eds.). *Behavior, aging, and the nervous system.* Springfield, Ill.: Thomas, 1965.
Weyl, N. Some genetic aspects of plantation slavery. *Perspectives in Biology and Medicine,* 1970, *13,* 618–625.
Wilkerson, H. R., Noble, C. E., & Skelley, C. S. *Sex and practice distribution in rotary pursuit tracking.* Paper presented at the meeting of the Southern Society for Philosophy and Psychology, New Orleans, March 1975.
Whipple, G. M. *Manual of mental and physical tests.* Part I: *Simpler processes.* Part II: *Complex processes* (3rd ed.). Baltimore: Warwick & York, 1921–1924.
Woodrow, H. The ability to learn. *Psychological Review,* 1946, *53,* 147–158.
Woodworth, R. S. The accuracy of voluntary movement. *Psychological Review Monograph Supplements,* 1899, Whole No. 13.
Woodworth, R. S. On the voluntary control of the force of movement. *Psychological Review,* 1901, *8,* 350–359.
Woodworth, R. S. *Le mouvement.* Paris: Doin, 1903.
Woodworth, R. S. Racial differences in mental traits. *Science,* 1910, *31,* 171–186.
Woodworth R. S. *Psychology* (Rev. ed.). New York: Holt, 1929.
Worthy, M. *Eye color, sex, and race: Keys to human and animal behavior.* Anderson, S. C.: Droke House/Hallux, 1974.
Wright, L., & Hearn, C. B., Jr. Reactive inhibition in normals and defectives as measured from a common performance criterion. *Journal of General Psychology,* 1964, *71,* 57–64.
Zeaman, D., & Kaufman, H. Individual differences and theory in a motor learning task. *Psychological Monographs,* 1955, *69*(6, Whole No. 391).
Zegoib, L., & Payne, R. B. Reminiscence in children as a function of sex. *Bulletin of the Psychonomic Society,* 1977, *9,* 173–175.

11

Epilogue: The Evolution and Variation of Human Intelligence

C. D. DARLINGTON
Oxford University, England

Our present understanding of human variation springs from many sources. One we find in Darwin's observations of men and animals all over the world. Another is the theory of selection he based on his studies not of man but of what men have done in domesticating plants and animals. Another we find in the work that grew out of Galton's confident, almost obsessive belief in his powers of measuring even immeasurable properties of living things.

A widely different kind of inquiry has led us forward from the experiments of Mendel, the observations of chromosomes by microscopists, and the inspired conjectures on their evolution by Weismann. Many utterly different lines of inquiry have related those chromosomes to the action of hormones, the differentiation of the sexes, and the development of the brain. All these independent pursuits of understanding are brought together in the present volume, some of them perhaps more closely than even their exponents recognize. For there is no aspect of our fast-developing knowledge of man's history and origins that does not concern his variation in intelligence.

The focus of this book is inevitably the great development of Galton's idea of the measurement of intelligence. Our starting point has to be the success of the methods of testing intelligence and their use in measuring the educability of individual Europeans and their capacity for useful employment in the European type of society, loosely stratified as it is in occupational classes. In achieving this strictly practical end, however, intelligence testing has thrown to us an intellectual *bonanza* that turns us back to the deepest problems that puzzled the four precursors whose names I have mentioned.

The reasons for this scarcely expected gift are that European society has common evolutionary origins. It is a complex population but homogeneous enough to demonstrate characteristic statistical properties of heredity and variation in the whole range of what is today physically and physiologically measurable, including the properties of intelligence.

Within the genetic continuum in space and time represented by European types of society, over a period of 70 years, investigators have discovered the normal curves of variation, the distinction between small and large gene discontinuities and whole chromosome defects, the correlations between kindred, the regular differentiation of the sexes, the greater variability of the heterozygous sex, that is, of the male rather than the female, and even the greater genetic discordance between the females than between the males among one-egg twins.

These refinements of analysis demand an assumption of a dominance of heredity over the environment that the analysts themselves have scarcely been prepared to make. When we look back at wider issues, however, we see why this is so. Throughout the evolution of the vertebrates, from the lancelet and the lamprey to man, there has been, as we have long known, a selection favoring the enlargement of the cerebral cortex. The enlargement has proceeded in jumps but with a direction that has persisted and seems never to have been reversed.

Such directional changes are quite conspicuous features of animal evolution (Simpson, 1951). They are due, in my opinion, as I explained in *The Evolution of Man and Society* (Darlington, 1971), not to any mysterious process but to natural selection operating under a novel principle, namely, that many genetic mutations, such as the lengthening of a horse's foot, inherently change the environment in which the animal lives. This is above all true of improvements in the memorizing, associating, and coordinating abilities of the brain. The development of human intelligence has, by its inventions, inherently improved the environments in which men have lived and to which they have had to adapt themselves. Every advance has positively fed back to favor further changes in the same general direction.

Darwin was well aware of the difficulty of separating heredity and environment in evolution. But his successors (often known as neo-Darwinians), dominated by the experimental methods of genetics, have attempted to enforce an artificial distinction between heredity and environment that does not exist for populations living and breeding in nature. The environment, for example, to which human populations adapt themselves is a social, cultural, and, indeed, genetic environment that they themselves, by their innovations, emotional and intellectual as well as material and mechanical, are continually changing. All men and every group of men have their own inherent characteristic differential environments.

At a certain point in evolution, a bifurcation occurred. As a result of a trivial mutation, or a shift of chromosome number from 48 to 46, or more probably both, our common ancestral stock split. Men, who became subject to this cultural and intellectual feedback, left behind the apes, who were shelved forever at a manifestly unchanging mental level.

For the last step in the evolutionary development of man's intelligence and his genetic environment we have now the historical evidence of the invention of agriculture and the growth of civilization. Are we then justified in supposing that European civilization and the culture it has created, and the tests that are derived from it, are measures of all human potentialities? Certainly not. But we are justified in asking, as many investigators have asked, just how far they take us in understanding the whole of human variation and human evolution.

Look first at the grounds for hesitation. In directed evolution, as we have seen, bifurcations occur. Blind alleys are reached. Success may bring its own downfall. We do not need to look far for grounds for this kind of misgiving. The European's failure to respect nature, his own environment, and other men's environments are obvious enough, as are the overwhelming consequences of this failure for the future of all mankind.

The European shares his failures, to be sure, with the more primitive people of Africa and India, but the evil results of the work of the European are magnified by his power. His responsibility is vastly greater since he is in a position to know better: He cannot claim, as others may, the excuses of either superstition or disease. He may therefore reflect on the proposition that the ability to understand nature and to protect posterity and our fellow men is a measure of his intelligence.

Again, looking around the world, we see that the Chinese and the Japanese are comparable with Europeans not only when measured individually and statistically by our intelligence tests but also when judged by their professional activity, social history, cultural achievement, and political organization. To all these, there is another kind of test to be added: that of success as an immigrant minority.

Wherever they have gone abroad, these Mongoloids have lived under social and political disadvantages from which Caucasoids have rarely had to suffer. They have endured unfavorable racial discrimination. The Chinese in Southeast Asia have been repeatedly expelled or massacred. Indeed, their experience has been closely parallel with that of the Armenians and Greeks in Turkey and the Jews in Europe, particularly Eastern Europe. It has not been unlike that of the Indo-Dravidians in Africa. The reason for this discrimination and suffering has been in all cases the same. The immigrant minorities have been economically successful.

What has given them this success? At first sight, it would seem to be simply their superior ability to assimilate and to exploit European civilization by a European type of intelligence. But when we look further at all the circumstances in such cases, we see that these intelligent people owe their success to many other social qualities that are not, and cannot be, measured merely by the process of intelligence testing. Long-range qualities of adaptability and foresight are evidently correlated with those short-range properties that can be quickly assessed. Nor is this surprising since (if we understand the lessons of Lorenz) a large part of human evolution has consisted in the progressive subordination of instinctive to rational processes, a subordination that has demanded a continual extension of the period of education and a continual readjustment of the relations of the individual to society. All these changes are seen developing in a well-documented history of civilized societies.

Conversely, they are seen *not* developing in primitive and peripheral societies cut off from the continual stimulus of invention of pottery and the wheel, of writing, of metallurgy and of engineering, all of them feeding back to favor the evolutionary selection of new mental abilities.

These developments we can see from the course of our own history. Even in the last 500 years, understanding of cause and effect has influenced the taking of evidence, the practice of the law, the displacement of credulity and superstition, the decay of magic, the extension of scientific inquiry. The absence of all these changes we can see in the operations of the primitive mind, so correctly diagnosed 50 years ago by Levy-Brühl.

The failure of the primitive mind to make use of what we call evidence represents an interruption or cessation of evolutionary development. One of its consequences is the failure to accept or demand the principle of individual responsibility, a failure that effectively blocks the further evolution of society. Another of its consequences is the failure of primitive language to convey the inferences and connections we require for expressing scientific or, indeed, civilized thought. All languages, of course, differ in the subtleties they can express. But they can be placed in an evolutionary series in an inevitable relation to the lives and experiences of the people who use them. A vast range, hardly to be measured in one dimension, may be found among those who speak some form of English, from the "pidgin" English of New Guinea or the "black" English of New York to our technical or poetic sophistications at the other extreme.

To understand this contrast between the civilized and the primitive, we must return to the starting point of civilization, the origins of grain-growing agriculture 10,000 years ago, above all in Southwest Asia. From this source, cultivators spread with their growing satellite industries in all directions. In Europe and in Asia they met fearful obstacles, but in the course of 5000 years they completed their journeys across the world. In Africa, however, and indeed in the tropical Old World as a whole, their expansion was blocked by an obstacle that defeated them. This obstacle consisted in the diseases inherent and endemic for all species of plants and animals in the region from which they take their origin. It is a principle whose discovery we owe to the Russian agricultural geneticist, Nickolai Vavilov.

Not only agriculture but civilization itself (contrary to the chauvinistic illusions of many nations) was carried by migrant peoples into Europe, China, and India. But from Africa it was excluded. This exclusion has had a complex history of direct and indirect causation. The exclusion of men by the mosquito and the diseases it carried was direct. So also was the exclusion of men mounted on horses by the tsetse fly and its diseases. Indirect was the genetic response to these diseases. The abnormalities of hemoglobin, which became a racial characteristic of all lowland Negroid populations, protected them from death by endowing them with disabilities that weakened them in their reaction to every other disease. In turn, they left the Negroids burdened with a second, no less spurious defense against disease, namely, witchcraft. For a belief in witchcraft, once established in a primitive community, is something that neither the science nor the religion of Europe has so far been able to exorcise.

When the Negroids were taken away from Africa by slave traders they were partly, and in the United States often wholly, rescued from the diseases that had infested them. But slavery did not rescue them from their genetic responses to disease, which have continued with them as an obstacle to their development individually, racially, and culturally. Nor did it rescue them from their primitive beliefs. These experiences teach us indeed that peoples cannot be separated from their history. They are bound down by their evolutionary antecedents, occupational, social, and medical.

Thousands of Negroids were removed from Africa, where the women bore the children and did the work while the men did the fighting and kept the population stable in numbers. On the North American and West Indian plantations, they all had to work. There was neither famine nor unemployment, and there was very little war. The environment was more favorable than anything they had experienced in Africa. As slaves, they improved in health and increased in numbers.

When the Negroids were liberated from agricultural slavery, they were thrown free to shift for themselves in largely urban Caucasoid societies. Discipline and protection having been withdrawn from them, they were offered the beguiling prospect of life first in American and then in European cities. These simple, unskilled rural people were suddenly offered irregular urban employment combined with the opportunities of drink and drugs, gambling and prostitution, and no reliable means of productive, creative, or congenial labor.

The intellectually well-endowed races, classes, and societies have a moral responsibility for the problems of race mixture, of immigration and exploitation, that have arisen from their exercise of economic and political power. They may hope to escape from these responsibilities by claiming an intellectual and, therefore, moral equality between all races, classes, and societies. But the chapters of this book, step by step, deprive us of the scientific and historical evidence that might support such a comfortable illusion.

European man is morally as well as intellectually responsible equally for the splendors and the miseries he has brought on mankind. For while creating his own environment, to which he is perhaps capable of adapting himself, he has also created for other peoples throughout the world environments to which they, especially in the tropical world, are disastrously ill-adapted.

How ill-adapted they are is shown most clearly by the painful records of governments established among certain African peoples with the aid of Western technology. When we compare the state of the Ashanti, Nigeria, and Uganda before and after European colonization, we see oppression made more oppressive. When we look at the century and a half of liberated Sierra Leone, Liberia, or Haiti, we see promises equally unfulfilled. In the absence of a large educable class, recognizable either on the IQ scale or by professional qualifications, we find a uniform incapacity for constitutional government—and in its place a variety of police states in which most people matter very little, while the natural resources of the country are being quickly extracted for the benefit of a few fortunate families.

We may now begin to see the world as a great field of natural experiment in the assessment of human variation. We have its history at our disposal marvelously revealed and open for comparisons with the tumultuous events of today. But the comparisons are loaded with emotion and charged with political interest. It is, therefore, all the more important that we can turn for guidance to the methods of measurement that psychologists, educators, and biologists have designed with the purpose of comparing the individual and inborn qualities of men and women with the practical results of the work they do.

The fruit of this enterprise has been to show us much more than we could ever have expected. For it has gone a long way toward revealing the connectedness of all human experience. And it has shown us in a new light the obligations that men and women owe to one another and to the earth that they hope to hand on to their posterity.

REFERENCES

Darlington, C. D. *The evolution of man and society.* New York: Simon & Schuster, 1971.
Simpson, G. G. *The meaning of evolution.* London: Oxford Univ. Press, 1951.

Subject Index

Q

R

A
B
C 8
D 9
E 0
F 1
G 2
H 3
I 4
J 5